HIGHWAY AND URBAN ENVIRONMENT

ALLIANCE FOR GLOBAL SUSTAINABILITY BOOKSERIES
SCIENCE AND TECHNOLOGY: TOOLS FOR SUSTAINABLE DEVELOPMENT

VOLUME 17

Aims and Scope of the Series

The aim of this series is to provide timely accounts by authoritative scholars of the results of cutting edge research into emerging barriers to sustainable development, and methodologies and tools to help governments, industry, and civil society overcome them. The work presented in the series will draw mainly on results of the research being carried out in the Alliance for Global Sustainability (AGS).
The level of presentation is for graduate students in natural, social and engineering sciences as well as policy and decision-makers around the world in government, industry and civil society.

For other titles published in this series, go to
www.springer.com/series/5589

Highway and Urban Environment

Proceedings of the 9th Highway and Urban Environment Symposium

Edited by

Sébastien Rauch
Water Environment Technology, Chalmers University of Technology, Sweden

Gregory M. Morrison
Water Environment Technology, Chalmers University of Technology, Sweden

Andrés Monzón
Universidad Politécnica de Madrid, Spain

Springer

Editors
Sébastien Rauch
Water Environment Technology
Chalmers University of Technology
Sweden
sebastien.rauch@chalmers.se

Gregory M. Morrison
Water Environment Technology
Chalmers University of Technology
Sweden
greg.morrison@chalmers.se

Andrés Monzón
Universidad Politécnica de Madrid
Spain
amonzon@caminos.upm.es

ISBN 978-94-007-3095-3 e-ISBN 978-90-481-3043-6
DOI 10.1007/978-90-481-3043-6
Springer Dordrecht Heidelberg London New York

Cover illustration:

Cover design:

Printed on acid-free paper

Springer is part of Springer Science+Business Media (www.springer.com)

THE AGS
The Alliance for Global Sustainability

Preface

The 9th Highway and Urban Environment Symposium (9HUES) was held on 9–11 June 2008 in Madrid, Spain. HUES is run by Chalmers University of Technology within the Alliance for Global Sustainability (The AGS). The symposium was hosted by Universidad Politécnica de Madrid (UPM).

HUES was initiated by Professor Ron Hamilton at Middlesex Polytechnic (now University) in the early 1980s and had the title "Highway Pollution". The initial aim was to measure and assess challenges in highway pollution, with a strong emphasis on urban photochemical smog, ozone formation and particle release. After the first symposium, the emphasis on air pollution issues continued through to Munich in 1989 where diesel particulate issues and the relevance to health through measurements of PM10 emerged. The focus on air quality issues was also strengthened by the co-organisation of the symposium with Professor Roy Hamilton at the University of Birmingham from 1986 to 1998. In parallel, the symposium started to receive an increasing number of scientific contributions from the area of urban run off, indeed to the extent that the title of the symposium was changed to "Highway and Urban Pollution". Also at this time the importance of science in support of policy was emerging as a key aspect of the symposium.

8HUES was held in Nicosia, Cyprus in 2006 and was hosted by the Cyprus Institute. For this 8th symposium, we decided to evolve the name of the symposium to "Highway and Environment" to provide a positive view of our common future looking to a positive environment. That said, paper addressing pollution issues in the highway and urban environment remain a central part of the symposium as they help to raise awareness around issues to be solved. 8HUES was also marked by an organizational change with Chalmers University of Technology taking over the organization of the symposium. For the first time, the proceedings of the symposium were published as a book in the AGS book series.

For 9HUES we worked with TRANSyT – Universidad Politécnica de Madrid, Spain to provide a professional and scientific forum on global examples of the science required to support pathways to a positive and sustainable future in the highway and urban environment.

Ana Botella Serrano, Delegada del Área de Gobierno de Medio Ambiente y Segunda Teniente de Alcalde del Ayuntamiento de Madrid opened the symposium with a keynote speech on the efforts done by the Municipality of Madrid to keep their environment sustainable. Plenary presentations were given by José María López from UPM, Spain; Stefan Norra from University of Karlsruhe, Germany; Thomas Pettersson from Chalmers University of Technology, Sweden; Maryse Labriet from CIEMAT, Spain and Alfredo Tolón Bercera from Universidad de Almeria, Spain. Presentations covered air pollution, soil and water contamination, pollution control technologies, land use and mobility, urban climate and climate change. Best poster awards was given to Ana Estela Barbosa from LNEC, Portugal and Adalgiza Fornaro from the University of Sao Paulo, Brazil.

The following facts provide a background for 9HUES:

Eighty-two delegates (37 female) from 26 countries, including all continents
One hundred and seven abstracts for papers and posters were received
Fifty oral presentations and 20 poster presentations at the symposium

We would like to take this opportunity to thank all who have contributed to the success of 9HUES. We would especially like to acknowledge Alexandra Priatna, Jonna Bjuhr and Yvonne Young at Chalmers' AGS office and Emilio Ortega and Patricia Rico at UPM whose organizational skills were essential to the success of this symposium. Maria Svane at Chalmers' AGS office is acknowledged for putting together the proceedings. Finally we would like to thank the delegates for the many valuable contributions and a highly enjoyable symposium. We would particularly like to acknowledge the continued participation of Paul Höglund of Transforsk-Transearch, Sweden.

The next symposium, 10HUES is planned for 9–11 June 2010 and will be held in Göteborg, Sweden.

Göteborg, Sweden — Sébastien Rauch
Göteborg, Sweden — Gregory M. Morrison
Madrid, Spain — Andrés Monzón

Contents

Part IV Urban Water Contamination

Part VI Urban Form and Climate Change

Part I
Sustainable Mobility and Management

Planning for Sustainable Mobility with Personal Rapid Transit in Small European Cities

Alexandra Lichtenberg, Patrícia Guimarães, and Heleen Podsedkowska

Introduction

The objective of this research is to introduce a new perspective to sustainable mobility strategic planning for small European municipalities – a framework for Strategic Sustainable Development (FSSD) – where back casting from a sustainable vision in the long-term future is central to the process. This vision portrays mobility in an efficient, safe and economic manner, where the use of private cars has been substituted by attractive and reliable integrated sustainable transportation modes. It is a holistic approach based on systems thinking, which provides a structure for organizing analysis and strategic planning in complex systems from the perspective of the mobility issue and the global socio-ecological system.

Continuing economic development and the consequent increase in atmospheric pollution from motorized transport emissions (particularly in urbanized areas) have transformed cities into one of the big contributors to the climate change issue around the planet.

Renewable energy for fuels and advanced technology mobility systems are already in place to mitigate undesirable effects of existent transportation systems. However, it has been proved that only mitigating the impact will not solve the problem, and the need for a long-term solution is urgent.

Besides its downstream impacts on the environment due to greenhouse gas emissions, car dependency has been causing several other problems to human society in the biosphere. The most impacting are time wasted in congestions, accidents, noise pollution, the need of raw materials to produce vehicles that do not have a suitable end of life destination, and the enormous infrastructure needed to support all of this.

In order to reach the vision of sustainable mobility in the future, offering excellent public transport capable to compete with cars was the primary target. Moreover, behavior change, integrated urban development and energy planning, and economics are also

A. Lichtenberg (✉), P. Guimarães and H. Podsedkowska
Department of Mechanical Engineering, Blekinge Institute of Technology, SE-37179 Karlskrona, Sweden
e-mail: alexandra@ecohouse.com.br

S. Rauch et al. (eds.), *Highway and Urban Environment*, Alliance for Global Sustainability Bookseries 17,
DOI 10.1007/978-90-481-3043-6_1,

important issues related to the transportation system which deserve careful systemic (integrated) analyses when planning for sustainable mobility. Personal Rapid Transit (PRT) systems were analyzed in this study as a suitable and attractive alternative transportation system, offering the advantages of private cars while avoiding its negative sides, leading urban areas towards zero car dependency contributing to a sustainable mobility vision in the future. Through Karlskrona municipality's case study a specific strategic planning method for sustainable mobility was defined: the Framework for Strategic Sustainable Development – Sustainable Mobility (FSSD-SM).

Throughout this paper the terms Personal Rapid Transit (PRT) and Podcar are used interchangeably and have the same meaning.

Method

Literature review on the issues of economics, behavior change, energy, land-use and transport modes related to sustainable mobility, as well as analyses of research projects (especially European Union funded and commissioned) and best practice cases on sustainable mobility and land use planning around the world.

The Framework for Strategic Sustainable Development (FSSD) approach, developed by Dr. Karl-Henrik Robèrt for planning in complex systems, is used here as a main tool informing the achievement of Strategic Sustainable Development in Karlskrona's Mobility System. The core elements of this methodology are the back casting process and the four sustainability principles which guide to a complete understanding of the unsustainable current reality as well as to serve as constraints for development of the vision of success in the future [1].

Compliance with these four principles was the basis for planning for a sustainable mobility society in Karlskrona in 2050, using the ABCD tool in order to perform its sustainability analysis, identifying its mobility current reality, a vision for its sustainable mobility in 2050 with zero car dependency, and compelling measures to reach this vision. Back casting from Karlskrona's mobility vision assisted the strat-egic planning process, ensuring that the actions will lead the municipality closer to its sustainable vision, bridging the gap between the current reality and the envisioned future. External advisors reviews and peer reviews completed the methods.

Results

Sustainable Mobility and Car Dependence

Car dependence is a persistent problem of our present society. The car is the dominant form of maintaining personal mobility in a very attractive manner. Its benefits are difficult to match: it is a door-to-door transportation system available at any time, the means to gaining access to life necessities and employment, and a source

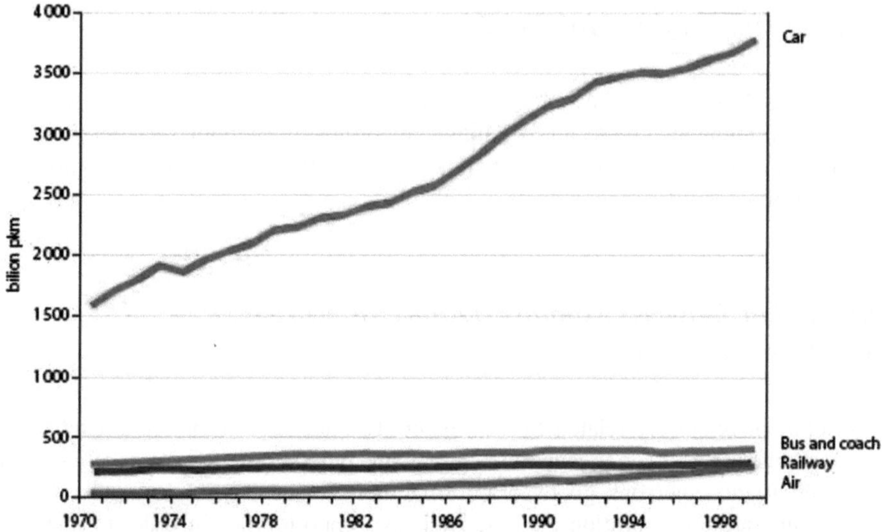

Fig. 1 Passenger transport – growth of traffic by mode of transport *EU-15: 1970–1999* [4]

of pleasure and social status. So are its disadvantages, including local air pollution, greenhouse gas emissions, road congestion, noise, mortality and morbidity from accidents, and loss of open space to roads, parking lots and urban sprawl. Many people are aware of the disadvantages but see no comparable substitutes to the automobile. The dilemma of an automobile owner is similar to that of a herd owner described in the classic case of the "tragedy of the commons"[2]. In the EU car use is on the rise (in Fig. 1).

As stated in a report by the Low Carbon Vehicle Partnership [3], voluntary agreements made by the European Commission in 1998 with automobile manufacturers in Europe, Japan and Korea to reduce average emissions to 140 g/km CO_2 emissions by 2009 are far from being achieved – and an EU regulatory framework has therefore been proposed, to be ready by mid-2008. There are a number of reasons as to why progress has been slower than anticipated: (i) consumer preference has been towards higher performance and larger cars; (ii) progress with tax incentives for low carbon vehicles across the EU has been mixed; and (iii) the three trade bodies have been unable to influence the CO_2 emissions of vehicles produced by their members and there has been no clearly laid out system of burden sharing.

Part of the analysis is to suggest ways to revert the car dependence trend, through evidencing its intricate relation to urban and energy planning, behavior change campaigns, and a shift in the perception of how to account for the externalities of road transport. Car-Free development encompasses a range of ideas that places *humans before cars* and stresses a new urban order based first on quality of life. To better understand sustainable mobility, causes and effects of transportation activities must be identified and its upstream issues dealt with accordingly.

What is Sustainable Mobility ?

Sustainable Mobility is the ability to meet society's need to move freely, gain access, communicate, trade and establish relationships without sacrificing other essential human or ecological values, today or in the future [5]. All of this must be achieved in a cost-effective manner.

Interaction Between Mobility, Land-Use, Energy and Behavior Change

Development is closely related to movement. To undertake commercial exchange, access public services, or engage in recreation and entertainment, society relies upon the ability to move persons, goods, or information from one location to another. Cities are a deliberate concentration of these exchange opportunities in order to increase both the diversity and accessibility of exchange opportunities.

Planners have long known that land use and transport are intertwined and correlate with energy consumption, particularly in urban areas. There is an inseparable link between establishing locations for human functions, the use of space, and the impact on the way people travel. Nevertheless, land use, transport and energy issues are too often viewed and addressed separately [6]. As a result, opportunities to *collectively address individual concerns* in each sector are lost. Planning approaches are often described as adversarial or lacking in proper representation [7]. Forrester states that "a systematic approach to analyzing and managing urban land use, transport, and energy systems requires *systems thinking*: the evaluation of interrelated and complex systems over time" [8].

Sustainability Principles

In the late 1980s Dr. Karl-Henrik Robèrt led a network of leading Swedish scientists from a variety of disciplines, which came up with a principle-based and systemic definition of sustainability. Over the years these principles have been successfully adopted by several organizations and communities. As basic or first order principles (SPs), they intend to be: (i) based on a scientifically agreed upon view of the world; (ii) necessary to achieve sustainability; (iii) sufficient to achieve sustainability; (iv) general to structure all societal activities relevant to sustainability; (v) concrete to guide action and serve as directional aides in problem analysis; (vi) non-overlapping or mutually exclusive in order to enable comprehension and structured analysis of the issues [9].

Sustainability Principle #1: IN A SUSTAINABLE SOCIETY, NATURE IS NOT SUBJECT TO SYSTEMATICALLY INCREASING CONCENTRATIONS OF SUBSTANCES EXTRACTED FROM THE EARTH'S CRUST.

> We dig up substances from the Earth's crust – various minerals, oil and gas, etc. – and use them in products and processes, and then release them into nature. We do this at a faster pace than nature re-deposits these substances back into the Earth's crust. And as a result, they accumulate in natural systems and eventually cause problems *if their concentrations get too high.* The first principle does not say a sustainable society requires that we not use any material at all from the Earth's crust. It does not say that there is no mining in a sustainable society. It does say that whatever materials we use from the Earth's crust, *we must use them in a way that prevents their accumulation in natural systems*. This means using them efficiently, and using them in products and processes where they can be recaptured and reused, rather than released into the atmosphere, water or soil [10].

There are thresholds beyond which living organisms and ecosystems are adversely affected by increases in substances from the earth's crust. Problems may include an increase in greenhouse gases leading to global warming, contamination of surface and ground water, and metal toxicity which can cause functional disturbances both in humans and animals.

Sustainability Principle #2: IN A SUSTAINABLE SOCIETY, NATURE IS NOT SUBJECT TO SYSTEMATICALLY INCREASING CONCENTRATIONS OF SUBSTANCES PRODUCED BY SOCIETY.

> Humans combine molecules into new, more complex molecules that nature has never seen before, and use these complex molecules in products and processes that eventually allow them to be released into natural systems. Because nature has never seen them before, it cannot break them down within its regular cycles, and so they accumulate. Note that the second principle does not say that there are no chemicals in a sustainable society. Rather, it says that a sustainable society will require that we be efficient in our use of them, and most importantly that we use them in ways that allow them to be captured and re-used, rather than dispersed into nature where they can accumulate [10].

Synthetic organic compounds such as DDT and PCBs can remain in the environment for many years, bio accumulate in organisms and cause harmful effects in many species. Freon and other ozone depleting compounds may increase the risk of cancer due to added UV radiation in the troposphere. The solution includes gradual substitution of certain persistent and man-made compounds with ones that are normally abundant or break down more easily in nature, and use all substances produced by society in an efficient manner.

Sustainability Principle #3: IN A SUSTAINABLE SOCIETY, NATURE IS NOT SUBJECT TO SYSTEMATICAL DEGRADATION BY PHYSICAL MEANS.

"We physically degrade nature's ability to run natural cycles by encroaching into natural areas, over-harvesting renewable resources, and eroding nature's ability to process our waste" [10]. Humans must avoid taking more from the biosphere than can be replenished by natural systems, safeguarding nature's ability to regenerate. Preserving the integrity of any living system requires that society understands that what effects one part of a system may ultimately impact the system as a whole. The downstream effects we witness nowadays – climate change, acid rain, deforestation, depletion of fish stocks, toxics in toys that accumulate in tissues – can be traced back to one or more of these three ecological mechanisms. They are all "downstream" symptoms of more fundamental problems in how our societies are designed. An “upstream” approach to planning anticipates and avoids problems before they occur, as opposed to a “downstream” approach, where resources are used to deal with the results of the problems.

Sustainability Principle #4: IN A SUSTAINABLE SOCIETY, PEOPLE ARE NOT SUBJECT TO CONDITIONS THAT SYSTEMATICALLY UNDERMINE THEIR CAPACITY TO MEET THEIR NEEDS.

In order to achieve the first three sustainability principles and have a thriving human society, we must also consider the barriers that exist to people meeting their needs worldwide. If human needs are not met then increasing poverty, conflict and other forms of social stress will result in more environmental degradation. All forms of political oppression need to be overcome and human rights respected for people to have any chance of meeting their own needs.

Backcasting × Forecasting

Far too much time and money have been invested to date in discussing and accommodating the movement of vehicles rather than the *movement of people*. Most of these projects use forecasting to solve mobility problems and address behavior change without much innovation in the services provided. Back casting starts with a desirable future situation, usually described by a set of goals or targets established by assumed events between the current and future situations. Therefore back casting is capable of highlighting discrepancies between the current and desirable future, and incorporating large and even disruptive changes.

A key obstacle to achieving a mode shift from private to public transport is that PT lacks the attractiveness and reliability. Low quality has a tendency to discourage users who have an alternative and very attractive option. It may be easier to deter people from using public transport via low quality than to attract them back via improved quality. Insufficient attention to improve the quality of public transport and raise awareness about these improvements could restrict the use of public transport to users who do not have a choice due to factors such as age or economic status [11].

PRT Systems Technology

PRT systems have been in development and proposed as an alternative to the use of private cars for over 30 years, against a wide range of vested interests in the automotive and public transport industries. They have the potential for contributing significantly to the solution of fundamental problems of modern mobility including congestion, harmful emissions, dependence on a dwindling supply of cheap oil, and most recently terrorism. It is a concept that can enable public transport to compete with the car in urban areas. Apparently, the main barrier to its implementation seems to be the imbalance between risks and financial support: No local authority buys a system, due to the development risks as long as it is an unproven technology; and no

developer invests fully in development, as long as there is no market for the end product [12, 13]. Only one system has been fully implemented (1972, in Morgantown, West Virginia). It has successfully demonstrated the viability of key PRT technology components, including off-line stations, on-demand service, nonstop origin-to-destination transport and fully automated control systems [14]. Since it began operating, the system has demonstrated very high reliability of 98% or greater.
The fundamental elements of a PRT system are:

On demand, origin to destination direct service: the service is *available on demand* rather than on fixed schedules. In most cases a pod will already be at the station, *available 24 h a day* at all stations on the network to *meet individual travel needs*. Its size allows for the passenger to choose to have his or her own ride. Non-stop travel service ensures *short trip times* (the stations are placed off of the main guide-way). In this system, traffic is adapted to the passenger, instead of the passenger having to adapt to traffic – and that is what makes the difference [15, 16].

Small, fully-automated vehicle: PRT vehicles are intended to operate under computer control and require no operator or driver, which considerably lower operating costs. The system will automatically route the vehicle to the desired destination within the shortest possible time and without the need for further user interaction with the system. Operations involve staff to support central control, network management and vehicle flow systems, maintenance, security, and station services [16]. Due to the need to provide a high degree of safety, speed and accuracy to effectively manage potentially large fleets of vehicles over complex networks, PRT networks have a number of unique operating and control system requirements.

Exclusive-use guide ways: designed as elevated systems with beams and support structures sized for lightweight two-four passenger vehicles. The light weight of each vehicle allows for economic construction of the guide-ways, and less land-use. PRT systems use networked and separated guide ways to permit higher transportation capacity and safety levels as the system will not interfere with current modes of transportation [17, 18].

Off-line stations: these are a major breakthrough, designed with a "siding" track or guide way. Vehicles not stopping at a particular station can bypass that station and are not delayed by other vehicles boarding and alighting passengers; the PRT station should be sized according to the necessary number of loading berths needed to handle the demand at each station.

Low-emissions and resource efficiency: electric power is used primarily for propulsion and vehicle amenities. Beyond the need for propulsion, a PRT vehicle needs power for on-board systems and passenger comfort. It should support limited HVAC services as well as vehicle control and communications. The electric power can come from many different, preferably renewable sources (Fig. 2).

Some power is used to move around empty seats. With PRT, empty vehicles are sometimes routed to another station with a higher demand.

Figure 3 shows that the power required per seat to move a PRT pod is very low compared to a bus or a car. The efficiency of the system depend on many other factors, most of all on the occupancy of the vehicle [19].

Vehicle and Passenger Accommodations: for urban environments, studies have shown that the typical private automobile has 1.63 occupants for all trips, 1.14

Fig. 2 Ultra vehicle, test site in Cardiff, Whales (Department of Rail and Public Transportation, 2008)

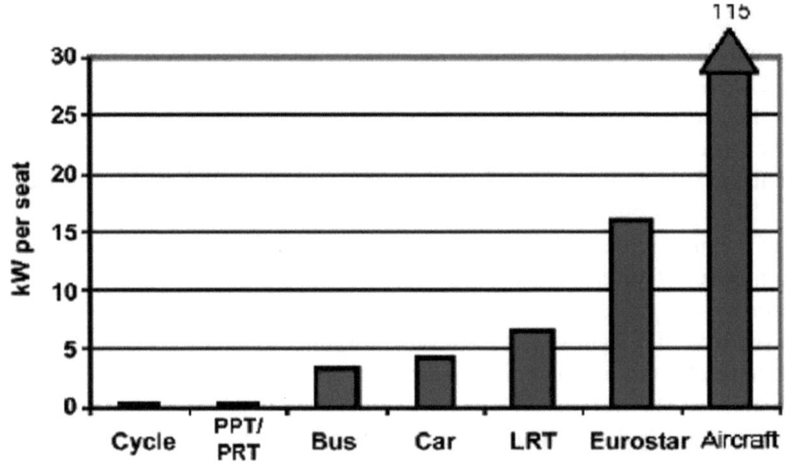

Fig. 3 Power per seat installed for various types of transport (kW/seat) [19]

occupants for work trips, and over 93% of urban trips have three or fewer co-travelers [18], *corroborating the need for individual PT services to compete with car use*. PRT vehicles are designed for accessibility, including wheelchairs or other assistive devices (Fig. 4).

Adapted Framework for Strategic Sustainable Development in Karlskrona Municipality's Sustainable Mobility System

Back casting from a desirable future is the core element of this methodology. "The major benefit of the back casting approach is that it provides a clear sense of how things need to be, instead of just aiming at making things better than they

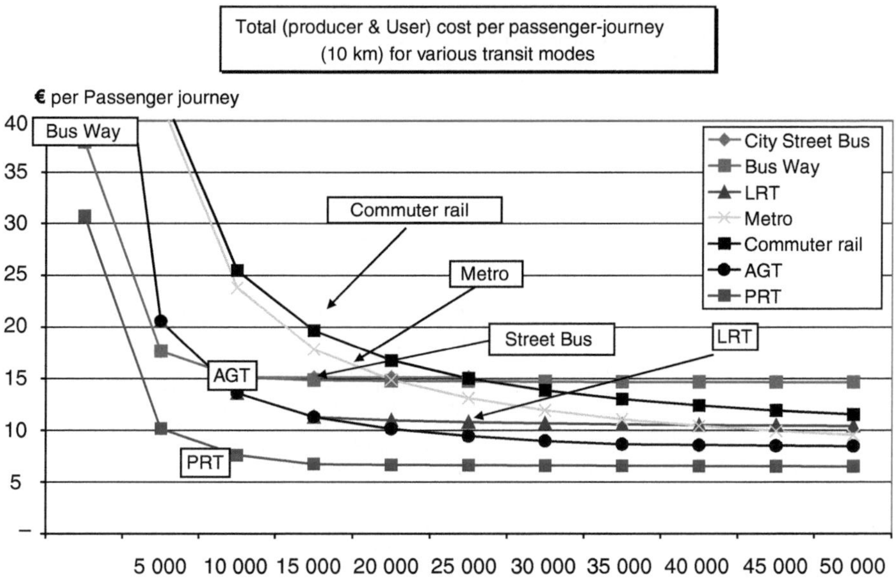

Fig. 4 Total cost per passenger–journey for transit modes [20]

are now. There is an emphasis on an upstream versus a downstream approach, which helps to address problems before they occur rather than after the fact."[21]. Table 1 show the working process to apply the FSSD to achieve Karlskrona municipality's sustainable mobility. The working process for developing the FSSD-SM consists in:

- Acknowledgments of the system
- Building a mobility vision
- Backcasting from Karlskrona sustainable mobility vision – strategy: analyses of Karlskrona mobility system – current reality, C step, strategic guidelines, D step
- Suggested actions; suggested tools

The research team facilitated an ABCD workshop with politicians and employees from the municipality of Karlskrona, which resulted in:

- A list of present barriers to sustainable mobility – B list
- The beginning of the construction of a vision for sustainable mobility in the municipality in 2050
- Brainstormed measures to get to this vision within the constraints of the 4 SPs – C list
- Thorough understanding of how complex systems work, the need for acknowledging the systems involved and for the creation of guidelines for prioritization of measures

Table 1 FSSD-Sustainable Mobility Framework for Strategic Sustainable Development for Sustainable Mobility in Karlskrona

SYSTEM	Karlskrona Municipality's mobility system within society, within the Biosphere: understanding the relationships between transport systems, land use, energy consumption, people behaviour, policy making, economical structuring and its impact on the social and natural environment
SUCCESS	• Vision – 2050 Fossil Fuel Free Municipality World class and ***integrated transport system*** that offers safe, efficient and ***sustainable*** travel ***for all***, where public transport, walking and cycling provide real and desirable alternatives to car travel; moreover, making the Municipality ***greener and more attractive.***
STRATEGY	• Backasting from the vision of a sustainable mobility society Structured actions must be guided by **strategic guidelines** leading towards eliminating car dependency – in order to achieve sustainable mobility the measure must meet: • Emissions reduction • Congestion elimination • Resource efficiency - • Recyclability • Semless modal integration • User satisfaction • Safety and sucurity Regarding to a mode system, this must be : • Comfortable, available on demand, accessible, fast trip time, affordable Actions must be prioritized according to the principles: • Right direction towards the vision? • Flexible platform to new developments? • Good return of investment (economical, social, etc)?
ACTION	Design the integrated systems network including walking, cycling and PT, according to prioritized measures – to achieve integrated implementation of the PRT system in Karlskrona Municipality.
TOOLS	Additional techniques, instruments, measurements, management tools, to assist and support the action: ABCD tool LCA CLD LRP planning system

Conclusion and Discussion

Sustainable mobility in a small municipality where car riding is rising can only be achieved with the supply of a new technology system, attractive enough to compete with the use of cars [I. Andrèasson, 2008, personal communication[1]]. PRT systems

[1] Adjunct professor in Transport and Logistics; Director for the Centre for Traffic Research at the Royal Institute of Technology in Stockholm Sweden.

were compared with the use of private cars, and not with other PT modal systems. The fact that it provides personalized on demand service 24 × 7, and is driverless (therefore allowing for the extended service hours with no extra cost) must be added to the careful step-by-step introduction of the system with the clear aim to *supplant car use*, where additional infrastructure for the private car will not be accepted. The Karlskrona case study highlighted several benefits which the implementation of PRT systems would ensure: (i) increased competitiveness for the municipality; (ii) attracting investment and visitors; (iii) direct and indirect employment effects; (iv) access benefits to citizens marginalized by poor PT and no access to cars; (v) congestions elimination; (vi) emissions and pollution reduction; (vii) accident and noise reduction. Thus, the municipality can and should go ahead to plan for the implementation of PRT systems in the medium term. Unless it decides to embrace this new technology and plans accordingly, it will go on investing time and resources into "dead-end" strategies, and not be prepared for issues like peak oil, ever more severe EU and National regulations concerning emission, etc. The next step would be to contract expert advice to develop a thorough feasibility study for the integrated implementation of a PRT system in the city.

PRT systems are not the only magical solution to sustainable mobility – they will only be an effective solution when properly integrated to the existing modal infrastructure. The integration will vary from region to region. In smaller cities it must be integrated to the existing bus (road) and train infrastructure, as well as to the shorter distance mobility systems like walking and cycling. The bus service must be reshaped to provide reliable and timely scheduling and routing to attract more customers. This might be more productive than trying to substitute the fuel the buses presently run on (if bus ridership raises, emissions per capita will lower considerably). Bus stops must be redesigned to provide necessary shelter; bike paths must be planned and implemented. Once planning is completed and implementation is under way, a marketing campaign can be launched in consonance with the implementation lead times.

References

1. Robèrt K-H, Broman G, Waldron D, Ny H, Byggeth S, Cook D, Johansson L, Oldmark J, Basile G, Haraldsson H, MacDonald J (2004) Strategic leadership towards sustainability: Blekinge Institute of Technology
2. Vergragt PJ, Brown HS (2007) Sustainable mobility: from technological innovation to societal learning. Journal of Cleaner Production 15:1104–1115
3. Kågeson P (2007) A European regulation on the fuel efficiency of new cars. Low Carbon Vehicle Partnership
4. European Commission (2001) White paper – European transport policy for 2010: time to decide
5. WBCSD (2002) The sustainable mobility project – July 2002 Progress Report. Geneva
6. Novem (2002) Local transport performance: the shortest route to a better living environment. Utrecht: Novem
7. Belaieff A, Moy G, Rosebro J (2007) Planning for a sustainable nexus of urban land use, transport and energy. Blekinge Institute of Technology

8. Forrester JW (1971) Counterintuitive behaviour of social systems. Technological Forecasting and Social Change 3:1–22
9. Holmberg J, Robèrt K-H (2000). Backcasting from non overlapping sustainability principles: a framework for strategic planning. International Journal of Sustainable Development and World Ecology 7 (4):291–308
10. Park C (2008) Remarks by Chad Park on Bill C-474. In House of Commons Environment and Sustainable Development. Ottawa, Canada
11. European Environmental Agency (2007) Transport and environment: on the way to a new common transport policy – term 2006: indicators tracking transport and environment in the European Union. Luxembourg
12. Tegnér G, Hunhammar M, Andrèasson I, Nowacki JE, Dahlström K (2007) PRT in Sweden: from feasibility studies to public awareness. Paper presented at 11th International Conference on Automated People Movers, Vienna, Austria
13. Anderson JE (2005) The future of high-capacity Personal Rapid Transit. PRT International LLC, Minneapolis, MN, November 2005
14. Raney S, Young S (2005) Circulation and driverless transit. Paper presented at Transportation Research Board Annual Meeting, Washington, D.C
15. Lincoln/Lancaster County. 2004. Multi-modal transportation plan study. Lincoln, NV: City of Lincoln Planning
16. Andrèasson I (2001) Innovative transit systems. VINNOVA
17. Anderson JE (2007) High capacity personal rapid transit: rationale, attributes, status, economics, benefits. Minneapolis, MN
18. Carnegie JA, Hofman PS (2007) Viability of Personal Rapid Transit in New Jersey. New Jersey Department of Transportation
19. Brand C, Preston J (2003) The Supply of Public Transport: A Manual of Advice. University of Oxford, Transport Studies Unit
20. Tegnér G (2005) PRT costs compared to bus, LRT and heavy rail –some recent findings. AATS European Conference, Bologna 7–8 Nov 2005"Advanced automated transit systems designed to out-perform the car"
21. Robèrt M (2007) Mobility management and climate change policies. Ph.D., Royal Institute of Technology

Financing Urban Transport Through Value Capture

Lucía Mejía Dorantes and José Manuel Vassallo Magro

Introduction

The European Union in the Green Book on Urban Transport notices that in order to foster urban and suburban sustainable mobility among citizens, they should be provided with transport services with quality that is efficient and reliable. However, to provide good services, it is essential to develop and adequately maintain and operate the necessary infrastructure facilities. The report intends to establish different strategies for sustainable urban transport, giving special emphasis in the need to rely on important economical resources to invest in transport infrastructure. Those strategies take shape in a group of measures related to budgets, regulations and different financial instruments such as specific local taxes, in order to accomplish the goals.

On the other hand, urban economics states that if a property has the opportunity to access to better benefits than the others, the demand for it will be higher, increasing its value.

A new station generates a benefit around it, which people living or working close to it take special advantage from. This benefit is noticed in the real estate market increase, especially in the areas close to the stations. The above is because people prefer to live close to a public transport station that will facilitate their commuting trips and lower their transportation costs. Likewise, retail stores and offices are benefited in its location by the number of people that uses this means of transport and by the facility that means of transport implies to the employees to commute.

L.M. Dorantes (✉)
TRANSyT – Universidad Politécnica de Madrid, 28040 Madrid, Spain
e-mail: imejia@caminos.upm.es

J.M.V. Magro
Universidad Politécnica de Madrid, Spain

S. Rauch et al. (eds.), *Highway and Urban Environment*,
Alliance for Global Sustainability Bookseries 17,
DOI 10.1007/978-90-481-3043-6_2, © Springer Science+Business Media B.V. 2010

It is convenient to comment that in some cases negative externalities are produced around the buildings close to the transport lines. It is important to differentiate between the externalities produced by being close to a station and the ones produced by being close to the transport line, especially when the line is at ground level. The negative externalities are produced principally by noise, vibrations and landscape affection, but nowadays due to new materials and new technology, these negative effects can be minimized and are topics taken into account when designing the infrastructure.

If the benefits produced by the infrastructure exist and can be accounted, it is necessary to question to whom theses benefits belong to. It is true that a new transport line provide a general benefit to the neighborhood but especially to the closest buildings. This value capture could be used to fund part of the transport infrastructure construction, operation or maintenance in order to diminish the contribution of public resources.

In the following part, some mechanisms used in different places are presented, that contribute to transport infrastructure funding.

Case Studies

There are many examples of value-capture due to a transport infrastructure. We discuss some of them in the following sections.

Los Angeles, California

To repay construction costs of the Los Angeles Metro Red Line, private businesses located half or one third mile close to the station, contribute by an assessment rate. This rate takes into account the benefited area, plus the last 3 years' delinquency rates. Furthermore, the price is fixed, therefore properties, independently of the type of business, pay the same for each square meter. It was a very long process to establish this mechanism, faced a lot of resistance but nowadays this rate is equal to the 9% total investment cost of this line [1].

France

This country has established a public transport funding system known as "*Versement Transport (VT)*". It is a specific tax paid by public or private companies with more than nine workers, located within a 10,000 habitant's urban transport perimeter. This tax is used to fund either operational costs or new transport infrastructure. The Urban Transport Authorities (AOTU) are in charge of establishing this tax as well as the rate, with a maximum contemplated by the law [2].

The VT represents around the 33% from all the urban transport's operational and investment costs without taking into account the Paris region. In this region, the companies also contribute with the 50% of the monthly ticket of their employees, therefore, the companies' final contribution increases to the 43% [3]. Hence, this tax is the main permanent funding source that allows vehicle modernization, to have new infrastructures been built and other operational costs.

Munich, Germany

The mechanism used is based on the interest of the urban developer. If he is interested in obtaining the license to construct a neighborhood, first he has to obtain the land value from a real estate experts' panel. After the land has the license and it has started to being constructed, the land value will increase. The difference between those two values is a measure that local authorities may claim, even till two third parts of this difference, as a source to construct public infrastructure. The rest of the value corresponds to the land developer [4].

Hong Kong

The train system in Hong Kong is known for its high quality and by its self-financing mechanisms without using public resources. To achieve so, they have implemented different mechanisms such as the "Railway-Property Development Model". This model gives the rail companies the right to build urban developments around the stations or even underneath or over the stations [5]. This mechanism allows maximizing the urban use around stations. Therefore, people could access very easily to the transport system and much more properties are benefited by the transport service.

The above implies that urban developments work as financial organisms of transport projects.

This kind of financial schemes have been used to develop 6 million square meters of housing and 5 million square meters of office and retail areas [5]. That explains in part their self-financing transport system.

Spanish Situation

Taxes and Tributes

In the next paragraphs there are briefly explained the most important property taxes as well as the Special Contribution tribute case.

Property Tax (IBI)

This tax is paid once a year; the value is established and paid to the city council as a percentage of the properties official value.

Urban Land Value Increase Tax

It has to be paid by whoever sells urban land, house, apartment, parking space, etc. for the land value increase it has got while being proprietary. Nonetheless, this tax does not take into account the reason of that value increase and its objective is not exactly to contribute to public transport funding. The taxes above, like all taxes in Spain, are not earmarked. On the other hand, there are as well other mechanisms to collect funds, such as fees and special contribution that may have self-financing purposes.

Special Contributions

The Spanish law contemplates this tribute as "those tributes where the taxable event consists in obtaining by the party who executes the event, a benefit or a value increase in his properties, as a consequence either of public works, or establishment or extension of public services" [6]. The Highways Law contemplates this financial mechanism as well, either for physical persons or with corporate identity that become especially benefited by the property value increase due to the public works carried out. The problem is how to justify this benefit, the benefit amount and the impact area. The above has made this tribute difficult to put into practice.

Spanish Experiences

Special contributions are covered in the legislation; however, most of the times are not used because it might take a long legal process. Nevertheless, there is an initiative to start using them. For example; Parla's light rail, where the city council invested 87 million euros to finance it. They look forward to recovering part of the investment by special contributions from private housing developers, around 10% of the invested [7].

Other simpler schemes have been put into practice, to fund transport infrastructures, most of the times through negotiations with developers. This is the case, long time ago, of the "*Cuatro Caminos*" metro station in Madrid. The transport company built a new neighborhood in order to collect money to construct the infrastructure. The same happened years latter with line 1 expansion in the "*Ensanche de Vallecas*" zone where 37% of tunnel total costs were paid by urban developers using the increase in property values generated to those properties close to the stations, approximately, 25 million euros.

Another interesting example is the metro line known as *Metronorte*, where the company "Telefonica" has funded part of the line to get to their new offices, in "Las Tablas" neighborhood, called "Ronda de la Comunicación". Through an agreement with the state government, the company contributed with 9 million euros to build the new station plus 5 million to construct the metro tunnel that crosses the building, around 260 m long.

Proposals for Spain

Developer's Fee

The simplest way to achieve value-capture, it saves time and it is based in looking for Transport Oriented Development (TOD) projects since the very beginning. It is recommended to legislate in this matter and specially to apply the existent mechanisms, like special contributions. If the developers are demanded to build part of the transport infrastructures, that will definitely improve the quality of the area, therefore increasing the benefits of the buildings, this implies a higher sale's price. It would have a high acceptability because society would not perceive this charge, it could be said that this is a win-win strategy.

It is a measure that has already been applied, either to developers or companies; however, until now it has depended on the good will of the benefited parts to contribute to the construction. As an example we have Telefonica Company in *Metronorte* facing the situation of "*La ciudad financiera*" of the Bank SCH, that did not have the willingness to contribute to the light rail construction so it could reach their new offices in order to benefit their 5,000 workers to encourage the use of public transport, in contrast, they built a parking lot for 3,500 places.

The main problem is that this scheme could be used to new housing developments but not in existent neighborhoods even though new transport infrastructure would be provided, for these cases additional mechanisms should be established.

A Modification to the Urban Land Increase Value Tax

This tax is established by the city council and they are the ones that manage it. The law says that the tax rate is based on the value increase between the value when the property was bought and the sale's price with a 30% maximum. Nonetheless, the collected taxes go to the general budget resources; hence it is impossible to differentiate which part belongs to transport value-capture.

It is proposed that an infrastructure fund is established by the city councils in order to capture the tax's proportional part due to transport infrastructure.

This is a very simple solution that would have neither acceptability, nor collecting problems. The main problem is that it has to be fulfilled that the city councils spend these funds only for public transit. It should be legislated in order to achieve this matter.

Office and Retail Fee

If retail and business centers are well communicated through different means of transport, the landlord as well as the tenant are benefited by the opportunities they have. Therefore, they should be asked to contribute to the infrastructure.

For that sake it is proposed to establish a fee where the companies or stores settled within a perimeter of 1,000 m from the transport station contribute to the transport infrastructure's operational costs or exchanging it by their employees' monthly transport ticket (100% or 50% of it). It would have a very good social acceptability because the fee would not be paid directly by the society and the companies located close to stations would receive a better amount of clients, they would contribute to improve the environmental quality of the cities, they would also diminish their employees' transportation costs, promoting the use of sustainable transport and improving their corporate image. Likewise, the benefits captured by this measure would be transferred directly to the transport authority and thereby they would be directly reinvested only in transport infrastructure.

In Madrid some private and public companies already offer this incentive, for example the *Transport Administration in Madrid*, most of the public administrations in Madrid city and in the region of Madrid as well, that provide to their employees the annual transport ticket. Some national ministries and public organisms like the *Universidad Complutense* give this incentive.

Finally, very few private companies provide this incentive, around five or ten and most of them are small companies.

Private Transport Fee

A similar measure has been proposed in London, Sydney and Perth, but through taxes to parking lots in business centers. These types of initiatives are very questioned although they may raise an important amount of money as well as may diminish the use of private transport.

It is proposed that if people that live in an area benefited by public transport still prefer to use private transport (at least that their condition turns it to be indispensable), they should be penalized through a private vehicle fee registered in that property. The objective is to reduce the number of cars per family, improving each home's mobility.

It seems to be a very complicated measure to establish because it has to be defined a border between the zones that are benefited by public transport from the ones that are not. It seems to be a measure with low social acceptability. Nevertheless, it would diminish each family's number of vehicles as well as promoting either public transport or a better mobility culture. Finally, it would induce people's relocation by their means of transport. Those who prefer private transport would move towards areas where this fee has not to be paid, giving priority to live close to the stations to the people that prefer to use public transport.

Conclusions

Nowadays, it can be said that value capture has not been taken advantage of to fund transport infrastructure. In order to achieve the goals set by the EU in the Urban Transport Green Paper, it is necessary to propose new financial mechanisms to reinvest in public transport projects in such a way that justice, feasibility and social acceptability are accomplished.

On the other hand, it is necessary to adapt the current fiscal system to capture the benefits produced by the transport infrastructures to use them to fund and maintain these infrastructures. Likewise, an effort should be made to apply the existent measures, especially those related to special contributions in Spain.

The presented proposals intend to promote the use of public transport as well as to penalize the use of private transport in areas close to train or metro stations, in order to promote that people that work or live close to the stations change to public transport.

It is important to notice the importance of urban densification especially in the closest zones to stations to attract as much people as possible to fulfill the optimum operational standards, where value-capture measures would be best exploited.

References

1. Rolon A (2008) An evaluation of value capture mechanisms from linkage capture to special assessment districts. Transportation Research Board, Annual Meeting. January 13–17, 2008. Cambridge Systematics, San Francisco, CA
2. Pascal V (2003) Les transports publics urbains en France. CERTU. Direction des transports terrestres, France
3. Varnaison Revolle P (2008) Le financement des transports publics urbains en France. Séminaire Régional sur les Déplacements Urbains en Méditerranée. Skhirat, Maroc. 22–23 Janvier 2008
4. Hass-Klau C (2006) Capture of land value premiums as a source of funding for public transport: evidence and practice in selected European metropolitan areas. EMTA Association. Final report March 2006
5. Jain A (2007). Railways-A catalyst for urban development in Hong Kong. Public Transport International No.5, pp. 17–19
6. Ballesteros A, Castelao J, Santos C (2003) Diccionario Enciclopédico El consultor. Diccionario Jurídico de Administración Local. Tomo I. 2ª edición. El Consultor de los Ayuntamientos y los Juzgados. Madrid, Spain
7. Delgado L (2007) Learning from Parla Tramway Experience. Transurban project workshop. Cambridge, April 16–17, 2007

Environmental Costs Account: A Base for Measuring Sustainability in Transport Plans

Andres Monzón, Alvaro Fernandez, and Pablo Jorda

Assessment of Environmental Costs in Urban Areas

Achieving sustainable mobility in urban areas means to reconcile environmental protection with social well-being and economic development. These targets include reducing the negative impacts from transport, including pollution, noise and congestion, while ensuring affordable, accessible and safe mobility, without affecting the living conditions of future generations [1, 2].

To measure how sustainable mobility is achieved we need to settle a number of indicators [3]. However, the assessment methods of transport policies more often use time and cost indicators rather than environmental ones [4].

The approach adopted to analyze transport sustainability is to develop the socio-environmental account of transport means for each specific city or region [5]. This integrated approach considers all transport externalities in a comparative way and is a good tool for planning taken into account social fairness [6–8]. We have used these concepts making the travelers transport account of Madrid. In this account, we have worked with social costs like a way for analyzing the sustainable transport.

This kind of assessment has been improved over time (decades) by including more factors [9, 10]. Thus, in Europe we can find several transport studies dealing with counts of general social cost or externalities at a national, regional or urban level [11, 12].

In this paper, we show the importance of urban and metropolitan environmental cost among various diverse transports modes. We will analyze the results of the Madrid case study and we will point to guidelines for a more sustainable development in cities.

A. Monzón (✉), A. Fernandez, and P. Jorda
TRANSyT-Transport Research Centre, Universidad Politecnica de Madrid, 28040 Madrid, Spain
e-mail: amonzon@caminos.upm.es

S. Rauch et al. (eds.), *Highway and Urban Environment*,
Alliance for Global Sustainability Bookseries 17,
DOI 10.1007/978-90-481-3043-6_3, © Springer Science+Business Media B.V. 2010

Social Transport Account: How to Measure

Environmental costs and externalities allocation are necessary to calculate the optimum social costs and sound basis to reduce global car mobility in cities [13].

Sustainable mobility is linked to the model for urban development in each city [14–17]. Suburban sprawl causes trips that are cheaper for user but generate more global mobility and less social welfare. The internalization of externalities can correct this process [18].

To deal with this kind of effects we need to know the social cost of all kind of trips, for each city zone and transport mode [19]. In the transport account presented here, we have used disaggregated data to provide detailed results for different geographical zones which influence the costs of trips: urban and metropolitan.

Transport social cost is the cost paid for an individual trip by the whole society. These costs have two components, one is internal and the other external. The cost structure is different for every transport mode, so we can use that cost account for comparing the performance of trips made with different modes. In this study we have taken into account the following modes: urban bus, suburban bus, metro, suburban train, car, taxi, and walking (in Madrid bikes are not representative nowadays).

The kind of costs we took into account include: infrastructure, operating cost, travel time, accidents, noise, air pollution, global warming and land taken. We have given special attention to externalities within social cost, although we have not taken into account some externalities such as downstream processes because they are still not significant [12] (Table 1).

We have made use of different cost methodologies, developed for previous studies, for each cost category [11, 12, 19, 20]. Then we had to adapt them to the specific case and values of the Madrid Metropolitan Area. Several previous studies have already been carried out in Spain on transport externalities costs in urban areas, both in Barcelona [21] and Madrid [22]. This study is based in the latter, but its methodology has been improved and applied to the 2004 Madrid Region mobility survey [23].

Table 1 Costs appraised in the study

Cost	Cost elements
Infrastructures	Capital costs of investments and replacement of assets, running costs for maintenance
Operating costs	Vehicle related costs, cost of personnel and administrative costs
Travel time	Time spent in traveling: access, waiting, trip and dispersion
Accidents	Risk value, human capital losses, medical care, administrative costs, damage to property
Noise	Human health impact, damage to property
Air pollution	Human health impact, environmental damage, damage to building structures
Global warming	Climate change, impact of energy use, impact of environment
Land taken	Loss of habitat, landscape charge

Madrid Metropolitan Area Case Study: Structure and Mobility Patterns

Madrid is the capital city of Spain. It is a city of 3.1 million inhabitants, surrounded by its Metropolitan Ring with a population of 2.3 million. Demographic density and job distribution varies considerably (see Table 2 and Fig. 1), where the core part (CBD) plays a big role. These differences have substantial effects on mobility patterns.

For the purpose of this study we have considered two different types of trips:

- **Urban trips**: origin and destination in Madrid City (either CBD or periphery)
- **Metropolitan trips**: origin and/or destination in the Metropolitan Ring

The use of different transport modes is rather different between Madrid City and its Metropolitan Ring (Table 3). Public transport (PT) is the most important transport mode in the denser central city districts whereas the car is dominant in the suburbs. The number of walking trips is on the rise in the city centre and is falling in favor of car use in the Metropolitan Ring. Very few trips are made by bicycle, just 0.1% in the urban area. Madrid City has a good mobility performance rating but car trips are increasing at a faster rate than trips by public transport or on foot. Car trips predominate in the Metropolitan Ring and are growing very rapidly, which is a clear threat for the future [24]. According to the 2004 mobility survey [23] PT

Table 2 Social Indicators. Madrid Metropolitan Area, 2004

		Inhabitants	Surface area (km^2)	Density (inh/ha)	Jobs	Jobs/1,000 inh
Madrid municipality	CBD	1,018,732	42	243	959,877	942
	Periphery	2,116,513	564	38	815,648	385
Metropolitan Ring		2,328,494	2,086	11	893,982	384
Total		5,463,739	2,692	20	2669,507	489

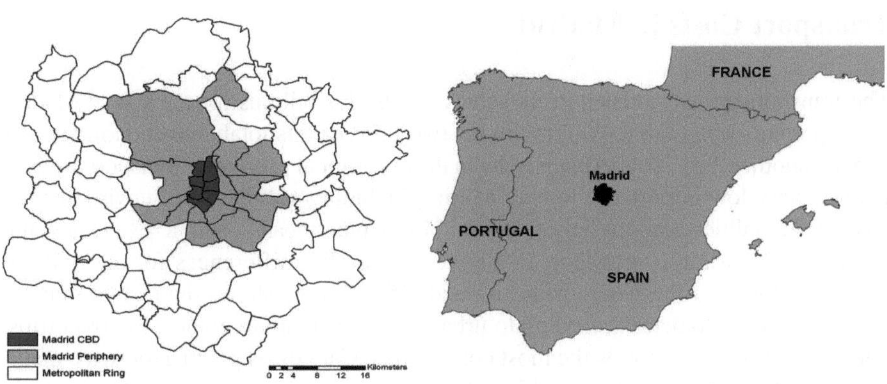

Fig. 1 Madrid Metropolitan Area zones

Table 3 Modal split in Madrid Metropolitan Area, 2004

Mode	Urban trips	Percent	Metropolitan trips	Percent
Urban bus	1,186,588	15.5	7,535	0.1
Metro	1,525,463	19.9	224,268	3.7
Suburban bus	33,835	0.4	735,463	12.1
Suburban train	173,052	2.3	513,909	8.5
Public transport	2,918,938	38.2	1,481,175	24.4
Car	1,716,583	22.4	2,700,828	44.5
Taxi	508,319	6.6	24,804	0.4
Motorbike	52,818	0.7	30,469	0.5
Walking	2,449,839	32.0	1,835,742	30.2
Total	7,646,497	100.0	6,073,018	100.0
Trips/inhabitants	2.44		2.61	
Trips/inhabitants (1996)	2.13		2.04	

accounts for 38% of trips in Madrid City and only 24% in the Metropolitan Ring; car patronage was 29%, and 45% respectively; while 32% are walking trips in Madrid City and 30% in the Metropolitan Ring. Comparing data from the two last mobility surveys 1996–2004, the first observation to make is that the mobility rate has been on the rise a lot. In Madrid City, it has increased from 2.13 to 2.44 (15%) trips per person over the 1996–2004 period [23, 25]. In the Metropolitan Ring the rate has grown even more: from 2.04 to 2.61 (28%) trips per person.

This is mainly due to the increase of car trips in the Metropolitan Ring by 75%. This change in mobility patterns has a big environmental impact and it is causing a vicious circle which will produce more car dependency and again more environmental costs.

Transport Costs in Madrid

The transport costs evaluated in this study are €21.7 billions for the whole Madrid Metropolitan Area (Table 4). Trip time is two third of this total, and economic costs explain another 28%. This explains why in the assessment process of transport projects is given very low importance to externalities, including accidents and environmental costs. Externalities account only €883 million, which means about a 4%. Analyzing trip costs according to transport mode we can see big differences for all kinds of costs and for the Area where trips are made [5]. Table 5 shows the results in costs per passenger–kilometer according to urban and metropolitan modes. In urban trips, collective public transport is the most competitive. Car cost is practically twice metro cost, and the metro is the most efficient transport mode in urban trips. However, in metropolitan trips, costs for different transport modes are more similar.

Table 4 Transport costs distribution (million $€_{2004}$)

	Economic costs			Externalities					
	Infra-structures	Operating costs	Travel time	Accidents	Noise	Air pollution	Global warming	Land taken	Total
Urban trips	229	3,431	7,325	124	56	137	33	1	11,335
Metropolitan trips	154	2,358	7,322	239	53	118	82	40	10,366
Total	382	5,789	14,647	363	109	255	115	41	21,701

Table 5 Costs by transport mode, 2004 ($€_{2004}$/100 passenger–kilometer)

	Urban trips				Suburban trips		
	Urban bus	Metro	Car	Taxi	Suburban bus	Suburban train	Car
Infrastructures	0.35	2.71	1.90	2.10	0.07	1.69	0.37
Operating costs	16.31	9.14	52.68	37.55	5.78	5.87	12.15
Travel time	56.71	44.26	42.79	48.60	23.29	23.68	23.18
Economic costs	73.37	56.11	97.37	88.25	29.14	31.24	35.70
Accidents	0.67	0.08	1.07	1.24	0.14	–	1.06
Noise	0.19	–	0.82	0.90	0.08	–	0.31
Air pollution	0.76	0.03	1.94	2.12	0.30	0.03	0.67
Global warming	0.09	0.06	0.52	0.56	0.05	0.06	0.52
Land taken	–	–	0.01	0.01	0.04	0.05	0.24
Externalities	1.71	0.17	4.36	4.83	0.61	0.15	2.80
Total	75.08	56.28	101.73	93.08	29.74	31.39	38.50

In other words, car is more competitive from the social point of view. Suburban train and bus trips are less costly from a social point of view, but differences are not big enough to influence a move in trips from car use.

We have to bear in mind also the average trips distances in each Area. Urban trips made by public transport have an average distance of 7.1 km while car trips are 7.9 km. This means that unitary costs in public transport are lower than in car, and also trips distances are more competitive. In the Metropolitan Area we find the opposite situation: average public transport trip is 21.4 km long, while in car it is 16.3 km.

The conclusion is that in urban areas it is possible to attract more trips towards public transport modes, but in the suburbs this is quite difficult. Public transport is a good competitor against cars in denser central areas both from the travelers and from the social point of view. On the contrary, it has clear disadvantages when it is operating in urban sprawl zones. In these Areas it is less competitive: longer trips distances and higher social and economic costs [15].

Externalities of Transport Modes in Madrid

If we focus our analysis only on the accidents and the environmental part of the total costs of transport, we see a very different picture both for urban and for metropolitan trips [26].

In both areas, rail modes (metro and suburban rail) present the lowest rate of externalities. Then buses came second, and the highest externalities correspond to cars.

In urban trips metro account only €1.7 per 1,000 passenger–kilometer (Table 6). Urban buses externalities are about €17.2 that is 10 times higher. Car and taxi reach more than €43.7, almost 25 times more. Among externalities, air pollution is the most important for urban trips. It accounts for about half of the car externalities and a high share of the buses.

In the case of metropolitan trips (Table 7), suburban rail accounts only €1.5 per 1,000 passenger– kilometer, the same cost as metro in urban trips. The other modes have lower unit costs than in the urban part, due to less congestion in the suburbs. However, trip distances are more than doubled: 18.1 km in the suburbs against 7.4 km for Madrid City trips. Comparing costs among suburban trips in the different modes, rail based trips have again the best performance in externalities. Bus externalities are four times higher and those related to car trip externalities are 19 times. The share of externalities corresponding to air pollution is still the most important, but lower than in urban Areas. On the contrary, accidents are more relevant, with a cost of €10 per 1,000 passenger–kilometer made by car.

Table 6 Urban trips. Externalities costs ($€_{2004}$/1,000 passengers–kilometer)

	Urban bus	Metro	Car	Taxi
Accidents	6.75	0.76	10.75	12.35
Noise	1.88	–	8.22	8.97
Air pollution	7.64	0.33	19.41	21.19
Global warming	0.87	0.62	5.16	5.63
Land taken	0.01	–	0.11	0.11
Total	17.15	1.71	43.64	48.26

Table 7 Metropolitan trips. Externalities costs ($€_{2004}$/1,000 passengers–kilometer)

	Suburban bus	Suburban train	Car
Accidents	1.39	–	10.64
Noise	0.79	0.02	3.07
Air pollution	2.96	0.32	6.69
Global warming	0.54	0.58	5.16
Land taken	0.37	0.54	2.44
Total	6.05	1.46	28.00

Recommendations for More Sustainable Transport Policy in Madrid

The results of this study have two main findings. The methodology developed to assess social and environmental costs of transport modes has proved to be useful. Firstly, it enables us to compare mobility impacts on different type of city developments: low and high density Areas, good and poor Public Transport supply, etc [8]. Secondly, its application to Madrid Metropolitan Area point out a number of policy guidelines to achieve more sustainable mobility patterns in each part of the city.

The urban development of Madrid City has a clear benefit both from the economic and environmental point of views. Denser developments are more PT oriented than those in the Metropolitan sprawl. Costs for the user, for society, and for the environment clearly fosters a higher concentration of activities in the city. The length of trips in dense city centers is about 2.5 times than those in the Metropolitan Ring. City centre trips could be made on foot, cycling or by PT, while trips in the Metropolitan Ring are more convenient for car.

Therefore we can conclude two urban policy guidelines. The revitalization of city centers brings economic and environmental benefits. They improve the life of the citizens, because they are less dependent on car trips. On the contrary low density developments bring more costs in the long term, resulting in less sustainable forms of city development [16].

Finally, big cities profit from the good supply of rail modes: metro in the central part and suburban rail in the suburbs. Although they require big investments, there are clear benefits in the long run. They produce much less operating and environmental costs because their capacity is higher and their technology is less harmful.

In summary, decision-makers in cities can use social and environmental transport cost accounts to help them with decisions in providing urban transportation. The socio-economic and environmental points of view provide complementary views on their duties. Technicians who help them also benefit from transport accounts. They provide a clear valuation of all transport costs in a comparative way [5, 13].

References

1. Åkerman J, Höjer M (2006) How much transport can the climate stand? Sweden on a sustainable path in 2050. Energy Policy, 34:1944–1957
2. May AD (2003) Developing Sustainable Urban Land Use and Transport Strategies: a Decision Makers' guidebook, PROSPECTS. Deliverable n°5. European Commission, 5th Framework
3. Black JA, Paez A, Suthanaya PA (2002) Sustainable urban transportation: performance indicators and some analytical approaches. Journal of Urban Planning and Development, 128:184–209
4. Monzón A (2006) Externalidades del transporte y ferrocarriles urbanos. Ingeniería y territorio, 76:32–39
5. Banister D (2008) The sustainable mobility paradigm. Transport Policy, 15:73–80
6. Gudmundsson H, Höjer M (1996) Sustainable development principles and their implications for transport. Ecological Economics, 19:269–282

7. Litman T, Burwell D (2006) Issues in sustainable transportation. Global Environmental Issues 6:331–347
8. Litman T (1999) Transportation Cost Analysis for Sustainability: Victoria Transport Policy Institute, Canada www.vtpi.org 1–17
9. Boiteux M (2001) Transports: choix des investissements et côut des nuisances:325. Commissariat General du Plan, Paris
10. Button K, Quinet E, Kageson P, Bleijenberg A, Rothengatter W, Banister D, Vreckem D, Bonnafous A (1994) Internalizing the social cost of transport: 191. European Conference of Ministers of Transport; Organization for Economic Co-operation Development, Paris
11. Nash C (2003) UNIfication of accounts and marginal costs for Transport Efficiency. Final Report: 60. Institute for Transport Studies, University of Leeds. Fifth Framework Programme. European Commission, Leeds
12. Schereyer C, Schneider C, Maibach M, Rothengatter W, Doll C, Schmedding D (2004) External cost of transport. Update study: 169. INFRAS. IWW Universitaet Karlsruhe, Zurich
13. De Borger B, Wouters S (1998) Transport externalities and optimal pricing and supply decisions in urban transportation: a simulation analysis for Belgium. Regional Science and Urban Economics, 28:163–197
14. Bertolini L, le Clercq F, Kapoen L (2005) Sustainable accessibility: a conceptual framework to integrate transport and land use plan-making. Two test-applications in the Netherlands and a reflection on the way forward. Transport Policy, 12:207–220
15. Chen H, Jia B, Lau SSY (2008) Sustainable urban form for Chinese compact cities: challenges of a rapid urbanized economy. Habitat International, 32:28–40
16. Cram J (2006) Urban form and sustainable transport. Journal of Transport Geography 14:226
17. Curtis C (2008) Planning for sustainable accessibility: the implementation challenge. Transport Policy, 15:104–112
18. Tabuchi T (1998) Urban agglomeration and dispersion: a synthesis of Alonso and Krugman. Journal of Urban Economics, 44:333–351
19. Maibach M, Schereyer C, Sutter D, Essen HP, Boon B H, Smokers R, Schroten A, Doll C, Pawlowska B, Bak M (2007) Handbook on estimation of external cost in the transport sector: 332. CE Delft, Delft
20. CERTU (2005) Etablir un compte déplacements à échelle locale: 128. CERTU, Paris
21. Robusté R, Margín J, Alsius F, Lorente A, Monzon A, Hormigo E and Subira E. (2000) Els Comptes del Transport de Viatgers a la Regió Metropolitana de Barcelona. Any 1998: 219. Autoritat del Transport Metropolitá, Barcelona
22. Guerrero MJ, Monzón A (2003) Cuenta económica y socio-ambiental del transporte terrestre de viajeros en la Comunidad de Madrid en 1996: 184. Consorcio Regional del Transporte de Madrid, Madrid
23. CRTM (2006) Encuesta Domiciliaria de Movilidad en día laborable de 2004 en la Comunidad de Madrid. Resumen de resultados. Consorcio Regional de Transportes de Madrid, Madrid
24. Bamberg S (2003) Does habitual car use not lead to more resistance to change of travel mode? Transportation 30:97–108
25. CRTM (1998) Encuesta Domiciliaria de Movilidad en día laborable de 1996 en la Comunidad de Madrid. Análisis y síntesis de la movilidad. Consorcio Regional de Transportes de Madrid, Madrid
26. Himanen V, Lee-Gosselin M, Perrels A (2005) Sustainability and the interactions between external effects of transport. Journal of Transport Geography, 13:23–28

A GIS Raster Model for Assessing the Environmental Quality of Spain Focused on SEA and Infrastructure Planning Procedures (LATINO Model)

Luis G. García-Montero, Santiago Mancebo Quintana, Miguel A. Casermeiro, Isaabel Otero Pastor, and Abdrés Monzón de Cáceres

Introduction

Strategic Environmental Assessment (SEA) research has made little progress in developing analytical methodologies to resolve technical problems and to incorporate new findings into the landscape planning process. Therefore, the complex and interdisciplinary nature of environmental problems requires the development of a new class of GIS, integrating mathematical models, databases and expert knowledge based on conceptual models [1–3].

Screening is defined by the European Commission as 'the process by which a decision is taken on whether or not SEA is required for a particular programme or plan'. Screening models have been proposed [4, 5], whose results have been effectively integrated into the decision-making process of two Spanish infrastructure plans: PIT 2000–2007 and PEIT 2005–2012 [6, 7]. These screening models are based on a prior GIS raster model containing an assessment model of environmental quality for the whole of a country, which classifies the Spanish territory.

The objective of the present work is to show the possibilities of this GIS model for assessing the value of the natural environment of a territory, for use in extensive geographic areas, and focusing on SEA and planning of infrastructure plans. We have selected as the main criteria 'the preservation of territorial environment quality'. This aims to 'conserve and improve the quality of the environment… based on the precautionary principle' (Articles 6 and 174 of the EU Treaty), and proposes 'the integration of conservation and sustainable use of biodiversity into the various plans

L.G. García-Montero
Dpto. Ingeniería Forestal, Universidad Politécnica de Madrid A.M.de Cáceres

S.M. Quintana, I.O. Pastor and A.M. de Cáceres (✉)
Transport Research Centre (TRANSyT), Universidad Politécnica de Madrid, 28040 Madrid, Spain
e-mail: amonzon@caminos.upm.es

M.A. Casermeiro
Dpto. Edafología, UCM, Ciudad Universitaria s/n 28040 Madrid, España

S. Rauch et al. (eds.), *Highway and Urban Environment*,
Alliance for Global Sustainability Bookseries 17,
DOI 10.1007/978-90-481-3043-6_4, © Springer Science+Business Media B.V. 2010

and programmes' (Convention on Biological Diversity 1992, Rio Earth Summit). [8, 9] also indicate that environmental quality is becoming recognised as a critical factor that should constrain landscape planning. The environmental quality of the territory may be defined as the 'degree of excellence or merit warranting the conservation of the essence and structure of the environment or of any of its elements', as proposed by [10]. We have chosen this concept, as this author has a markedly practical focus.

However, the author proposes environmental quality assessment models focused on landscape planning projects applied to territories which range from a small extension (a municipal district) to an average extension (region). Nevertheless, SEA procedures require methodologies which enable the assessment of plans affecting large territories (on a national scale). For this reason we propose a model which evaluates the environmental quality of large extensions of land; we have called this model LATINO (LArge Territory Integrated eNvirOnmental model).

Methods

The proposed model is used to compare territorial units [10], and it is combined with a multi-criteria method which uses six forest structure evaluation criteria to assess the quality of forest landscapes [11]. We also examine other evaluation criteria based on previously existing environmental information on a national scale. We consider that with 12 variables it is possible to assess the environmental quality of the Spanish territory. However, this model is an open system that allows continuous incorporation of new quality scores.

The map projections, GIS and mathematical software, scale and level of detail used are described in [4]. As enormous amounts of information are analyzed, it is impossible to use vector-type GIS operations, and we thus chose to use a GIS raster-type model. We use GIS models based on logical operations of reclassifying and combining the raster layers, and numerical vector operations based on the vector module. [12, 13] propose that the vector module is a suitable method for integrating multiple indicators into a single index, for practical operating reasons in planning procedures.

Inventory of 12 Environmental Qualities on a National Scale in Spain

Based on the available digital environmental information on a national scale, we use five digital maps: Corine Land Cover 1990 [14]; Spanish Landscape map [15]; Soil map [16]; Habitats map [17]; and Spanish Forestry Vegetation map [18]. These five maps allow us to generate, either directly or by deduction, a set of 12 raster layers (pixel: 100 m): naturalness and singularity of the 50 units in the Corine Land Cover map; naturalness, singularity and fragmentation of the units in the Habitats map; quality and singularity of the units in the Landscape map; quality and singularity of the units in the Soil map; total vegetation and forest cover; and forest fragmentation.

A basic panel of 15 experts (from the Technical and Complutense Universities of Madrid) was consulted in order to obtain a set of four raster layers: naturalness of the units in the Corine Land Cover and Landscape maps; and quality of the units in the Soil map. A raster layer of naturalness of the units in the Habitats map was generated using the naturalness value for each polygon on the map, which were previously assigned by the experts who participated in the European Habitat Project.

We assess the territorial basic singularity of the different categories or classes in these maps, using an objective index of basic singularity Eq. 1 [10].

$$S = Ln\left(\left(1-\left(\frac{Max-x}{Max-Min}\right)\right)\times 100+1\right) \tag{1}$$

where S = territorial basic singularity index; Max = ha of the map's largest category; Min = ha of the map's smallest category; x = ha of the map category being evaluated.

This continuous scale is then transformed into a discrete scale of five classes (rounding each decimal value up to the next whole number). We thus obtain a higher singularity value for the least represented classes in the territory, in order to safeguard biodiversity.

We evaluate the degree of territorial basic fragmentation of the different categories in the Forestry and Habitats maps, using an objective classification calculated with the GIS [10]:

1. We calculate the surface area of each of the polygons on these maps.
2. We assign a scale of four discrete values which correspond to each of the four percentiles of the numeric distribution of frequencies of surface sizes: Fragmentation class 1 = those sizes of polygons corresponding to the first percentile (0–25%)… fragmentation class 4 = those sizes of polygons corresponding to the fourth percentile (75–100%). The greatest value of the fragmentation variable is assigned to the polygons in the fourth percentile.

Normalisation of the 12 Variables

The 12 variables are normalised (transformed into a common final continuous scale from 0 to 1) to avoid overlapping during their subsequent integration into the model. Formula 2 is used to convert the discrete scales.

$$Xn = \left(\frac{x-0.5}{Max}\right) \tag{2}$$

Equation 3 is used for the transformation from a continuous scale into another normalised continuous scale.

$$Xn = \left(\frac{x - Min}{Max - Min} \right) \quad (3)$$

Integration of the 12 Variables into the Model

The 12 normalised raster variables are integrated using GIS combine operations. Each pixel of 1 ha of territory is assigned a vector with the 12 natural variables valued. We obtain n vectors distributed among the 50 million 1-ha pixels in Spain.

The next step is to order the n vectors using the modulus, to assign a synthetic value of theoretical environmental quality. The vector modulus equation (Eq. 4) is used to order the n vectors obtained based on their components.

$$v = \sqrt{v_1^2 + ... + v_i^2 + ... + v_{12}^2} \qquad i = 1...12 \quad (4)$$

where v is the vector modulus; and ^{v}i is a vector component. Then the values obtained for each of the n Euclidean distances are normalised into m equivalent classes of theoretical territorial environmental quality required (from 1 to m), applying the formula (5).

$$\text{Theoretical natural quality class} = ((\nu - \nu_{MIN}) / (\nu_{MAX} - \nu_{MIN})) * (m + 0.5) \quad (5)$$

where v is the vector modulus of each of the n vectors obtained; v_{MIN} is the minimum vector modulus obtained; v_{MAX} is the maximum vector modulus obtained. In our example, we use five environmental quality classes (m = 5), therefore, the higher territorial environmental quality class is obtained when $4.5 < v < 5.5$. The uses of 5 to 100 classes are considered sufficient to assist the SEA and screening models of the two Spanish infrastructure plans: PIT and PEIT (to distinguish clearly extreme cases of environmental quality in territory).

Verification of the Model

The goodness of fit is checked against the real model of the five classes of environmental quality by integrating our environmental quality map with a map of protected natural areas in Spain (30% of Spanish territory). We apply a contingency table with the corresponding chi squared test, which enables us to analyse whether there is a statistically significant distribution of squares with greater territorial environmental quality inside the protected areas.

Results and Discussion

The common premise governing environmental assessment and landscape planning procedures should be "to devote the greatest possible effort to the most significant problems". Therefore, [10] recommends using evaluation criteria to identify critical environmental areas in order to limit the area involved in the landscape planning. We use the criteria of environmental quality to identify critical areas in Spain.

In our example, a total of 102,240 different vectors are obtained with 12 components, assigned to each of the 50 million 1-ha grid squares for Spain. The classification of the vectors (vector modulus) provides a raster map with 100 classes of territorial environmental quality (Fig. 1). The raster layers and the combining and reclassifying operations are able to deal with enormous amounts of information with no difficulty at all.

The distribution frequency for each of the quality classes is very suitable for the development of landscape planning procedures (Table 1). The model shows that the planning of land-use activities would be acceptable in 48.28% of Spanish territory,

Fig. 1 Assessment model of territorial natural quality in Spain (LATINO model – 100 natural quality classes). Metadata: European projection standard Lambert Equal Area and Datum ETRS89. One hundred metres pixel-raster maps. Model scale: 1:500,000. Colour codes: low natural quality classes 1 to 39 = light grey; natural quality classes 40 to 59 = grey; natural quality classes 60 to 79 = dark grey; greatest natural quality classes 80 to 100 = black

Table 1 Territorial natural quality in Spain. Application of a new GIS raster model (LATINO model)

Classes of territorial natural quality	Area (ha)	Area (%)
Quality 1 to 19	9,110,329	18.27
Quality 20 to 39	14,962,451	30.01
Quality 40 to 59	15,522,632	31.13
Quality 60 to 79	9,673,891	19.40
Quality 80 to 100	588,741	1.18
Total	49,858,044	100.00

Table 2 Distribution frequency of grid squares with different territorial natural quality (1 to 5 classes) inside and outside the protected natural spaces in the whole of the Spanish territory

Classes of territorial natural quality	Quality percentage outside protected natural areas	Quality percentage inside protected natural areas
Quality 1	87	13
Quality 2	85	15
Quality 3	70	30
Quality 4	51	49
Quality 5	33	67

as this would affect grid squares with low environmental quality values (classes 1 to 39). However land-use activities should be minimised in 20.58% of the territory, as this would affect grid squares with high values (60 to 100).

Finally, a check against the environmental quality map shows that these results have a good fit with the actual situation (in relation to the scale and level of detail used). Thus a second test showed that there is a high frequency of squares with high environmental quality in protected natural areas, and that this is significantly greater than in the rest of the Spanish territory ($p < 0.0001$) (Table 2).

In our example of an environmental assessment model focused on SEA and infrastructure planning procedures (LATINO model), we analyse the environment in the Spanish territory as a whole. However, this objective contrasts with the availability and actual capacity, for handling environmental information in a country with an area of some 506,000 km^2. Spain has a low population density (81 inhabitants/km^2 compared to the EU 15's average of 115 inhabitants/km^2) and many largely undisturbed natural areas. These factors make Spain an exceptional and unique place in Europe for natural resources. There are, for example, more endemic plant species in Spain than in the rest of Europe. Moreover, due to its geographic situation and the complexity of its climate, geology, geomorphology and history, the Spanish territory is an interrelated mosaic of rich and diverse landscapes and ecosystems, of great natural value and quality [19].

Spain thus represents a good starting point for developing this type of environmental assessment model – which would also be valid in other countries – and this model must take into account Spain's vast biodiversity and natural richness.

In summary, the GIS raster model developed (LATINO model) has proved to be a simple and effective tool which provides a useful environmental quality assessment for a large territory, based on panels of experts and on objective GIS calculations, and focusing on SEA and infrastructure planning procedures. This GIS model is an open system that allows continuous incorporation of new environmental data.

Acknowledgements We would like to thank all the personal of TRANSyT (UPM) and Soil Science Department (UCM) for their invaluable contribution to the assessment models of this study. We thank A.P. Esplugas, M. Navarra, A. García-Abril and J. Sierra for their teachings and assessment models, and M. Montero, L. García, P. García, M. García and C. Pascual for their support. We thank Prudence Brooke-Turner for her English translation. We thank the Spanish Ministry of Defence and DGCONA for their maps and their contribution to this study. The research dealt with in this article forms part of the Projects entitled "Evaluación económica y ambiental del Plan de Infraestructuras 2002–2007" (Ministerio de Fomento, granted in 2003) and "Evaluación de los efectos del Plan de Infraestructuras 2000–2007 sobre la movilidad, el territorio y la socioeconomía en el contexto de la U.E. ampliada" (CyCIT, granted in 2004).

References

1. Polichtchouk Y (1998) Geoinformation systems and regional environmental prediction. Safety Science 30:63–70
2. Noble BF (2004) A state-of-practice survey of policy, plan, and program assessment in Canadian provinces. Environmental Impact Assessment Review 24:351–361
3. Nilsson M, Bjorklund A, Finnveden G, Johansson J (2005) Testing a SEA methodology for the energy sector: a waste incineration tax proposal. Environmental Impact Assessment Review 25:1–32
4. García-Montero LG, Otero Pastor I, Mancebo Quintana S, Casermeiro MA (2008) An environmental screening tool for assessment of land use plans covering large geographic areas. Environmental Science and Policy 11:285–293
5. Monzón A, García-Montero LG, Otero Pastor I, Mancebo S, Casemeiro MA (2005) Metodologías de EAE: el caso de los planes de infraestructuras de transporte. Paper presented at the 3rd Congreso Nacional de Evaluación de Impacto Ambiental, Universidad de Navarra, Pamplona, 6–9 April 2005
6. Ministerio de Fomento (2002) Plan de Infraestructuras de Transporte PIT 2000–2007. Madrid: Ministerio de Fomento
7. Ministerio de Fomento (2004) PEIT Plan Estratégico de Infraestructuras y Transporte 2005–2020. Madrid: Ministerio de Fomento
8. Barker A, Fischer TB (2003) English regionalization and sustainability: towards the development of an integrated approach to strategic environmental assessment. European Planning Studies 11:697–716
9. Schumaker NH, Ernst T, White D, Baker J, Haggerty P (2004) Projecting wildlife responses to alternative future landscapes in Oregon's Willamette Basin. Ecological Applications 14:381–400
10. Ramos A (1979) Planificación física y ecología. Modelos y métodos. Madrid: EMESA
11. Norris WR, Farrar DR (2001) A method for the natural quality evaluation of Central Hardwood forests in the Upper Midwest, USA. Natural Areas Journal 21:313–323
12. Martínez-Falero JE, González S (1995) Quantitative Techniques in Landscape Planning. Boca Raton, FL: CRC Lewis Publishers
13. Tran LT, O'Neill RV, Smith ER (2006) A generalized distance measure for integrating multiple environmental assessment indicators. Landscape Ecology 21:469–476
14. EEA European Environmental Agency (2003) Corine Land Cover Map, 1990 version. Brussel: European Environmental Agency

15. MMA Ministerio de Medio Ambiente (2004) Mapa del paisaje de España. Madrid: Ministerio de Medio Ambiente
16. FAO (2000) Soil map. Roma: FAO
17. DGCN Dirección General Conservación Naturaleza (2004) Mapas del Proyecto Hábitat. Madrid: Ministerio de Medio Ambiente
18. DGCN Dirección General Conservación Naturaleza (2004) Mapa Forestal Español. Madrid: Ministerio de Medio Ambiente
19. MMA Ministerio de Medio Ambiente (2000) Guía para la elaboración de estudios del medio físico. Contenido y metodología. Madrid: Ministerio de Medio Ambiente

Sustainable Mobility in Metropolitan Areas

María Eugenia López-Lambas

This paper presents some experiences around Europe: Holland, Spain, France, UK, etc., and their legislative and policy framework. These good practices show how to combine the right level of provision of infrastructures to attend mobility demand, but taking care for quality of life and environmental standards in new urban and business developments. The paper points out which are the key elements of a mobility plan, its phases, measures and the importance of a well designed public participation procedure.

Introduction

Some facts [1]:

- Europe is one of the most urbanized continents. Approximately 75% of its population lives in urban areas; by 2020, this will be 80% (EEA).
- EU-27 greenhouse emissions would have fallen (1990–2005) by 14% instead of 7.9% (TERM). Short term projections indicate that 2010 greenhouse gas emissions will be the same as in 2005.
- Urban transport contributes to 40% of transport related CO_2 emissions and is responsible for 70% of emissions of other pollutants arising from road transport.
- The number of road traffic accidents in cities is growing each year: one in three fatal accidents happens in urban areas.
- According to the World Health Organisation, about 100,000 deaths a year could be linked to ambient air pollution in cities in Europe.
- NO_2 and PM_{10} concentrations at selected traffic monitoring stations indicate that cities are at or above European quality air limits.

M.E. López-Lambas (✉)
Universidad Politécnica Madrid (UPM), Research Transport Centre TRANSyT-UPM

S. Rauch et al. (eds.), *Highway and Urban Environment*, Alliance for Global Sustainability Bookseries 17, DOI 10.1007/978-90-481-3043-6_5, © Springer Science+Business Media B.V. 2010

With this panorama, it is not surprising to consider transport as the *villain of the movie*, guilty of almost every wrong things regarding environment and, all in all, quality of life. But the transport sector may be seen from different perspectives, each on them, obviously, depending on which side are we. So, from an economic point of view, transport contributes for more than 10% of the GDP of the European Union, and gives employment to more than ten million people in the same context. On the other side, as a social cohesion and land integration factor transport means other interesting and by no means insignificant key aspect for economic competitiveness of the different countries, fostering a number of associated activities, such as auto-propulsion industry, assets, etc.

This said, transport has a dark but not hidden side, linked to the sustainable development of our societies, focused on one of the three pillars of what we have called "sustainability": the environment, in such a way that being the results positive in the other two – social and economic –, not the same can be said regarding this third one. In other words: transport, and most of all, urban transport, is one of the main responsible of pollution and climate change, not to mention the high energy consumption, noise, dead and injuries, illness, congestion [2], etc. Only road traffic congestion amounts to 0.50% of EU GDP, and *business as usual*, will increase to 1% for 2010 [2]. In fact, in Spain road traffic congestion meant 0.2% of the GDP [3].

Since the aim of this paper is not to provide detailed information on issues that are well known – emissions, climate change, etc., – we will focus on how some cities have faced this problem through integrated strategies which take into account the interrelations among land use patterns, transport supply and the role of the different modes of transport, individual and collective, benefiting from the co-modality promoted by the European Union in one of its most recent documents, the Green Paper: Towards a new culture for urban mobility that, obviously, includes walking and cycling as well [4].

That integrated approach, searching the balanced coordination of land use and mobility, is well expressed through the Local Transport Plans (LTP, UK), Plans de Déplacements Urbains (PDU, France), Piano de Mobilitá Urbana (PUM, Italy), Planes de Movilidad Urbana Sostenible (PMUS, España), and the generic umbrella for them all: Sustainable Urban Transport Plans (SUTP) [5]. This apparently tongue twister (PDU, LTP, PUM, PMUS, etc.), has a common set of features through which many cities have proven that, simple solutions are the most effective:

- Covers the whole urban area
- Aims to reduce the negative impacts of transport and tackle the rising volumes of traffic and congestion
- Should be linked to national/regional strategies
- Cover all modes of transport
- Include people and goods
- Modal shift in favor of "soft" and alternatives modes

Furthermore, those strategies involve the coordination and efficient use of the different modes of transport, to enhance the shift to more energy efficient transport modes, to foster the reduction of transport demand and the implementation of technological measures.

From PDU to PUM and PMUS, Going Through LTP: The Tongue Twister of Urban Mobility

France. French PDU's are in the origin of the urban mobility planning with sustainability criteria. They are based on three laws, the first one from 1982 (Loi d'Orientation des Transports Intérieurs (LOTI)-Inland Transport Law); the second, from 1996, was environmentally oriented: the Loi sur l'Air et l'Utilisation Rationnelle de l'Energie (LAURE) and make PDU compulsory for cities with more than 100,000 inhabitants. The last one, the Loi relative à la Solidarité et au Renouvellement Urbain (SRU), from 2000, was issued by the Transport Ministry as the first one, and in the framework of the territorial coherence scheme, obliges to set road safety objectives and to be compatible with the land use planning [6].

PDUS – time horizon, 10 years – are elaborated, passed and implemented by the Public Transport Authority, after a public survey process. Local Authorities inside the performance perimeter of the Transport Authority are responsible for the funding (except for Île de France (Paris). Nowadays, the Regions and Departments fund the plan.

As main results, PDU's have put the urban transport plans on the political agenda and that has been translated into the implementation of a lot of tramways, the construction of exclusive bus lanes, a decisive parking policy), the deep involvement of the stakeholders and the increasing in the number of travel plans, among others important outcomes [7]. Nevertheless, is always difficult to assert to what extent the good effects on urban mobility are only due to PDU's, since population ageing or price of fuel, for instance, can also influence [8, 9].

More recently, in France have been also launched the so-called "micro PDU", made of specific proposals around the local PDU, such as areas 30, pedestrianisation, parking schemes, bike lanes, etc. Sometimes those micro plans realized in green commuter plans or company transport plans, usually through soft measures such as public transport, on foot and bicycle.

United Kingdom. LTPs are landmark documents that require authorities to plan a 5 year programme for managing transport services within a comprehensive and multi modal strategy to achieve local and national objectives [8], that is, must be consistent with the national plans and goals, both in transport and other issues. After the 5 years period, the plan is evaluated by the Local Transport Authorities [10].

The first round of LTPs were launched for the period 2000/2005, following the Transport Act issued in year 2000, that gave to the local authorities the competencies to implement those plans and, in order to help the authorities to do so, the Government launched a guide that sets the following objectives: to protect the environment, to improve passengers safety, to promote the sustainable growth of the economy, to provide global accessibility, especially for those without a car and to integrate land use and transport policies [11]. The objectives must be coherent with the national goals. Time horizon: 5 years. One of the key aspects of the LTPs is the existence of a set of indicators to evaluate the outcomes against the objectives. An annual progress report is required as a basis for the national funding that could be increased if the objectives are reached.

Funding: National Government, if the annual report is positive. Then, the Local Authority can be eligible for 75% of the investment, plus 25% depending on the quality evaluation and level of achievement. A new guide for the second round of plans (2006/2011) has been launched.

Italy. Italian PUM's [12] starts in 1996, with a Law from the Public Works Ministry on Urban Traffic Plans for cities with more than 30,000 inhabitants. The law 340/2000 and the National Transport Plan defined the methodology for its implementation. Stricto sensu only the traffic plans are compulsory (i.e., appropriate disciplinary measures are taken), since the mobility plans are responsibility of those cities and regions with more than 100,000 inhabitants which can obtain funds from the National Government (up to 50% of the investment costs of the measures planed in the PUM).

There are a set of national guidelines for the implementation, even if the region produce its own. These guidelines define a set of indicators to measure the effectiveness of the programme. Anyway, the goal is to integrate the PUM at sectoral levels, that is, with the PUT and with the urban and environmental planning. Time horizon: 10 years, with a bi-annual review.

Funding: National Government 60% of the investment (as a maximum), and the rest between municipalities and regions.

Objectives: to satisfy the citizen's mobility needs, reduce pollution, noise, energy consumption and private car use, to increase the road safety, to foster car-pooling and car-sharing as mechanisms to reduce congestion in urban areas.

Spain. The Sustainable Urban Mobility Plans (SUMP) [13] were launched in the framework of a strategic and well defined plan, made up of the Master Plan for Infrastructures and Transport and the Energy Saving and Efficiency Strategy. Apart from that, and with the remarkable exception of the Mobility Law issued by the Catalan Government in 2003, there is not any compulsory rule to implement them.

But in 2006 a guide for the elaboration and implementation of Sustainable Urban Mobility Plans (SUMP) was launched. The guide explains in a clear and accessible way the main characteristics that a plan must have; measures, implementation methodologies, stakeholders, public participation process, good practices, etc. There is national funding foreseen to foster its implementation as well.

The guide recommends to those municipalities with more than 50,000 inhabitants the adoption of a SUMP, within a time horizon depending on the kind of measures to be implemented, 2 to 8 years.

Regarding the objectives, the guide does not provide a list beyond those that the word "sustainable" suggests: it will depend on each case, since the needs of each city differs. It recommends keeping the plan within a regional strategy, coordinating both municipal and regional levels.

Very schematically, those are some measures to be implemented by means of a SUTP, grouped by intervention areas [13]:

Traffic management and control (TMC)
Parking management (PM)

Fostering public transport (PT)
Recovery urban quality (UQ)
Mobility management through specific measures (MM)
Improvement mobility for impaired people (MIP)
Urban freight transport (UF)
Integrated mobility and land use policies (MLU)
Environmental quality improvement and energy saving (EQ/ES)
Mobility plans for large companies (MPC)
Safety improvement (SI)

Some Strategies and Measures

There are four basic policy categories in SUTP: (1) Reducing the need for motorized transport, (2) enhancing modal shift, (3) developing clean and silent transport systems, (4) improving transport efficiency (Sustainable Urban) [14] (Table 1).

The lack of space prevents this paper from showing the analysis of cases which would result on a more detailed benchmarking exercise. So, let's say that each country has developed its legislation, rules or guidelines having in mind the different needs of their municipalities. The monitoring process carried out by and in each one of them, shows different results against the specified objectives of the plan. It is important to remark that, despite the long list of cities analyzed; only a few provide data regarding the quantitative reduction of emissions (Table 2).

Almost all of the plans give information about the impact of the measures on congestion and modal shift, which obviously have positive effects on urban environment, but it is not enough, since detailed consideration of local circumstances would be needed.

Results and Discussion

Each country names its urban mobility plans in a different manner: PDUs, PTUS, PUM, LTP are all different names given to the same reality. But, as "head document" above all of them the UE is remarkable the recent Green Paper Towards a new culture for urban mobility (COM 2007), that claims for integrated urban mobility approaches, expressed in Sustainable Urban Transport Plans (SUTP). This document proves that it is time to act and, so, the European Commission committed itself to produce guidelines to implement those plans, not yet published. Indeed, if in 2001 the White Paper European Transport Policy first came out under the title "time to decide", maybe this Green one should be called as "time to act".) These guidelines should be the umbrella that covered the different national plans, suggesting a minimum content for each one in such a way that all of them had a common evaluation framework, with a set of indicators and goals that allowed, through a benchmarking exercise, the dissemination of good practices. Anyway, it is indifferent

Table 1 Possible measures of a sustainable urban mobility plans (Own, from Guía práctica para la elaboración e implementación de PMUS)

TMC	Traffic calming, speed limits 30 km/h, etc.	Junctions regulation: bus priority and trams	Ring roads: only when connecting routes		
PM	Blue areas, more parking spaces	Park and ride			
PT	Interchanges (design, location integration in the city)	Exclusive lanes, HOV lanes	Network enlargement, frequency, fleet renewal, interlinks, etc.	Integrated tariff system	New technologies: information, ITS, etc.
UQ	Pedestrian zones	Cycle lanes	Bicycle loans	Bicycle parking	
MM	Park and ride	Transport on request	Carpooling/Carsharing	Urban tolls	
MIP	Road accessibility	Adapted PT stops and vehicles			
UF	Traffic control for heavy vehicles	Limited timetables	Transportcentres		
MLU	Pedestrian zones	Integrated public transport/urban planning	City and neighbourhood friendly design		
EQ/ES	Fleets: electrical, gas, bio diesel; ZBE	Tax incentives for buying or renewal	Cycle lanes and pedestrian itineraries		
MPC	Placement	Company buses (shuttle/routes)	Specific PT lines; cycles parking; cycling facilities (showers, lockers, etc.)	–Tele working –Flexible or compressed timetable	Company transport season ticket
SI	Roadsigns improvement	Junctions improvement	Flows division: one space for mode		

Table 2 Measures of a sustainable urban mobility plan (Own)

City	Population	Main objectives	Measures	Funding	Monitoring
Lille Metropole (FR) Scheme: PDU	1.2 million 87 Communes	Double PT use 2000/2015	Increase PT offer and network, interchanges	Communauté urbain, region, department, UE	PT increased 30% 98/06) Private car reduce
Darlington (UK) Scheme: LTP	100,000	Accessibility, congestion, air quality(00/06)	Integrated transport (PT, cycles, pedestrian), road safety; maintenance	£ 3,062(05/06) National, regional, local	On track
Burgos (SP)	160,000	Reduce emissions, PT, recover public space (05/09)	Restricted access, parking strategy, PT improvement, bio-diesel, public bike scheme	UE (Civitas), municipality, region	90%pedest., 100%public fleet bio-diesel, increase number of cyclists, three off-street parking, PT =
Genoa (IT) Scheme: PUM	622,000 (MA)	PT, reduce pollution, renewable, reduce traffic in city centre, road safety	Access control, urban toll, collective taxis, parking information, PT clean vehicles	UE (Civitas), national, Region Municipality,	Increase renewable 17% (PT), increase PT users, less traffic in the centre, increase pedestrian areas
Grenoble Campus (FR) Scheme: MicroPDU (07)	Univ. area; 50,000 students	Foster soft modes, reduce private car (40%) and parking space	Tramway enlargement, bike parking + lanes, improve accessibility soft modes, car pooling	PDU	On track
Apeldoorn (NL) Scheme: Traffic and Transport Plan (99/10)	156,000	Access to city centre, bicycles, Mobility plans for big areas	Pedestr, bikes priority, areas 30, parking information, PT + P&R, special buses for schools, hospitals and industrial areas	Municipality, Province, public and private institutions	Accidents decrease 50% (99/04), road congestion improved (citizen satisfaction index increased from 7 to 8)
Aalborg Traffic/ Environment Action Plan (1994–)	192,000	Reduce traffic, energy consumption, emissions	Parking management, bus priority, company bikes, car sharing	Municipality, national gov.	On track[a]

[a]*Ad exemplum*: regarding environmental and energy impacts of the parking system information, calculations of energy reductions were made on the basis of an average speed of 30 km/h. All 'searching' traffic is car. 20% of all cars are equipped with catalytic exhaust purifiers. Based on this the reduction in energy consumption will be 3,521 MJ or 971 L of petrol daily. On an annual basis this is 980,000 MJ or 29,400 L of petrol. The equivalent reductions in emissions on an annual basis are: CO_2: 7 t, NO_x: 0.5 t, CO: 7.7 t, HC: 0.57 t, particles 5 kg. This reduction equals approximately 0.3% of the total annual emissions from traffic in Aalborg City.

the use of the term "transport" or "mobility" as long as the plan reflect the concept of mobility regarding objectives, policies and measures (Expert Working Group on SUTP, Rupprecht Consult, 2004).

On the other hand, if is true that there is nothing new under the sun, we could ask then which could be the added value of the EU initiative derived from the Green Paper. The answer is written on it, adopting different forms: promoting the exchange of good practice at all levels (local, regional and national); underpinning the establishment of common standards and the harmonization of standards if necessary; offering financial support to those who are in greatest need of such support; encouraging research the applications of which will make it possible to bring about improvements in mobility safety and environmental; simplifying legislation and, in some cases, repealing existing legislation or adopting new legislation. Not certainly minor questions to put into practice, although none of them are a big novelty.

The cases analyzed in this paper show that regardless the existence or not of a specific law, and the different names adopted, there are some countries where successful urban mobility plans are implemented. In Spain for instance, these plans are not compulsory. Many municipalities that have started to implement them benefiting from the national subsidies, what is, at last, the main problem: funding.

Regarding funding schemes, it is very important to link it to the achievement of the objectives (as in the UK). Closely related is the importance of a good set of homogeneous key indicators, available and clear that allow calculating the effectiveness and level of achievement of the measures implemented. This in turn, will allow the design of alternatives and become a key support of the decision making process.

Another important aspect is the need of coordination between the different levels of the administration, both hierarchical and horizontal, that is, National Government/Region/Department/Local, and between departments of the same administration, beyond informal cooperation agreement. The coordination is especially visible in France and the UK, where the urban plans must be coherent and consistent with the national and regional land use planning, and with the air quality legislation. In the UK, integration with sectoral policies (disabled, social inclusion, noise) is required as well.

Plans must be realistic, since in some cases it seems obvious that the objectives are clearly Utopian, and public institutions seems more interested in "sell" it than implement (at last, are the citizens who vote). Sometimes, results may be better acting on a *small* area, such as universities, whose mobility is half a commuter plan and half an urban mobility plan.

Finally, it is highly recommended to avoid the temptation brought about by the language tricks: pompous names do not turn into better what is not and, all in all, it is the same a Sustainable Urban Mobility Plan that a Local Development Strategy … as long as it is sustainable, of course.

References

1. European Environment Agency (2008) TERM 2007: Indicators tracking transport and environment in the European Union Report 1/2008
2. European Commission (2001) White Paper European Transport Policy for 2010: time to decide COM 370 final
3. INFRAS-IWW (2004) External costs of transport
4. European Commission (2007) Green Paper: Towards a new culture for urban mobility COM 551
5. European Commission (2006) Communication from the Commission to the Council and the European Parliament on Thematic Strategy on the Urban Environment. SEC 16
6. Loi d'Orientation des Transports Intérieurs (LOTI) (1982) Loi sur l'Air et l'Utilisation Rationnelle de l'Energie (LAURE) (1998) Loi Solidarité et Renouvellement Urbain (SRU) (2000)
7. Centre D'etudes sur les reseaux, les transports, l'urbanisme et sur les constructions publiques (CERTU) (1996) Les Plans de déplacements Urbains. Guide Méthodologique
8. Thevenon J (2007) (CERTU) The urban travel plan in France: 25 years on. From UTP to SUTP? IDAE, Espacios Urbanos Espacios Humanos…hacia una movilidad sostenible Madrid
9. Benoît T (2001) Local public transport organisation in France: A new deal? Seventh THREDBO Conference, Molde (Norway)
10. Department for transport (DfT) (2004) Full Guidance on Local Transport Plans (online) Second Edition. Draft for consultation. London http://www.dft.gov.uk/pgr/regional/ltp/guidence/fltp/fullguidanceonlocaltransport3657
11. Ministerio de medioambiente (2005) Planes de Movilidad Urbana. Tecnologías de Reducción de Emisiones en el Transporte. Madrid
12. Expert group in group on sustainable urban transport plans (2004) Final Report
13. Instituto para la diversificación y el ahorro de energía (2006) Guía práctica para la elaboración e implementación de Planes de Movilidad Urbana Sostenible, IDAE Madrid
14. Sustainable urban transport plans and urban environment: Policies, effects and simulations (2005) Final Report

Bases for Building a Sustainability Indicator System for Transport

Alfredo Tolón-Becerra, Isabel Otero-Pastor, Pedro Pérez, Alejandra Ezquerra-Canalejo, and Xavier Lastra-Bravo

To achieve these objectives, we recommend using an iterative cycle of the indicator: generation and selection, technical construction, application, development and finally, communication, transfer and use. As examples of the methodology, we show:

- An example of two blocks of indicators of a social nature
- The development of a subset of road accident indicators, which have been applied in several geographic areas throughout Europe and Spain

Introduction

In recent years the range of transport systems, infrastructures and services available in Spain has grown in response to social demand, within a climate of economic growth. This increasing trend in transport use – both for passengers and freight – has been a contributing factor to the rise in energy use and carbon dioxide (CO_2) emissions. The current transport system is based on transport modes with high energy intensity such as road transport. It has a serious impact on the environment, including air pollution and noise, as well as contributing to global warming [1].

Both transport demand and greenhouse gas emissions are constantly increasing, and at a greater rate than the GDP. On the other hand, there has been some decrease in air pollution, although to a lesser degree [2]. Unless we can dissociate transport activity from income, the use of energy in the transport sector will continue to grow [3]. However, we have already reached the maximum elasticity of greenhouse gases in relation to the GDP, and this implies that future increases in GDP will entail more minor changes in emissions [4].

A. Tolón-Becerra and X. Lastra-Bravo
Department of Engineering Projects, University of Almería

I. Otero-Pastor, P. Pérez, and A. Ezquerra-Canalejo
TRANSyT (Transport Investigation Centre), Polytechnic University of Madrid

S. Rauch et al. (eds.), *Highway and Urban Environment*,
Alliance for Global Sustainability Bookseries 17,
DOI 10.1007/978-90-481-3043-6_6,

Technological innovations can create a transport system that is sustainable from an environmental point of view, and there is evidence of an improvement in energy intensities, which has drastically reversed pollution trends [5]. However, it is not enough to rely on technology to obtain a sustainable transport system, and far-reaching shifts in policy and current regulations are required [6].

The direct usefulness of a system of indicators is that it can provide information from processes where major mobility policies are formulated and implemented [7]. A direct link between environmental indicator systems and policies will provide a solid mechanism for integrating the criteria of environmental sustainability into the decision-making process within the area of transport. This link is not only important for following up trends, but also for designing the measures themselves [8].

The general increase in transport activity has gone hand in hand with a rising concern of the potential threat that our current lifestyle poses to the environment and to human health [9]. Therefore, a sustainable transport system needs to address the balance between economic, social and environmental considerations in order to define transport policy objectives. In particular, this system must be cost-efficient while respecting the environment (pollution, land use, energy consumption and natural elements), but also safe [10] and provide society with sufficient quality of service throughout the whole territory (fairness, social integration, etc.).

Transport Indicator Systems

The European Environment Agency's TERM (Transport and Environment Reporting Mechanism) indicator system is a tool which responds to the European transport policy and to the concept of sustainable mobility. It currently comprises 40 indicators, which are primarily economical and environmental. Only four of these indicators, however, involve social aspects, and deal with accident rates, access to the service sector and to modes of transport.

The transport and environmental indicator system in Spain (TRAMA) is designed to measure the degree of sustainability of the Spanish transport system, and specifically analyzes the demand for transport in relation to environmental quality indicators. This makes it possible to analyze trends and to formulate transport policies over the long term. It has the declared objective of reducing external effects. Furthermore, it is based on key sustainability indicators; and it follows the guidelines and methodology of the TERM [11]. TRAMA provides a model of transport and trends, and a photograph of the environmental pressures and impacts in the sphere of transport, thus enables identification of causal factors (basically technological and socio-economical). This system measures the effectiveness of the policies, commitments and objectives adopted with regard to transport and the environment.

Finally, the Spanish Observatory for Sustainability (OSE) compiles indicators for the transport sector which deal almost exclusively with economic and environmental implications of sustainability [12]. The 2007 report has a structure which mirrors the European Union's indicator system and it dedicates one section to the transport sector. The report includes 17 indicators, but only one of them deals with the social dimension (fatal road accidents by age group), which is a response to the European Commission's proposal to reduce by half the number of fatalities from traffic accidents during the period 2000–2010 [13].

Generally speaking, we can identify three stages in the process of generating and applying sustainability indicator systems. This have led to first, second and third generation systems [14]. First-generation systems originated in the 1980s from the work done by the OECD, and are characterized by being very theoretical, and exclusively environmental. They include an environment-based approach (air, water, earth and biodiversity), an objective-based approach (in response to legal and administrative requirements, Agenda 21), and a sector-based approach (transport, tourism, industry, etc.). Second-generation systems were developed in the 1990s on a national scale, and pursue a multidimensional approach (economic, environmental and social) to sustainable development. In recent years, the need to link aspects of development together with its indicators has given rise to systems where the indicators are grouped, transversally and systematically, into themes or multidimensional areas (third-generation systems).

In summary, the transport indicator systems developed presently (TERM, TRAMA, OSE, ….) basically address economical aspects (demand, growth, investments, …), environmental aspects (emissions, clean technologies, …) and the interaction between economy and the environment (eco-efficiency…). However, a balanced solution needs to be found to the conflicting objectives of economic, environmental and social sustainability. So far, social aspects have been very superficially addressed.

Moreover, the proliferation of sustainability indicator systems has led to a certain disparity and confusion among the different methods which makes them difficult to integrate. Greater homogeneity and uniformity is required when building indicator systems, in order to improve comparability, adjust to the structure of the current observatories, and make it easier to integrate the various sectorial systems (transport, urban planning, energy, social well-being, …).

Methodological Bases for Building a System of Transport Indicators

We establish the methodological bases for building an indicator system in phases with which to monitor the sustainability of the transport sector in its three aspects: economic, environmental and social. We recommend using the following iterative cycle:

1. Context of the indicator system. This must include at least a definition of the geographic scope of application; political-administrative context; and the key time periods, given that sustainable development involves constant change [15].

2. Generating and selecting the indicators. This will be done on a structure classified into areas, subsystems, blocks or levels, in order to facilitate their analysis by specialization [15]. The indicators will be selected based on their previous characterization according to their suitability (significant, mature, not redundant, with adequate geographical coverage and integrable), the quality of the data required (available at a reasonable cost, reliable, and methodologically consistent), and their social impact (interest, resonance, whether they are understandable and easy to communicate).
3. Technical construction and application of the selected indicators. This will allow them to be quantified and compared on different scales for specific years (synchronic comparison) and to analyze trends over time (diachronic comparison). In doing this, it is important for the selection of relative indicators to include not only trends (percentage of variation over time) and the relativization of the indicator by population and by area, but also other data and variables which help to provide a better understanding of the indicator, and to qualify its complexity.
4. Establishing critical and desirable threshold values, and objectives for realistic progress towards the desirable objectives. This will be done using reference values which allow a preliminary estimation or approximation in terms of distance.

   The threshold values are considered limit; this means that in principle, there could be a negative limit or critical value, and another optimal or desirable value. The critical value indicates the minimum or maximum value that an indicator may have, below or above which there is clearly a situation of unsustainability, which is therefore the direct opposite of the desirable values corresponding to the optimal – albeit utopian – situation to be attained. As there is no official operative measurement of sustainable development, in most studies this is estimated based on the best existing situation in the scope of the study, or greater [16].

   The objective values must establish some pragmatic progress towards the desirable values; this progress will be more accelerated the further they are from the desirable values in order to reach a rapid convergence. Greater efforts will be required towards the end, as we come closer to reaching the desirable values. These values or intervals are the ones we wish to attain as the ultimate objective of the policy to be applied, and they are estimated in terms of distance and convergence.
5. Communicating and using the indicator system. This involves transferring the indicators to the users (politicians, public managers, society, …) to enable the results to be used in the decision-making process, thereby legitimizing the indicators. In order to ensure their usefulness, they must be expressed and then submitted to debate at different levels of complexity (the most complex for experts, through to the simplest). There should be a limited number of indicators per debate (about five), with one predominant and other secondary indicators.
6. Generating new indicators. These will become progressively more complex, dense and transversal between the sectors (e.g., accessibility and social well-being) and between different scales.

Results

To illustrate the methodology, we show part of the results from the beginning of the stage for generating and selecting a non-exhaustive list of indicators grouped into two blocks: accessibility to social services (Table 1), and road accident rate (Table 2). These two blocks correspond to the initial structure of the sub-system of social indicators for the transport sector ordered into a framework of pressure-state-response.

The "accessibility to social services" block shows the beginnings of integration of two blocks belonging to two indicator subsystems of different nature: transport and social well-being. Some of the proposed response indicators recommend improving transport infrastructures (road network, public or private transport, ...) whereas others opt to build more service centers and to reduce average distances. This highlights the need to integrate the indicator systems. As an example of the application of indicators, priority has been given to the "road accident rate" block, because this is the means of transport that causes most deaths.

Table 1 "Accessibility to social services" block

Pressure (causes)	State (social)	Response
Distance to social centers. Quality of transport infrastructures: distance to depots for each service (train, bus, service area, taxi)... Quality of public and private transport services: no. of lines, no. of services per day, waiting time between services, schedules,...	Average travel time to social centers, hospitals, health centers, schools, universities and colleges, administrative centers...	Investment in transport infrastructures, investment in public transport systems, building new social centers closer by...

Table 2 "Road accident rate" block (European Road Safety Action Programme" (2003). "Mid-term review of the European Commission's 2001 Transport White Paper" (2006))

Pressure (causes)	State (social)	Response
Quality of the road network: curves, state of repair, signposting... Causes attributable to drivers: traffic offences; driving while under the influence of alcohol, drugs or medicines; driving licence obtained by fraud; age-related problems; and physical condition of driver... Causes deriving from vehicles: age of the vehicles, design defects...	No. of accidents involving victims, without victims, with fatalities... Total no. of fatalities broken down by sex, age groups... No. of black spots: per area, length of road network... Time taken by emergency services to reach victims...	Improvement in vigilance and control: no. of police officers, frequency of official technical inspection of vehicles, licenses incurring penalty points, fines... Improvements and new designs in vehicle safety... Correct use of legal drugs and medicines, rehabilitation programmes for those involved in accidents... Publicity: campaigns to encourage good driving practices... Improvement of emergency services...

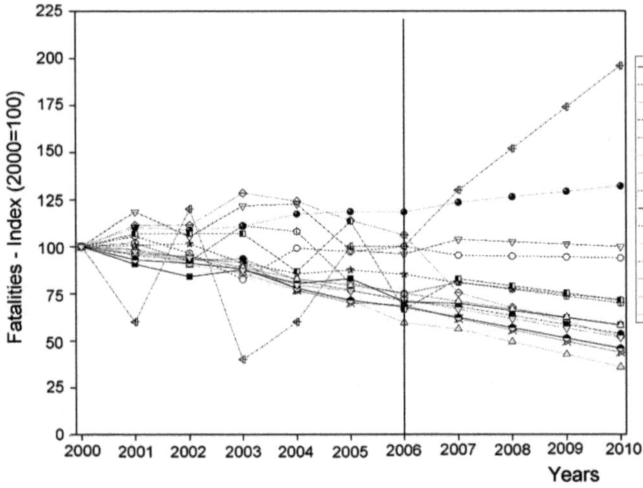

	2000	2006	2010
Alicante	100	70,62	58,39
Almeria	100	100,00	93,80
Barcelona	100	59,59	35,89
Cádiz	100	100,00	196,00
Ceuta y Melilla	100	95,88	99,81
Madrid	100	68,80	43,79
Málaga	100	106,03	52,16
Sevilla	100	75,00	70,01
Germany	100	67,85	45,99
Italy	100	85,26	71,53
Lithuania	100	118,41	132,13
Malta	100	66,67	71,43
Spain	100	71,05	53,73
EU25	100	75,13	58,20
EU15	100	71,97	51,70

Fig. 1 Number of fatalities in road accidents updated to 2000 (%). Evolution 2000–2010 (CARE 2008, DGT 2007)

We first approach the indicators of state before addressing the indicators of pressure and response. We have selected the number of fatalities in traffic accidents as the main indicator, as one of the priorities within the EU is to decrease this figure to half between 2000 and 2010. The values have been updated to 2000, and figures for 2010 have been estimated by quadratic minimum regression (Fig. 1). It can be seen that the objective is being met, although generally more satisfactorily in the EU-15 (72 in 2006 and 51.7 in 2010) than for the EU-25 (75 and 58), and that Spain is following a similar trend (71 and 54). However, this progress is not homogeneous, either in European countries (68 and 46 in Germany, compared to Lithuania, where fatalities increased to 118 and 132), nor in Spanish cities. Barcelona or Madrid (with 60 and 36, and 69 and 44, respectively) shows greater progress than Almería or Cadiz (with 100 and 94, and 96 and 99, respectively).

In addition to measuring the rate of progress, it is also necessary to measure the current situation, and to compare areas on different scales. For this reason we consider that this primary indicator should be complemented with other secondary indicators, expressed in relative terms. The first of these relativizations is by population (Fig. 2).

In this case it can be seen that although there has been a similar rate of progress, this progress has been more irregular, and particularly there is considerable difference between the different areas. If we take 2006 as the year for synchronic comparison, we can see that the EU-15 value (76) is also lower than the EU-25 value (85), but here it is more disperse with both extremely low values (Malta, 25) and extremely high values (Lithuania, 223). In Spain, the value is slightly higher (92) than for the EU, and widely dispersed by province (from values around 36 in Vizcaya, through to shockingly high figures, 330 in Cuenca, with an ongoing upward trend).

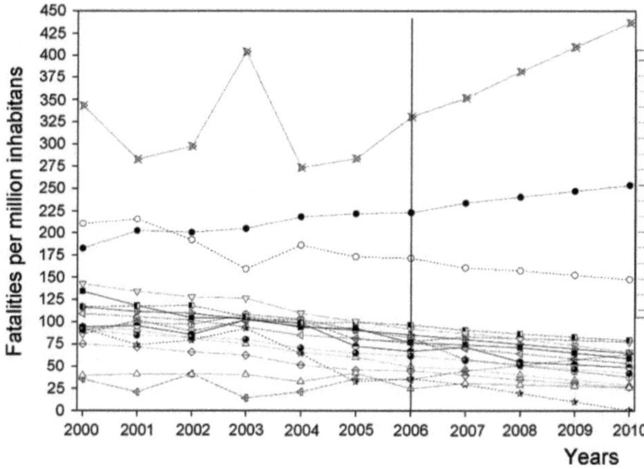

	2000	2006	2010
Alicante	134,24	76,81	59,14
Almería	210,33	171,42	147,62
Barcelona	93,53	49,72	28,47
Cádiz	86,21	77,89	77,70
Ceuta y Melilla	35,33	35,03	63,75
Cuenca	343,19	330,75	436,65
Madrid	75,11	44,77	26,35
Málaga	90,71	82,48	37,74
Sevilla	94,53	67,03	49,22
Vizcaya	92,70	35,97	0,59
Germany	91,32	61,76	42,01
Italy	116,81	96,49	79,24
Lithuania	182,51	223,02	253,81
Malta	39,45	24,73	25,91
Spain	142,62	91,79	66,39
UE15	109,20	76,13	53,49
UE25	116,53	85,39	64,93

Fig. 2 Number of fatalities in road accidents per million inhabitants. Evolution 2000–2010 (CARE 2008, DGT 2007, EUROSTAT 2008, INE 2008)

Other relative indicators worth mentioning are the percentage of victims, when broken down into urban and country roads, in comparison with the total number of traffic victims. These indicators underscore the fact that the value is always higher for country roads than for urban roads, although the values are heterogeneous (57% in Barcelona and 72% in Madrid, compared to 97% in Avila and 94% in Cuenca, for 2006). Also worth noting is the relative indicator regarding number of fatalities for every 1,000 accidents with victims (fatal and non-fatal), which is decreasing in Spain (41 in 2006 compared to 57 in 2000) and even more disperse (14 in Barcelona, 17 in Madrid and 25 in Vizcaya, compared to 99 in Almería and 83 in Ciudad Real).

Conclusions and Discussion

Generally speaking, the accident rate is decreasing, albeit unevenly, and with very heterogeneous values. A certain more or less pronounced negative correlation can be seen, depending on the relative indicator analyzed, between the accident rate and the characteristics of the geographical area considered, according to whether it is densely-populated, developed, urban, flat, coastal, … But the great dispersion of the results highlights the need for micro-investigations which would make it possible to detect local situations (black spots…), and to break down the indicators by sex, age, proximity to major urban areas etc. and to determine their causes. This would enable non-linear and de-concentrated decisions to be taken by provincial delegations of the Interior Ministry, regional governments and local councils, among others.

In any case, the desirable value for all road accident indicators should be zero, while the critical and objective values may vary according to the geographical area considered. The critical values must be obtained in terms of distance, and the objective values in terms of distance and convergence, also expressed in absolute terms, or percentage of improvement.

Given that the objective value has been fixed at a reduction coefficient of 0.5 (fatalities in 2010/fatalities in 2000) for the EU, greater progress should be seen in more local areas, as the distance is greater with regard to the value of the highest scope (in this case the EU).

Therefore the objective reduction coefficient should be modulated using a normalized multi-criteria function depending on the population of the area considered and the distance of the relative accident rate compared to the average value for the EU. This coefficient may vary between 1 (hypothetical case of an initial accident rate of practically zero), and 0 (theoretical case of total initial accident rate for the population).

References

1. Sperling D (2004) Environmental impacts due to urban transport. Urban Transport and the Environment. An International Prospective, Elsevier, Oxford, pp. 99–189
2. Pérez PJ, Monzón A (2007) Mobility and environment in Spain. In: G Morrison, S Rauch (eds.) Highway and Urban Environment: Proceedings of 8th International Symposium on the Highway and Urban Environment, Springer Science, Dordrecht
3. OECD Environment Directorate (2003) Analysis of the links between transport and economic growth. Project on decoupling transport impacts and economic growth. Caid N. ENV/EPOC/WPNEP/T (2003) 4/FINAL, Paris, France, 94 pp
4. Pérez-Martínez PJ (2006) Mobility and environment in Spain. Are we tending towards sustainability? In: R. Joumard (ed.) Environment and Transport – Transport and Air Pollution, 2nd Conference, Reims, 12–14 June 2006, INRETS, Les Collections de LInrets, Bron Cedex, France, pp. 60–65
5. Lakshmanan TR, Xiaoli H (1997) Factors underlying transportation CO2 emissions in the U.S.A.: a decomposition analysis. Transportation Research D 2:1–15
6. Rodenburg CA, Ubbels B, Nijkamp P (2002) Policy scenarios for achieving sustainable transportation in Europe. Transport Reviews 22:449–472
7. Acutt MZ, Dodgson JS (1997) Controlling the environment impacts of transport: matching instruments to objectives. Transportation Research D 2:17–33
8. Gudmundsson H (2003) Making concepts matter: sustainable mobility and indicator systems in transport policy. International Social Science Journal 55:199–217
9. Lenz HP, Prüller S, Gruñen D (2003) Means of transportation and their effect on the environment. In: D. Gruden (ed.) Traffic and Environment, Springer Verlag, Berlin, pp. 107–173
10. Sinha KS (2003) Sustainability and urban public transportation. Journal of Transportation Engineering 129:331–341
11. Pérez PJ, Monzón A (2005) Informe sobre transporte y medio ambiente. Trama 2005. Centro de Publicaciones Secretaría General Técnica Ministerio de Medio Ambiente, Madrid, España, 108 pp
12. Observatorio de la Sostenibilidad en España (2007) Sostenibilidad en España 2007. Ministerio de Medio Ambiente, Fundación Biodiversidad, Fundación Universidad de Alcalá. 2005. Mundi-Prensa Libros, S.A
13. European Commission (2001) White Paper – European transport policy for 2010: time to decide. Office for Official Publications of the European Communities. Luxembourg, 119 pp

14. Gallopín G (2006) Los Indicadores de Desarrollo Sostenible: Aspectos Conceptuales y Metodológicos. Ponencia realizada para el Seminario de Expertos sobre Indicadores de Sostenibilidad en la Formulación y Seguimiento de Políticas. Chile
15. Bossel H (1999) Indicators for Sustainable Development. Theory, Method, Applications. International Institute for Sustainable Development. Canada
16. Mega V, Pedersen J (1998) Urban Sustainability Indicators. Office for Official Publications of the European Communities. Luxembourg, 49 pp

Part II
Air Pollution

Biodiesel from Waste Olive Oil: Transesterification Kinetics, Exhaust Emissions and Fuel Consumption

Laureano Canoira, Ramón Alcántara, Nikolaos Tsiouvaras, M. Jesús García- Martínez, Juan García Galeán, Alberto Llamas Lois, Evripidis Lois, and Dimitrios M. Korres

Introduction

Biodiesel is defined as "a fuel comprised of mono alkyl esters of long chain fatty acids derived from vegetable oils or animal fats" [1]. Transesterification (also called alcoholysis) of triglycerides [2] and jojoba oilwax [3] for biodiesel manufacture has been studied in our research group in the past few years. The European Union has issued Directive 2003/30/EC, which mandates the use of biofuels in a percentage ranging from 2% in 2005 to 5.75% in 2010 (calculated on the basis of energy content), for all transportation fuels marketed within the member states. It is expected that a significant portion of this amount will be biodiesel. However, although the literature concerning the production of biodiesel has increased dramatically in the last 5 years [4, 5], data related to the kinetics of transesterification are rather scarce: [6] presented the kinetics and final state of methano-, ethano- and butanolysis of cottonseed, peanut, soybean and sunflower oils catalyzed by sodium hydroxide and methoxide or sulphuric acid, and [7] studied the kinetics of methanolysis of sunflower oil catalyzed with KOH. The effect of water on the kinetics of cotton oil ethanolysis catalyzed by KOH [8], and the kinetics of the non-catalytic transesterification of soybean oil at 220°C and 230°C [9], was described. The kinetics of transesterification of rapeseed oil to biodiesel fuel in supercritical methanol without any catalyst was described [10], and [11] described the kinetics and mechanism of the KOH catalyzed methanolysis of rapeseed oil for biodiesel production. Moreover, the kinetics of methoxide base-catalyzed methanolysis of

L. Canoira, R. Alcántara, N. Tsiouvaras, M. Jesús García- Martínez, J.G. Galeán, and A.L. Lois
Department of Chemical Engineering and Fuels, School of Mines, Universidad Politecnica de Madrid, Ríos Rosas 21, 28003, Madrid, Spain

N. Tsiouvaras, E. Lois, and D.M. Korres
School of Chemical Engineering, National Technical University of Athens, Iroon Polytechniou 9, Athens, 15780, Greece

S. Rauch et al. (eds.), *Highway and Urban Environment*, Alliance for Global Sustainability Bookseries 17,
DOI 10.1007/978-90-481-3043-6_7, © Springer Science+Business Media B.V. 2010

soybean oil at 40°C was studied [12]. To the best of our understanding, any data have been published to date on the kinetics of the transesterification of used olive frying oil catalyzed by sodium methoxide at 60°C to produce biodiesel, which is the first aim of this paper.

In Spain, edible vegetable oil consumption is approximately 600 ML/year. Most of this oil (70%) is olive oil that is mainly used for deep-frying processes. According to the Spanish National Institute of Statistics, about 74 ML of waste olive oil are collected per year, which is an approximate value since most of the household waste frying oil is thrown through the drainage. Transesterification of waste olive oil to produce biodiesel could decrease the waste disposal problem. Studies concerning the exhaust emissions produced by the combustion of pure waste olive oil biodiesel have been published recently [13], but no data exist to date on the emissions produced by the combustion of blends of waste olive frying oil biodiesel in 5% vol and fossil diesel fuel. The Spanish Government allows petroleum product operators to blend biodiesel up to 5% vol with conventional diesel fuel without the need to legally declare this blending.

Nitric oxide (NO) and nitrogen dioxide (NO_2) are very important in polluted air. Collectively designated NO_x, high NO_x concentrations can cause severe air quality impoverishment. NO_x enters the atmosphere as a result of the combustion of fossil fuels in internal combustion engines. Like carbon monoxide, NO attach to haemoglobin and reduces oxygen transport efficiency. Particulate matter (PM) exhaust emissions are also of great concern, since they are involved in acid rain processes and respiratory diseases. Substantial reduction in PM emissions can be obtained by addition of biodiesel to diesel fuel. The PM reduction appears to be related to the amount of oxygen content in the fuel blend [14]. This paper also reported the exhaust emissions from ternary blends of waste olive oil biodiesel (5% vol), ethanol (5% vol) and fossil diesel fuel, the so called BE-diesel.

Although some studies have been published recently on exhaust emissions of BE-diesel [15], no data exits on the use of the blends described in our paper, especially with the use of waste olive oil biodiesel. Ethanol is a low cost oxygenate of vegetal origin with high oxygen content. However, there are many technical barriers to the direct use of ethanol in diesel fuel due to the low cetane number of ethanol, and the poor solubility in cold climates. In fact, diesel engines cannot operate normally on ethanol-diesel blends without special additives [16]. Biodiesel is known to act as an emulsifier for ethanol, and blending biodiesel and ethanol into a conventional diesel fuel greatly improved the solubility of ethanol in diesel fuel over a wide range of temperatures [17].

Experimental

Methanol, tetrahydrofuran (THF), sodium, hydrochloric acid and anhydrous sodium sulphate were all commercial grade and used without further purification. The oil transesterified in all the experiments was waste olive frying oil of domestic origin.

For the production of biodiesel from waste olive oil, the method of transesterification with methanol and basic catalyst (sodium methoxide) was used in the following conditions: molar ratio methanol/oil: 7.85/1; 60°C; 600 rpm, amount of catalyst from 1.4 to 2.8% wt of the oil. The total quantity of the oil 760 mL (700 g) was placed in the reactor with part of the methanol, 155 mL (120.9 g, 3.78 mol). The stirring and the heating were switched on. The rest of the methanol, 100 mL (78 g, 2.44 mol) was used for the preparation of the catalyst by reaction with 4.35 g (0.189 mol) of metallic sodium. After total dissolution of the sodium, the catalyst was added to the reactor were the desired temperature was already achieved. A slight increase of temperature to 64°C was observed after the addition of the catalyst which was soon controlled to 60°C. The reactor was left for the time specified for the reaction, and then the heating and the stirring were switched off. One hour after the refrigerating water was also switched off, and the two phases were decanted overnight. After the distillation of the methanol in both phases, the glycerine phase was neutralized with 14.5 mL of concentrated (37% wt) hydrochloric acid. The biodiesel phase was neutralized in a decanting funnel with two washings with diluted aqueous solution of hydrochloric acid, and afterwards it was washed with diluted aqueous solution of sodium chloride and three times with water. The biodiesel was then dried with anhydrous sodium sulphate and filtered. For the analysis of the content in glycerol and mono-, di- and triglycerides according to the European standard EN 14105 [18], a gas chromatograph Hewlett Packard HP 6890 series GC system was used, in connection with a mass spectrometer Hewlett Packard 5973 Mass Selective Detector, including an automated system of sample injection. The results of these analyses are summarized in Table 1.

The kinetics experiments have been carried out with the same amounts of reagents, using the same procedure. 10 mL samples, taken at 2, 5, 10, 15, 25, 40 and 60 min

Table 1 Properties of the biodiesel of waste olive oil

Analysis	Biodiesel	Max	Min	Units	Standard
Free glycerol	0.03	0.02		% Mass	EN 14105
Monoglycerides	1.29	0.80		% Mass	EN 14105
Diglycerides	0.17	0.20		% Mass	EN 14105
Triglycerides	<0.05	0.20		% Mass	EN 14105
Total glycerol	0.36	0.25		% Mass	EN 14105
Water and sediments	<0.05	0.05	–	% Vol	ASTM D 1796
Density at 15°C	880	900	860	kg/m^3	ASTM D 4052
Sulfated ash content	<0.005	0.02		% Mass	ASTM D 482
Copper strip corrosion	1a	Class 1	Class	ASTM D130	
Total contamination	24.0	24.0		mg/kg	EN 12662
Flash point	141.9	–	120	°C	ASTM D 92
Cloud point	5			°C	ASTM D 2500
Cold filter plugging point	–5.0			°C	EN 116
Kinematic viscosity at 40°C	5.44	5.00	3.50	mm^2/s	ASTM D 445
Higher calorific value	9748.6			cal/g	ASTM D240
Cetane number	57.9		51		ASTM D 613

were immediately neutralized with 1:1 hydrochloric acid (2 mL). Methanol was evaporated from the sample using a rotary evaporator. 10 mL of THF were added, and this solution was dried with anhydrous sodium sulphate. 200 mg (exactly weighed) were used for the analysis of free glycerol, and mono-, di- and triglycerides according to the standard EN 14105 [18] (instead of the 100 mg to balance for the addition of the THF). The Fatty Acids Methyl Esters (FAME) content was calculated from the analyses and the mass balance in each kinetic sample.

Some physicochemical properties of the biodiesel prepared according to this recipe were determined in a preliminary study of the feasibility of use this biodiesel as a fuel in diesel engines. The higher calorific value was measured after the ASTM standard D 240.

The cetane number was calculated from the time needed for the rise of the pressure by 2 mbar in the fuel ignition chamber, (ASTM D 613). The equipment was from Fuel Tech. Co. and the cooling water circulation (Circulating System 253) from the Precision Co.

The fossil diesel was mixed with the waste olive oil biodiesel. Blends containing 5% vol biodiesel, were chosen in order to prepare samples to fulfill the requirements of EN 590 specification (max. biodiesel content 5% vol) and because it was a reasonable percentage within the requirements of Directive 2003/30/EC (maximum 5.75% based on lower heating value). The engine tests were made in a single cylinder diesel Petter AVL-LAB engine; operating characteristics are given in Fig. 1. Apart from the engine, the system consists of an electric dynamometer, some heat exchangers for the lubricants and a control panel. The dynamometer was connected with the rotor and was originally moving the same way in order to help the engine to start, but later the dynamometer was used as a power generator transforming the mechanical power of the engine to electrical power. In this way we were able to adjust the charge of the engine by changing the potential in the dynamometer. The parameters controlled

Fig. 1 Test AVL-LAB engine used in the emissions and fuel consumption experiments. Type: single cylinder direct injection, speed: 1,500 rpm, compression: 19:1, volume: 533 cm^3, maximum power: 5 hp (3.8 kW)

from the control panel are the charge (resistance) of the dynamometer, the temperature of the air at the inlet and the outlet, and the temperature of the inlet and outlet of the lubricant and the cooling water. The experiment consisted of measurement of NO_x and PM emissions and fuel consumption for different charges of the engine from 0% to 100% (3.8 kW).

The analyzer (MEXA-754 GE, Horiba) was used to calculate the amounts of NO_x in the exhaust gases. The most important aspects of this analyzer were: exhaust gases measured, NO, NO_2; units, ppm; range, (0–5,000) ppm; accuracy, ±1%. The analyzer was connected to the engine through heated lines of exhaust transport from Signal Instruments Co (type 530/540) and a pre-filter unit (Signal Instruments Co, Pre-filter Unit 333).

For the analysis of particles in the exhaust gases, the equipment of Joy Manufacturing Co.; recommended by the Western Precipitation Division was used. The filters where of Whatman type 934-AH.

Results and Discussion

Biodiesel Quality

The composition of biodiesel after the EN 14214 [1] specification is summarized in Table 1. The transesterification reaction rate is much higher than it was originally estimated. After the first 10 min of the addition of the catalyst to the reactor, an almost quantitative conversion of TG to FAME and glycerol takes place. All the time that the reactor is left running after this initial time, the reactions are reversible reactions, and the reaction mixture is slowly moving towards equilibrium. The chromatograms of the kinetic samples show an initial quantitative conversion of TG to FAME but after the first minutes the monoglycerides MG (1.29% wt) and diglycerides DG (0.17% wt) reappear. However, only MG are out of the specification (1.29% wt, max. 0.80% wt), and consequently also the total glycerol G (0.03% wt, max. 0.02% wt). The biodiesel produced meets most of the specifications required by the Technical Regulation EN 14214.

Kinetic Experiment

The % wt of TG, DG, MG, G and FAME are plotted against time in Fig. 2. The transesterification reaction is very quick at 60°C and almost all TG were converted after the first 5 min; MG and DG were transient intermediate species, always at very low concentrations.

The transesterification reaction could be considered as an irreversible second order reaction (first order in TG and first order in methanol) which had been solved as follows. The transesterification reaction had the general form:

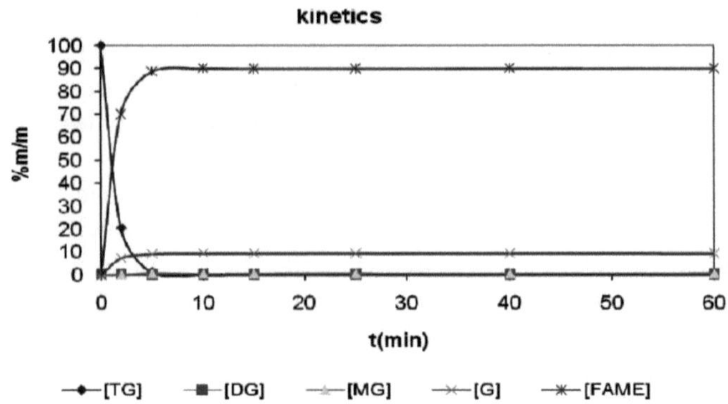

Fig. 2 Concentrations of mono-, di-, triglycerides, FAME and glycerol vs. time in the kinetic experiments

$$A + bB \xrightarrow{k} \text{products}$$

Here the function that expresses the rate of reaction vs. time was [19]:

$$k \cdot t \cdot b \cdot C_{A_0} (M-1) = \ln \frac{C_B}{b \cdot M \cdot C_A}, \text{where } M = \frac{C_{B0}}{bC_{A0}}.$$

The simplified reaction was:

$$TG + 3CH_3OH \xrightarrow{k} G + 3FAME$$

Here C_A was the concentration of TG, C_B the concentration of methanol, b = 3. 760 mL (699.2 g) of waste olive oil (mean mw 885.43 g/mol), and 255 mL (201.45 g) of methanol (a total of 1015 mL) was introduced to the reactor. Thus, the mol number *n* and the initial concentrations C_{I0} are: $n_{A0} = 0.79$ mol; $C_{A0} = 0.78$ mol/L; $n_{B0} = 6.3$ mol; $C_{B0} = 6.20$ mol/L. M = 2.65.

Consequently, the second order reaction equation remains as follows:

$$k \cdot t \cdot 3 \cdot 0.78 \cdot (2.65 - 1) = \ln \frac{C_B}{3 \cdot 2.65 \cdot C_A}$$

This second order kinetic equation was solved using the mass fractions of TG and methanol. The mass fraction of methanol was calculated in every point from the difference of the measured mass fractions of G, MG, DG, TG and FAME. The kinetic plot of $\ln(C_{MeOH}/aC_{TG})$ vs. time gave a straight line (Fig. 3). The 95% confidence limits for the intercept and slope of the straight line are:

Slope: 0.2245 ± 0.0521

Intercept: −0.8571 ± 0.1621

The resulting rate constant *k* is 0.2245 $L \cdot mol^{-1} \cdot min^{-1}$. The transesterification reaction of used olive frying oil with methanol catalyzed with sodium methoxide, follows the equation:

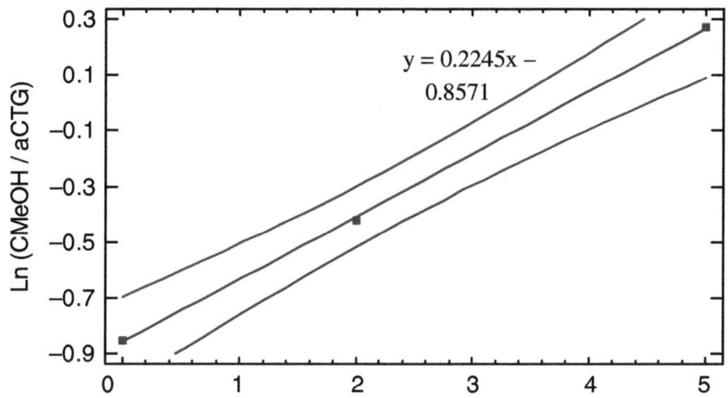

Fig. 3 Kinetic plot of Ln ($C_{MeOH/}$a C_{TG}) vs. time for the second order kinetics transesterification reaction

$$-r_{TG} = 0.2245[TG][MeOH](\text{mol}\cdot\text{L}^{-1}\cdot\text{min}^{-1})$$

Exhaust Emissions and Fuel Consumption

The fossil diesel was mixed with the waste olive oil biodiesel, and blends containing 5% vol biodiesel were prepared. Also, samples were prepared that contained 5% vol of ethanol to compare the performance of BE-diesel fuel with fossil diesel and biodiesel-fossil diesel blends. The results of these tests are summarized in Tables 2 through 4.

The effect of oxygenated fuels blends on NO_x emissions is complex and not conclusive. Cetane number, fuel density and aromatic fuel composition can influence NO_x emissions. Many studies indicate that oxygenated fuel blends could cause slight increases in NO_x emissions [17]. However, in our work the reduction of NO_x exhaust emissions was significant when using blends of fossil diesel and waste olive oil biodiesel (–11.7%), and in good agreement with previous results [13]. It was very interesting with ternary blends of BE-diesel (–32.7%) although previous

Table 2 NO_x Exhaust emissions reduction (%)

Sample	Base fuel	Base fuel + 5% waste olive oil biodiesel
Diesel	0	–11.7
Diesel + 5% vol ethanol	–15.3	–32.7

Table 3 Particulate matter (PM) exhaust emissions reduction (%)

Sample	Base fuel	Base fuel + 5% waste olive oil biodiesel
Diesel	0	–6.1
Diesel + 5% vol ethanol	–13.1	+2.0

Table 4 Mass consumption (g/h) and energy consumption (MJ/h) vs. engine charge (kW) of blends of diesel and diesel + 5% vol ethanol with waste olive oil biodiesel

Sample		0 kW	0.08 kW	0.95 kW	1.90 kW	2.85 kW	3.80 kW
Diesel	g/h	485.9	472.4	556.1	638.4	710.1	783.4
	MJ/h	21.7	21.1	24.9	28.5	31.8	35.0
Diesel + 5% vol waste olive oil biodiesel	g/h	405.2	408.1	512.8	571.0	617.4	713.1
	MJ/h	18.1	18.2	22.9	25.4	27.6	31.8
	%	−16.6	−13.7	−8.0	−10.9	−13.2	−9.1
Diesel + 5% vol ethanol	g/h	482.6	473.9	558.0	663.8	704.3	826.1
	MJ/h	21.7	21.4	25.2	29.9	31.8	37.3
	%	0	+1.4	+1.2	+4.9	0	+6.6
Diesel + 5% vol ethanol + 5% vol waste olive oil biodiesel	g/h	426.1	418.8	504.3	588.4	624.6	688.4
	MJ/h	19.1	18.8	22.7	26.4	28.0	31.0
	%	−12.0	−10.9	−8.8	−7.4	−11.9	−11.4

results with BE-diesel fuel showed an increase in NO_x emissions, albeit for a different ethanol-biodiesel-diesel fuel ratio [15]. Oxygenate diesel fuel blends are known to reduce total PM emissions, although the mechanisms are not clear. Soot formation mainly takes place in the fuel-rich zone at high temperatures and pressures, specifically within the core region of the fuel spray.

It is commonly assumed that oxygenates blended with diesel fuel effectively deliver oxygen to the pyrolysis zone of the burning diesel spray, resulting in reduced PM generation. The blends of fossil diesel + biodiesel and fossil diesel + ethanol shows good results in the reduction of PM (−6.1% and −13.1%, respectively), but the ternary blends of BE-diesel shows an increase of PM (+2.0%), in contrast to previous reported results [15]. This difference could be explained by the lower amount of biodiesel used in our experiments that described in the literature [15]. From a practical point of view, the advantage of reducing NO_x emissions by one third should be balanced against a slight increase in the PM emission when using this ternary BE-diesel fuel blend.

The binary blends of fossil diesel and biodiesel show a decrease in the consumption from −16.6% at the lowest engine power to −9.1% at the highest power. On the contrary, the binary blends of fossil diesel and ethanol always show an increase in fuel consumption. However, the ternary blends of BE-diesel fuel show a reduction in the fuel consumption at all engine powers tested. This, coupled with the reduction in the NO_x emissions, makes this fuel a good promise for diesel cars in a near future.

Conclusions

The transesterification reaction of waste olive oil with methanol, catalysed with sodium methoxide, to produce biodiesel follows a first order kinetics in methanol and in triglycerides, with a rate constant of $k = 0.2245$ L mol^{-1} min^{-1}. The reaction time should be optimized to avoid the reversible reactions that increase the amount of monoglycerides in the final products.

When added in a 5% vol to conventional fossil diesel, biodiesel fuel reduces NO_x and PM emissions, especially when using ternary blends of fossil diesel, biodiesel (5% vol) and ethanol (5% vol), BE-diesel fuel. It also significantly reduces the fuel consumption of the engine.

References

1. European Committee for Standardization (DIN) (2003) Automotive fuels. Fatty acid methyl esters (FAME) for diesel engines. Requirements and test methods. European Standard EN 14214
2. Alcantara R, Amores J, Canoira L, Fidalgo E, Franco MJ, Navarro A (2000) Catalytic production of biodiesel from soy-bean oil, used frying oil and tallow. Biomass and Bioenergy 18:515–527
3. Alcantara R, Canoira L, Garcia-Martinez MJ Carrasco J (2006) Biodiesel from jojoba oil-wax: transesterification with methanol and properties as a fuel. Biomass and Bioenergy 30:76–81
4. Fangrui Ma, Hanna MA (1999) Biodiesel production: a review. Bioresource Technology 70:1–15
5. Meher LC, Vidya Sagar D, Naik SN (2006) Technical aspects of biodiesel production by transesterification: a review. Renewable and Sustainable Energy Reviews 10:248–268
6. Freedman B, Butterfield RO, Pryde EH (1986) Transesterification kinetics of soybean oil. Journal of the American Oil Chemist Society 63:1375–1380
7. Mittelbach M, Trathnigg B (1990) Kinetics of alkaline catalyzed methanolysis of sunflower oil. Fat Science Technology 92:145–148
8. Bikou E, Louloudi A, Papayannakos N (1999) The effect of water on the transesterification kinetics of cotton oil. Chemical Engineering and Technology 22:70–75
9. Diasakou M, Louloudi A, Papayannakos N (1998) Kinetics of the non catalytic transesterification of soybean oil. Fuel 77:1297–1302
10. Kusdiana D, Saka S (2001) Kinetics of transesterification in rapeseed oil to biodiesel fuel as treated in supercritical methanol. Fuel 80:693–698
11. Komers K, Skopal F, Stloukal R, Machek J (2002) Kinetics and mechanism of the KOH-catalyzed methanolysis of rapeseed oil for biodiesel production. European Journal of Lipid Science and Technology 104:728 737
12. Boocock DGB, Konar SK, Mao V, Sidi H (1996) Biomass and Bioenergy 11:43–50
13. Dorado MP, Ballesteros E, Arnal JM, Gomez J, Lopez FJ (2003) Exhaust emissions from a diesel engine fueled with transesterified waste olive oil. Fuel 82:1311–1315
14. Shi X, Yu Y, He H, Shuai S, Wang J, Li R (2005) Emission characteristics using methyl soyate-ethanol-diesel fuel blends on a diesel engine. Fuel 84:1543–1549
15. Shi X, Pang X, Mu Y, He H, Shuai S, Wang J, Chen H, Li R (2006) Emission reduction potential of using ethanol-biodiesel-diesel fuel blend on a heavy-duty diesel engine. Atmospheric Environment 40:2567–2574
16. McCormick RL, Parish R (2001) Technical barriers to the use of ethanol in diesel fuel. Milestone report to NREL MP-540–32674
17. McCormick RL, Michael S, Graboski TL, Alleman A, Herring M (2001) Impact of biodiesel source material and chemical structure on emissions of criteria pollutants from a heavy-duty engine. Environmental Science and Technology 35:1742–1747
18. Fat and oil derivatives. Fatty Acid Methyl Esters (FAME) (2003) Determination of free and total glycerol and mon-, di-, triglyceride content. Reference method EN 14105
19. Levenspiel O (1999) Chemical Reaction Engineering. 3rd ed. Wiley, New York

Intake Fraction for Benzene Traffic Emissions in Helsinki

Joana Soares, Miranda Loh, Ari Karppinen, Leena Kangas, Kari Riikonen, Matti Jantunen, and Jaakko Kukkonen

Introduction

Intake fraction (iF) is defined as the integrated incremental intake of Bz released from a source (or source category) and summed over all exposed individuals during a given exposure time, per unit of emitted pollutant [1].

$$iF = \frac{\sum_{people,\, time} \text{mass intake of pollutant by an individual } (mass)}{\text{mass released into the environment (mass)}} \tag{1}$$

The intake fraction concept has been developed as a metric which can be useful in both screening-level order-of-magnitude estimates and more detailed policy modeling [2]. For instance, an intake fraction of 1 in a million (10^{-6}) means that for every ton of a pollutant emitted, 1 g is taken in by the exposed population. Intake fraction takes into account the dispersion of pollutants, locations and activity of population and human breathing rates. The calculated intake fraction for benzene is directly applicable to any other inert substance emitted by the traffic, for example CO, NO_X, so the calculations also provide a ready-to-use tool for health effect studies concerning other pollutants and emission scenarios. However, for chemically active urban pollutants like NO_2 and O_3, the presented intake fraction results are not applicable, thus a model run accounting correctly for the most important chemical mechanism must be performed for these substances.

Intake fractions have been most frequently calculated for whole populations, rather than for individuals. Population intake fraction is the sum of individual

J. Soares (✉), A. Karppinen, L. Kangas, K. Riikonen, and J. Kukkonen
Finnish Meteorological Institute, Erik Palmenin aukio 1, P.O. Box 503 FI-00101, Helsinki, Finland
e-mail: Joana.Soares@fmi.fi

M. Loh and M. Jantunen
National Public Health Institute, Neulaniementie 4, P.O. Box 95FI-70701, Kuopio, Finland

S. Rauch et al. (eds.), *Highway and Urban Environment*,
Alliance for Global Sustainability Bookseries 17,
DOI 10.1007/978-90-481-3043-6_8, © Springer Science+Business Media B.V. 2010

intake fractions, and understanding the spatial and temporal variability across a population for individual intake fractions provides useful information about which exposure factors dominate the inter-individual intake fraction variation.

Methodology

This study examines spatial distributions of benzene from local vehicular traffic, in the Helsinki Metropolitan Area (HMA), using different methods of intake fraction calculations. The first method incorporates both spatial and temporal information on population activity patterns as spatial concentration distribution for 1 year, and subsequently examine the spatial distribution of intake pattern in an urban area, using the exposure model EXPAND model [3]. This model was further developed to calculate iF for several substances. The spatial benzene concentration distributions were obtained by using dispersion models: CAR-FMI [4] and OSPM [5]. For HMA CAR-FMI was used for the emission calculations and the results were compared with previous studies conducted in HMA by [6]. OSPM was used for the street canyon calculations in a segment of Hämeentie Street. The activity patterns available for EXPAND are for HMA working age population (25–55 years old), representing 46% of the whole HMA population [3]. The microenvironment activity data in EXPAND is divided into four categories: home, workplace, traffic and other activities. Note however, that EXPAND does not calculate an individual's personal intake fraction, but rather the intake fraction per unit area, averaged over the number of people that are located within that area for all or some portion of the averaging time. Although several activities are comprised in the EXPOLIS database, the EXPAND model accepts a single breathing rate only, so we assumed a constant rate of 1 m^3/h, based on the U.S. Environmental Protection Agency's (USEPA's) Exposure Factors Handbook [7]. In order to include the air exchange rates between outdoor and indoor, an infiltration efficiency of 0.8 for buildings [8] and 4 for vehicles [9] was used to calculate intake fractions for indoor environments, acknowledging that concentrations inside vehicles are higher [10].

The second method uses data from Air Pollution Exposure Distributions of Adult Urban Populations in Europe (EXPOLIS) project to estimate intake fractions for individuals in different environments. The data describes the adult population between 25 and 55 years old when exposed to different pollutants, including benzene, in different micro-environments: home, workplace and outdoor. Detailed descriptions of the EXPOLIS study design and methods have been published previously [11–13]. We assume that the residual personal benzene exposure, not accounted for by residential indoor, outdoor, and workplace exposures is due primarily to exposure while commuting. Since EXPOLIS did not measure in-vehicle benzene concentrations, individual intake fraction while commuting was estimated. The method to calculate individual iF while commuting is described in [14]. An individual's intake fraction is computed from:

$$iF = \frac{\sum_{i=1}^{m} C_{v,k} t_k Q_k}{E_v} \tag{2}$$

where Cv,k = the benzene concentration in microenvironment k due to vehicular sources (μg/m³), tk = the time spent in microenvironment k (h), Qk = the breathing rate while in microenvironment k (m³/h), and Ev = benzene emissions (μg) over the time evaluated. Thus, each EXPOLIS participant's intake fraction (iFi) is:

$$iF_i = \frac{C_b t_c Q_l + pC_o t_{hi} Q_r + pC_o t_w Q_l + pC_o t_{oi} Q_l + C_o t_o Q_m}{E_{v,y}} \tag{3}$$

where C_b = residual benzene concentration; C_o = the concentration measured outside the participant's home while the participant was at home; t_c = time spent commuting (h); t_{hi} = the time spent indoors at home; t_w = the time spent at work; t_{oi} = time spent in other indoor microenvironments (h); t_o = time spent in all outdoor microenvironments; p = the infiltration efficiency for benzene from outdoor air into the indoors; Q_l = light activity breathing rate; Q_r = resting breathing rate; Q_m = moderate activity breathing rate; $E_{v,y}$ = emissions from vehicles in the Helsinki metropolitan area in y = year 1996 or 1997, depending on when the participant was sampled [15]. The terms on the right-hand-side of Eq. 3 represent exposure while commuting; at home, at work, in other indoor microenvironments and in all outdoor microenvironments. We assume here that the outdoor contribution is predominantly from vehicular emissions and that other sources (including long-range transport) are negligible. The times spent in each microenvironment were taken from participants' time-activity diaries and breathing rates were from the U.S. EPA's Exposure Factors Handbook [7].

Results and Discussion

The results obtained for 2000 and 2005 show the same spatial distribution pattern, though a decrease of concentration for the year 2005 was noticed (~factor 0.5). This might be a result of the European legislation to reduce benzene content in fuel (Directive 98/70/EC) and setting limit values for benzene ambient air concentrations (Directive 2000/69/EC).

The modeled traffic contribution to benzene concentrations in the in HMA in 2000 vary between 0 and 2.7 μg/m, as depicted in Fig. 1. Previous studies in HMA in 2000 [6] show traffic benzene concentrations varying between 0.2 and 2.2 μg/m³.

The higher concentrations are situated closer to the line sources, where the benzene is emitted, and decreasing for receptors further away from the streets. The centre of Helsinki, due to a denser and highly trafficked road network, including main highways, has the highest concentrations ranging between 1.5 and 2.5 μg/m³. For the street canyon calculations the differences between the calculated yearly average

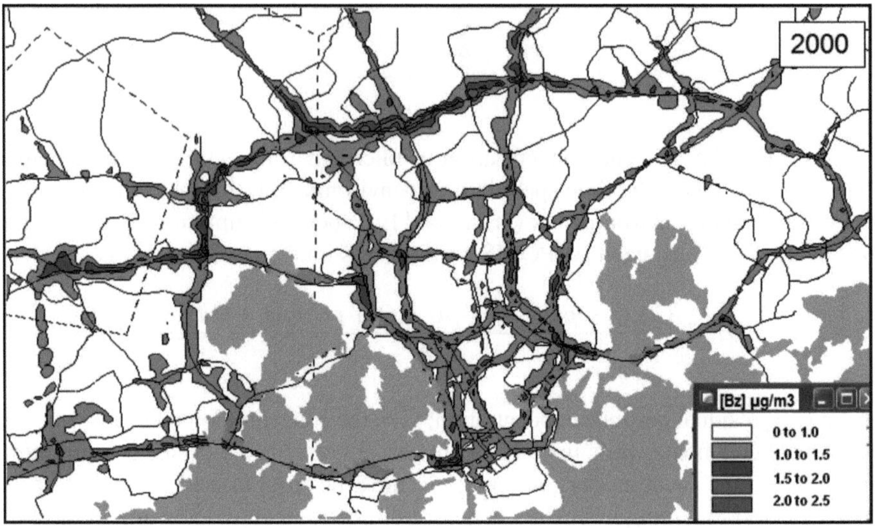

Fig. 1 Spatial distribution of benzene emissions from mobile sources in 2000

benzene concentrations were not substantially high, varying from 0.2 to 0.5 µg/m^3. These concentrations are considerably lower than the results for the HMA which is mainly attributable to the low height/width (0.66) ratio of the studied street canyon.

The annual average benzene intake results for the working-age population within HMA during 2000 and 2005 were calculated using the EXPAND model. Only results for 2005 are shown in Fig. 2, as the results for both years were almost identical.

Higher values of total intake fraction, ranging between 10^{-10} and 10^{-7}, are concentrated in residential and commercial areas, where people spend most of their time. Therefore, there is an increase of receptors but also sources (traffic amount increases) in these specific grid cells. The average intake fractions due to exposure in home, work, traffic, and other microenvironments for the working-age population within the HMA are 2.5×10^{-6}, 1.4×10^{6}, 5.9×10^{-6} and 0.2×10^{-6}, respectively. The total intake fraction that corresponds to the exposure of the working-age population summed across these different microenvironments in the HMA is 1.0×10^{-5}. Population exposure was higher during commuting, because of the relatively higher concentrations in the vicinity of the main roads and streets, and subsequently higher concentrations of benzene inside vehicles. People are also more exposed to benzene traffic related emissions in the home and work environments, since these account for around 90% of people's time. About 20% of the subjects have an intake fraction of zero from combined exposure in traffic and outdoor microenvironments. These null values can be attributed to either no reported time spent in these microenvironments or nonexistent traffic related benzene concentrations, according to the dispersion computations. Additionally, a total intake fraction of 3.7×10^{-5} was calculated for the street canyon. Using measured data from EXPOLIS as the average, individual intake fraction of benzene from traffic sources over a 48-h period was 8.0×10^{-11},

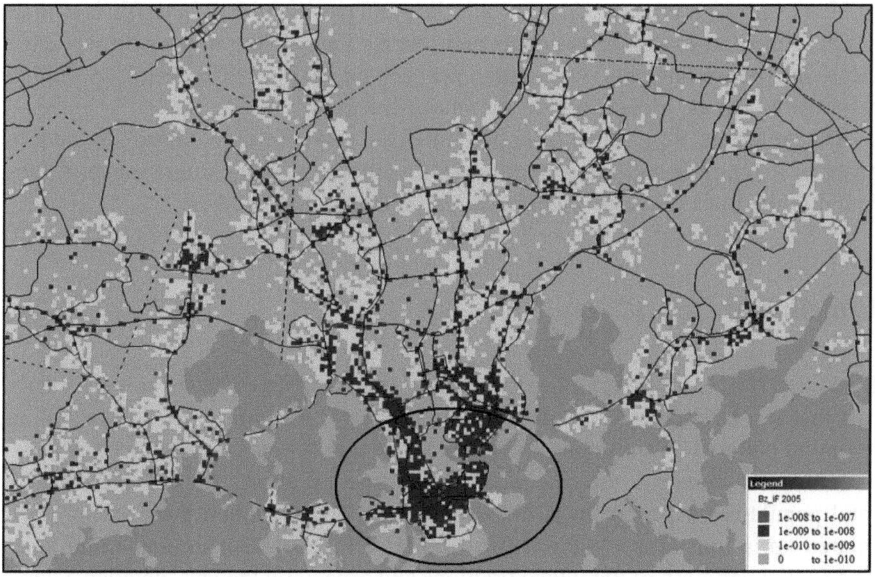

Fig. 2 Spatial distribution of total intake fraction for benzene from mobile sources in 2005. Helsinki city centre is marked with the *black ellipse*

Table 1 Intake fraction by inhalation in Hämeentie Street (2000)

Groups	Breathing rate (m^3/day)	No. of people	Time of exposure (days)	iFi
Inhabitants	1	176	0.56	1.6 E–05
Working and costumers	1	113	0.37	2.0 E–05
In traffic	1	20,380	1.04 E–04	1.1 E–06
Total iF				3.7 E–05

with a total intake fraction for these 129 people of 1.0×10^{-8}. If we extrapolate from the 129 EXPOLIS people to the working-age population (i.e., by multiplying this iF with the ratio of the total working-age population within the area to the EXPOLIS population, [3]), we estimate a population intake fraction of 3.9×10^{-5}. The intake fraction values computed using the EXPAND model and based on EXPOLIS data are not directly comparable, as the location of the populations and the years considered are different, and due to methodological differences. However, they both provide useful information about the patterns of exposures across space and individuals, and the results obtained by the two methods are fairly similar.

In a densely trafficked street canyon, such as the segment of Hämeentie considered in this study, about 40 μg/g of benzene emitted on that road is inhaled by the people living, working, and passing through that road segment. It can be seen that due to the longer amount of time spent in the area, the intake fraction for residents is largest and the in-traffic intake fraction is smallest (Table 1).

However, the intake fraction per unit time and per individual in traffic is actually larger compared to that of residents and others. Thus, if the time spent in each of these microenvironments was equal per person, exposure in traffic would be responsible for the greater part of the intake fraction.

The intake fraction results from both modeled and measured data are in the same order of magnitude and are generally within an order of magnitude range of those found in the most previous mobile source intake fraction studies.

Conclusion

Intake fraction is useful measure to evaluate and compare different sources categories that can be used a health decision support tool. Assuming that the spatial and temporal patterns of emissions and population activity remain unchanged in different emission scenarios, the calculated intake fraction allows for direct estimate of total population exposure in all these scenarios. We evaluated the benzene traffic related emission intake fraction for the MHA for different domains and years. The spatial variation of the intake fractions and especially the very detailed calculations for the smallest domain, street canyon, demonstrate clearly how crucial it is to have access to detailed information on traffic patterns and locations and activities of the people in order to get a reliable estimate on the real burden of pollutants on human exposure and health.

The highest values of 100×100 m^2 grid cell iF's (10^{-10}–10^{-7}) are concentrated in residential and commercial areas where the density of receptors and sources is higher. In other words, between 0.1 ng and 0.1 μg are inhaled in each 100×100 m^2 grid cell per gram of benzene emitted from mobile sources in the whole HMA. The intake fraction of the same population summed across these different microenvironments in the HMA is 10^{-5}. Population exposure to benzene was higher during commuting due to higher concentrations in the vicinity of the streets as well as inside vehicles. The same population is also exposed at home and workplace as a result of spending a large fraction of time indoors.

References

1. Bennett D, McKone T, Evans J, Nazaroff W, Smith K, Margni M (2002) Defining intake fraction. Environmental Science and Technology 36:206A–211A
2. Bennett D, Margni M, McKone T, Jolliet O (2002) Intake fraction for multimedia pollutants: a tool for life cycle analysis and comparative risk assessment. Risk Assessment 22(5):905–918.
3. Kousa A, Kukkonen J, Karppinen A, Aarnio P, Koskentalo T (2002) A model for evaluating the population exposure to ambient air pollution in an urban area. Atmospheric Environment 36:2109–2119
4. Karppinen A, Kukkonen J, Elolähde T, Konttinen M, Koskentalo T (2000) A modelling system for predicting urban air pollution: comparison of model predictions with the data of an urban measurement network in Helsinki. Atmospheric Environment 34:3735–3743

5. Berkowicz R (2000) A simple model for urban background pollution. Environmental Monitoring and Assessment 65:259–267
6. Hellén H, Kukkonen J, Kauhaniemi M, Hakola H, Laurila T, Pietarila H (2005) Evaluation of atmospheric benzene concentrations in the Helsinki Metropolitan Area in 2000–2003 using diffusive sampling and atmospheric dispersion modelling. Atmospheric Environment 39:4003–4014
7. U.S. Environmental Protection Agency (USEPA) (1997) Exposure Factors Handbook. Office of Health and Environmental Assessment. Volumes 1 and 2. EPA/600/P-95/002Fa and EPA/600/P-95/002Fb
8. Joint Research (JRC) (2003) Draft EU Risk Assessment Report on Benzene (Human Health Part March 2003). http://ecb.jrc.it/existing-chemicals/
9. Marshall J, Teoh S, Nazaroff WW (2005) Intake fraction of nonreactive vehicle emissions in US urban areas. Atmospheric Environment 39:1363–1371
10. Marshall JD, Behrentz E (2005) Vehicle self-pollution intake fraction: children's exposure to school bus emissions. Environmental Science and Technology 39:2559–2563
11. Edwards RD, Jantunen MJ (2001) Benzene exposure in Helsinki, Finland. Atmospheric Environment 35:1411–1420
12. Jantunen MJ, Hänninen O, Katsouyanni K, Knöppel H, Künzli N, Lebret E, Maroni M, Saarela K, Sram R. Zmirou D (1998) Air pollution exposure in European cities: the "EXPOLIS" study. Journal of Exposure Analysis and Environmental Epidemiology 8:495–518
13. Saarela K, Tirkkonen T, Laine-Ylijoki J, Jurvelin J, Nieuwenhuijsen MJ, Jantunen M (2003) Exposure of population and microenvironmental distributions of volatile organic compound concentrations in the EXPOLIS study. Atmospheric Environment 37:5563–5575
14. Loh MM, Soares J, Karppinen A, Kukkonen K, Kangas L, Riikonen K, Kousa A, Asikainen A, Matti J (2008) A comparison of intake fraction distributions for benzene from vehicles in Helsinki Metropolitan Area using different methods (submitted to Atmospheric Environment) Intake fraction distributions for benzene from vehicles in the Helsinki metropolitan area Atmospheric environment 2009, vol 43 pp 301–310
15. Myllynen M, Aarnio P, Kostkentalo T, Malkki M (2006) Air quality in Helsinki Metropolitan Area year 2005. Helsinki: Helsinki Metropolitan Area Council (YTV)

Modelling of Air Pollutants of Highway Roads in Hungary

Györgyi Baranka

Introduction

As new European Community member Hungary is obliged to harmonize its transport sector according to the European rules. The harmonization process concerns among others road network development, transport security as well as environmental issues. Due to the lack of suitable technical tools the difficult contradictory issues related to the environmental matters were not rationally managed. For the accomplishment of such concept a sophisticated environmental model is necessary considering all traffic related atmospheric pollutant emission of major roads. Implementing such an environmental model Hungary could optimize its transport sector in conformity with the EC rules.

The present work aims to provide decision makers the necessary technological support and scientific information on air quality of highway roads. To achieve this goal a road dispersion model will be used to estimate the possible effects of road traffic on ambient air.

Emission

The expressway network in Hungary was in 2004 comprised of 650 km of motorway, 205 km of clearways and highways (which are comprised of highways serving as expressways, highways with expressway features later converted to expressways, and highway sections with expressway features) as well as 188 km of junction point sections, 52 km of roads leading to rest areas, and 106 rest areas (see Fig. 1).

The Hungarian CO and NO_x emissions from mobile sources have been investigated by Merétei [1]. Emission of nitrogen oxides from road traffic continuously

G. Baranka (✉)
Hungarian Meteorological Service

S. Rauch et al. (eds.), *Highway and Urban Environment*,
Alliance for Global Sustainability Bookseries 17,
DOI 10.1007/978-90-481-3043-6_9, © Springer Science+Business Media B.V. 2010

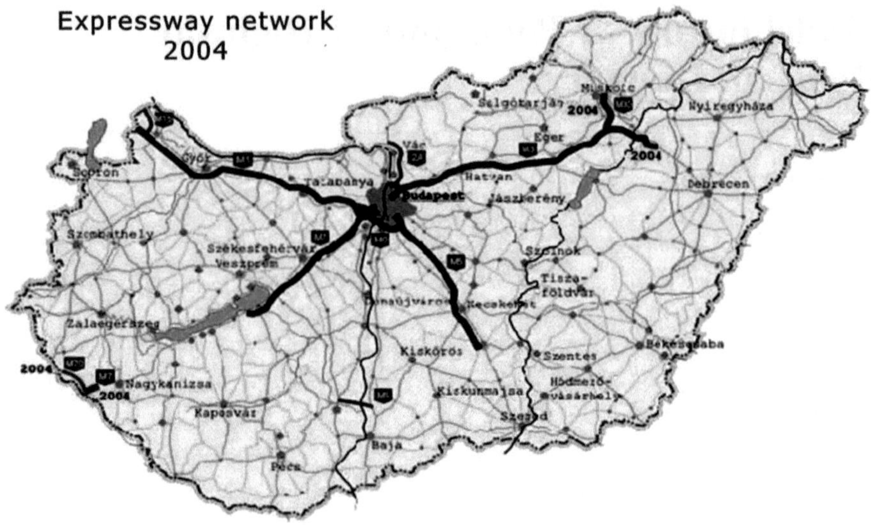

Fig. 1 Expressway network in Hungary in 2004

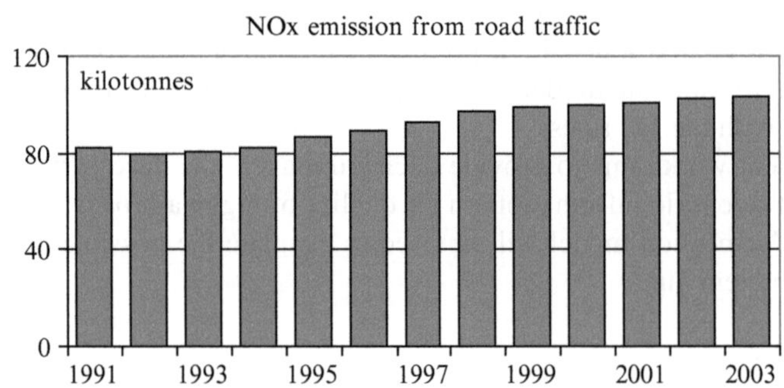

Fig. 2 NO_x emission from road traffic in Hungary, determined by [1]

increased (Fig. 2), while the emission of carbon monoxide from the same sources decreased during the last 20 years (Fig. 3). The NO_x emissions from road traffic is growing (growing rate is 20%) due to the increase in traveled kilometers by cars and duty vehicles. In the same time NO_x emissions from the European member countries decreased by 20% due to the application of three steps catalysts in new cars and the changes of the compound of the fuel.

Source emissions are a function of both the vehicle emission factors and the vehicle activity (usually measured in vehicle kilometers traveled). The vehicle emission factors can differ greatly depending on the type of road, vehicle fleet mix, and traffic flow encountered. As a result, emission factors were determined specifically for each roadside location to ensure that the appropriate input was used for the

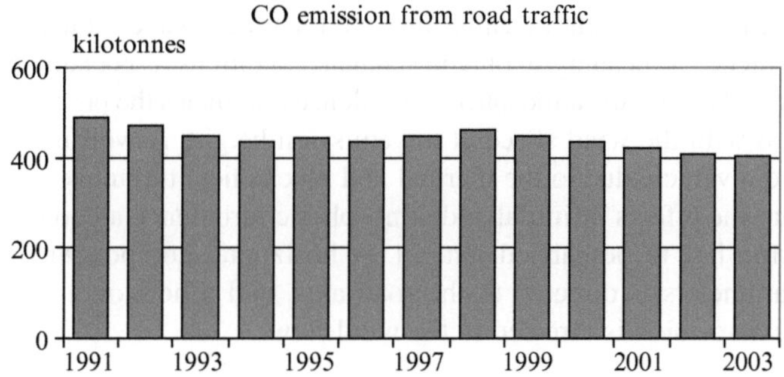

Fig. 3 CO emission from road traffic in Hungary, calculated by [1]

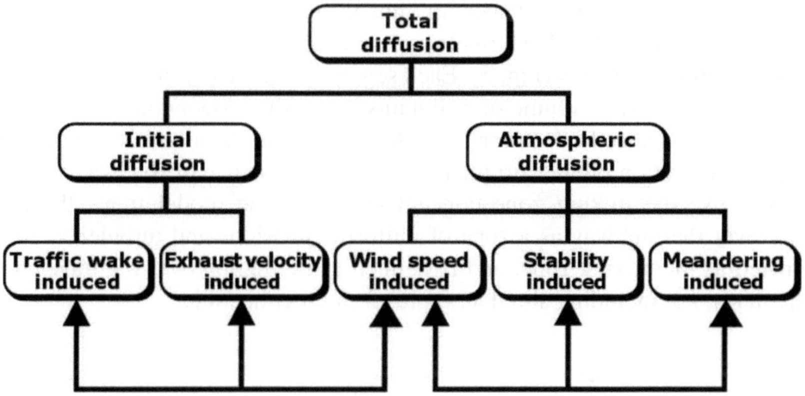

Fig. 4 Diffusion processes in road traffic modeling. *Arrows* represent interactions

modeling of subsequent roadway emissions. These site-specific emission factors were then combined with the vehicle kilometers traveled data to give a total emission rate for the road.

Diffusion Processes in Road Modeling

The total diffusion is divided into primary diffusion affected by the traffic and diffusion caused by ambient meteorological conditions (see Fig. 4). Contaminants outcome from the tailpipe take part in the primary process, on one hand by the thermal turbulence caused by differential surface heating and by eddy movements created by the buoyant exhaust plume, on the other hand by mechanical turbulence from wind flow and the traffic wake induce turbulence. Thermal influences interact with mechanical effects. There are interactions between the ambient wind speed, exhaust velocities of the tailpipe emissions and the traffic wake-induced turbulence. Primary turbulence is meaningful only in stagnant conditions (especially in the case of cold-running) otherwise the atmospheric turbulence dominates the process. The plume is transported with the wind speed at the emission height. A well-mixed zone can occur behind a vehicle due to the thermal and mechanical turbulence.

To sum up the effects of initial and atmospheric turbulence a coordinate system transformation has to be carried out [2]. A horizontal component of the initial diffusion parameters is directed to the road axis, and a horizontal component of atmospheric dispersion is directed to the wind flow.

Description of CALINE4 Model

CALINE4 is a Gaussian type dispersion model for predicting air pollution concentrations near roadways [3]. Each roadway link analyzed in the model is treated as a sequence of short segments (Fig. 5). Each segment is treated as a separate emission source, which produce a plume of pollutants dispersing downwind. Pollutant concentrations at any specific location are calculated using the total contribution from overlapping pollution plumes originating from the sequence of roadway segments. CALINE4 use the mixing zone concept (Fig. 6). The model treats the region directly over the highway as a zone of uniform emissions and turbulence. This is designated as the mixing zone, and is defined as the region over the traveled way (traffic lanes; not including shoulders) plus 3 m on either side.

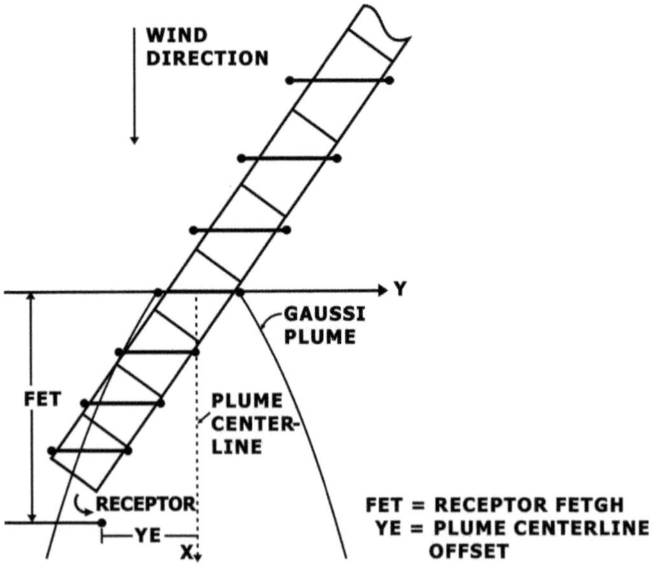

Fig. 5 Gaussian plume and roadway segments

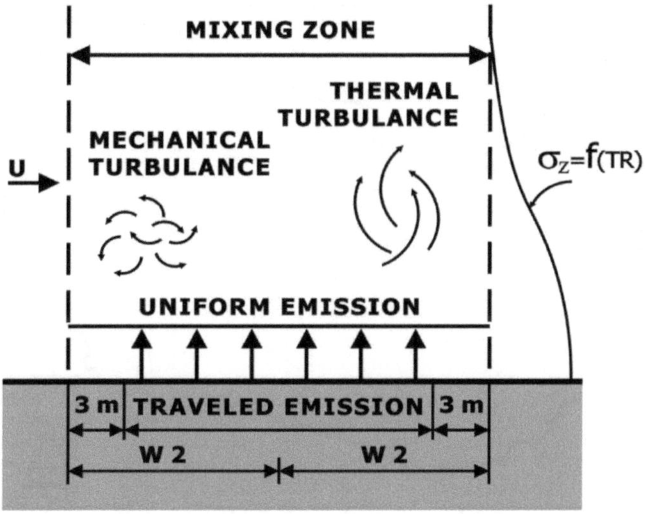

Fig. 6 Main characteristics of the mixing zone concept

The additional width accounts for the initial horizontal dispersion imparted to pollutants by the vehicle wake. Within the mixing zone, the mechanical turbulence created by moving vehicles and the thermal turbulence created by hot vehicle exhaust is assumed to be the dominant depressive mechanisms.

The chemical transformation is modeled by the so-called receptor-oriented discrete parcel method. The model includes the basic reactions of nitrogen oxides, oxygen and ozone, but the influence of other compounds is neglected. The influence of plume dilution has been accounted for according to the receptor oriented Discrete Parcel Method [4], which is a modified version of the original method.

The CALINE4 requires relatively minimal input from the user. Input data are traffic volume, emission factors, roadway geometry, wind speed and direction, ambient air temperature, mixing height, atmospheric stability class and coordinate of receptors.

Model Experiments and Results

By using CALINE4 dispersion model, hourly and 8-h average concentrations can be obtained. The CO and NO_2 concentrations have been calculated at 100 m away from the road axis for the Hungarian motorway network. NO_2 concentration is shown in Fig. 7 for worst-case meteorology, where input is the combination of the worst wind speed, wind direction, and stability class. The highest concentrations occur around Budapest, which is explained by arterial roads structure in Hungary and by urban sprawling.

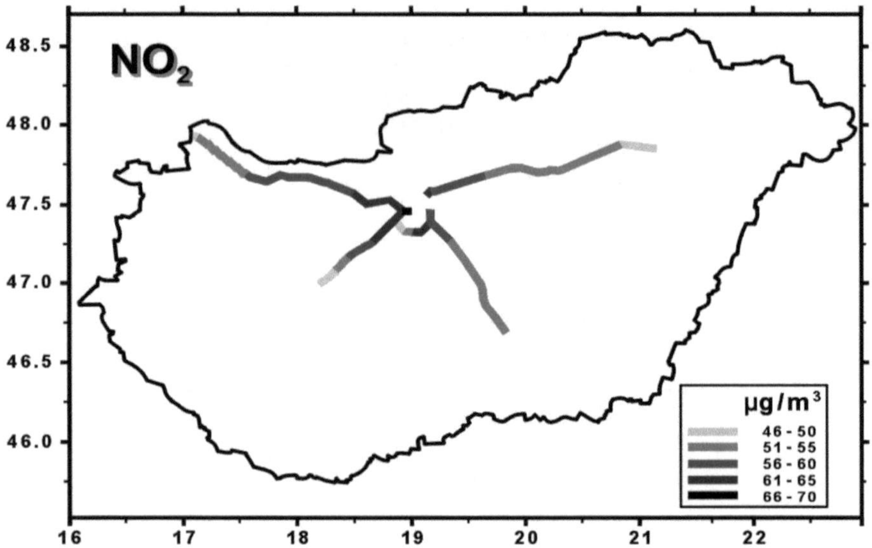

Fig. 7 Worst-case 1-h NO_2 level at 100 m away from road axis

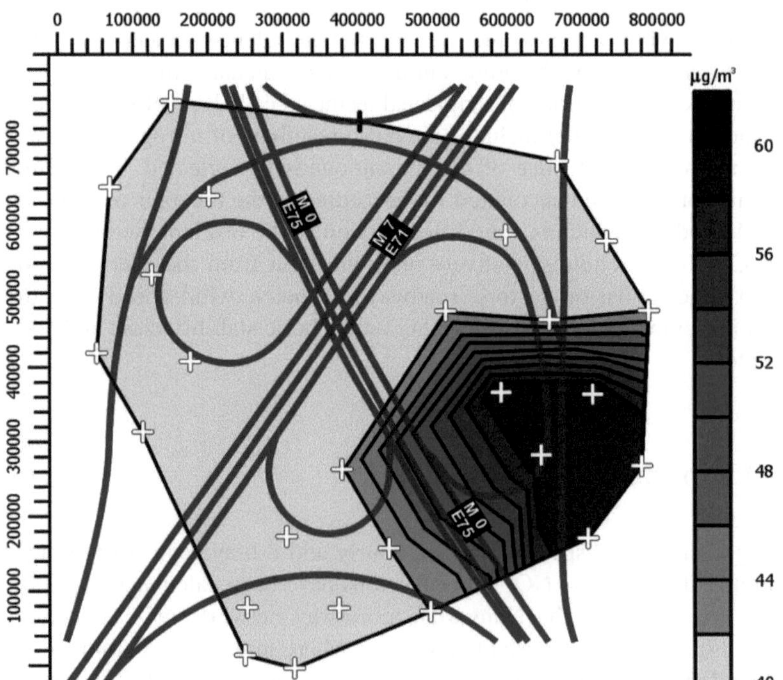

Fig. 8 Hourly average NO_2 concentration around an intersection near Budapest

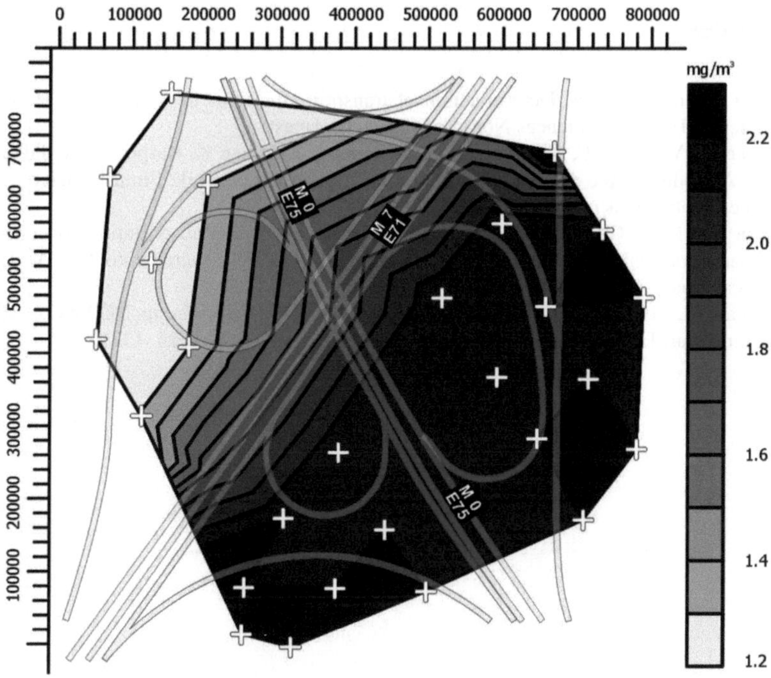

Fig. 9 Hourly average CO concentration around an intersection near Budapest

A sensitivity analysis of CALINE4 model was also carried out. CO concentration was estimated to the distance from a roadway for typical daytime conditions. Wind was perpendicular to the roadway. It was established that good results could be obtained at receptors located within 500 m from the roadway.

CALINE4 model provides a possibility to predict air pollutants near intersections. Spatial distributions of hourly averaged CO and NO_2 concentrations are shown in Figs. 8 and 9 for an intersection near Budapest. The distance between receptors (indicated by crosses in the figures) is 100 m. Air pollutants are transported by the most frequent airflow to the SW direction. At this intersection the traffic of motorway M7 is dominant.

Conclusion

Examples simulated by a road dispersion model showed that acceptable results could be obtained in order to consider detailed effect of traffic flow. Using a dispersion model connected to an emission model is a good tool to estimate air quality where ground based measurements are not available or not possible.

References

1. Merétei T (editor) (2006) Determination of transport emissions in Hungary (in Hungarian) Institute for Transport Sciences Non-profit LTD, Budapest
2. Härkönen J, Valkonen E, Kukkonen J, Rantakrans E, Lahtinen K, Karppinen A, Jalkanen L (1996) A model for the dispersion of pollution from a road network. Finnish Meteorological Institute, Publications of Air Quality 23, Helsinki, 34 pp
3. Benson PE (1989) CALINE4 a dispersion model for predicting air pollution concentrations near roadways. California Department of Transportation (1988), Sacramento, CA. Air Quality Technical Analysis Notes, Sacramento, CA
4. Hanrahan PL (1999) The plume volume molar ratio for determining NO2/NOx ratios in modelling. Part I: Methodology. J. Air Waste Manage. Assoc. 49:1324–1331

Seasonal Trend of Water-Soluble Ions at One TSP and Five $PM_{2.5}$ Sampling Sites in Beijing, China

Nina Schleicher, Stefan Norra, Fahe Chai, Yizhen Chen, Shulan Wang, and Doris Stüben

Introduction

Approximately 15 million people currently are living in Beijing, the capital of China. The rapid economic growth comes along with a rapid growth of the urban population and a huge increase in energy consumption. Beijing faces severe anthropogenic air pollution caused by an intense increase of traffic, coal combusting power plants, heavy industry, huge numbers of household and restaurant cookers, and domestic heating stoves. Additionally, each year especially in spring, dust storms carry particulate matter over long distances from the deserts of Gobi and Takla Makan and other semi-arid areas in the northwest of Beijing towards the city. Other sources of particulate matter such as bare soils, coal heaps and construction sites occurring in and around Beijing contribute to the air pollution. The single shares of the specific aerosol sources are not known, neither are the mixing processes of these aerosols within the area of Beijing.

Currently, the air pollution due to particulate matter in Beijing exceeds by far the pollution in cities of Western Europe as well as national and international threshold values. The high atmospheric pollution is reflected in a severe reduction of visibility at many days in Beijing. This situation also affects the public health with regard to respiratory diseases [1]. Water-soluble ions are easily released to the environment and are known to have negative effects on human health [2, 3]. Furthermore, anions such as nitrate and sulphate, cause acidification of rain. Therefore, this study focuses on water-soluble ions in Beijing dust samples.

N. Schleicher, S. Norra, and D. Stüben
Institute of Mineralogy and Geochemistry, Universität Karlsruhe, Germany

F. Chai, Y. Chen, and S. Wang
Chinese Research Academy of Environmental Sciences, Beijing, China

S. Rauch et al. (eds.), *Highway and Urban Environment*,
Alliance for Global Sustainability Bookseries 17,
DOI 10.1007/978-90-481-3043-6_10, © Springer Science+Business Media B.V. 2010

Sampling and Methods

Five $PM_{2.5}$ (particulate matter ≤2.5 μm) sampling sites were positioned along a transect from NW to SE through the city of Beijing (Fig. 1). Sampling was carried out with mini-volume-samplers at a flow rate of 200 L/h separately for day and night on a weekly basis. Day-time samples were taken from 7 a.m. to 7 p.m. and night-time samples from 7 p.m. to 7 a.m., respectively. Site 1, 4 and 5 were installed at 1.5 m above ground – a height especially relevant for the exposure to humans. Site 2 and 3 were set up on building roofs at heights of about 20 and 7 m, respectively.

At site 4 TSP samples (total suspended particles ≤100 μm) were taken additionally with a TSP-Sampler at a flow rate of 1 m^3/h. TSP samples were also collected weekly, but without distinction between day and night. All samples were collected on quartz fiber filters (MN QF 10, Macherey-Nagel, d = 50 mm). The sampling period presented within this paper comprises 1 year, from January to December 2006.

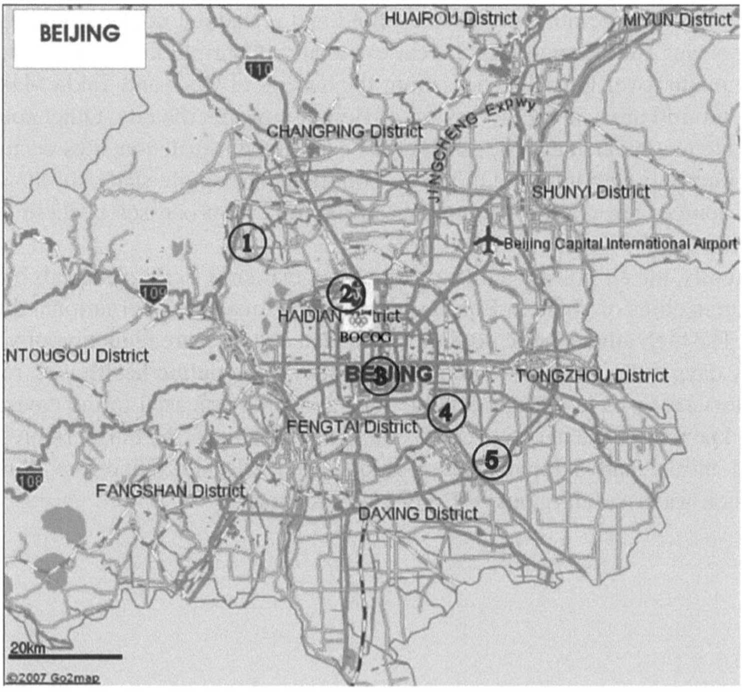

Fig. 1 City map of Beijing with the locations of the five sampling sites (from beijing2008.go2map.com)

Analysis

For determination of total mass concentration, the filters were weighted using a microbalance (Sartorius SE 2-F, readability = 0.1 μg) before and after sampling. In each case the mean value of five weightings was used for calculation. The standard deviation was kept smaller than 0.1 (in most cases ≤0.05). Additionally, the weighing was controlled by blank filters.

Water-soluble ions were analyzed by ion chromatography (IC). For this analysis, one quarter of each of the collected filters was shaken for 45 min in 15 mL ultrapure deionized water (Milli-Q) and additionally extracted ultrasonically for 30 min in order to release the water-soluble parts.

For anions, a Dionex ICS-1000 with ASRS-ULTRA suppressor, Ion Pac AS4A-SC column and a Na_2CO_3/$NaHCO_3$-eluent was used. Cations were analyzed with a Dionex DX-120 with CSRS-ULTRA suppressor, Ion Pac CS12, using a 0.1 M H_2SO_4-eluent. Standard solutions of both anions (Multi ion anion IC standard solution Specpure, Alfa-Aesar) and cations (Multi-component Cation Mix 2, Acculon) were used in different concentrations for quality control.

Results and Discussion

Particle Mass Concentration

In 2006 the average total mass concentration for $PM_{2.5}$ samples from all five sampling sites was 90.7 ± 57.3 μg/m^3 for day samples and 91.6 ± 67.9 μg/m^3 for night samples (all mean values in this paper are expressed plus/minus standard deviation). Site 4 showed highest annual $PM_{2.5}$ concentrations with 114.7 ± 79.1 μg/m^3 (average for day and night). The average TSP concentration (sampled only at site 4) was 371.4 ± 141.9 μg/m^3 in 2006. These values exceed even the Chinese class III standard for annual TSP concentrations. China has three classes for air quality standards which are the following for TSP: class I – 60 μg/m^3, class II – 120 μg/m^3, class III – 150 μg/m^3. No threshold value exists for $PM_{2.5}$ in China until now. The yearly average $PM_{2.5}$ concentrations for all five sites are shown in Fig. 2.

The mass concentration varied considerably over the course of the year. Highest concentrations, both in $PM_{2.5}$ and TSP, were found in winter with 113 ± 72 and 448 ± 165 μg/m^3, respectively. Summer values showed lowest particle concentrations; 72 ±44 and 235 ± 60 μg/m^3, respectively. Spring and autumn concentrations were also quite high (spring: 94.3 ± 53.8 and 405.3 ± 126.3 μg/m^3, autumn: 76.3 ± 38.4 and 396.5 ± 103.0 μg/m^3).

In this study, the seasons were defined as follows: winter – Dec, Jan, Feb, spring – Mar, Apr, May, summer – June, July, Aug, autumn – Sept, Oct, Nov.

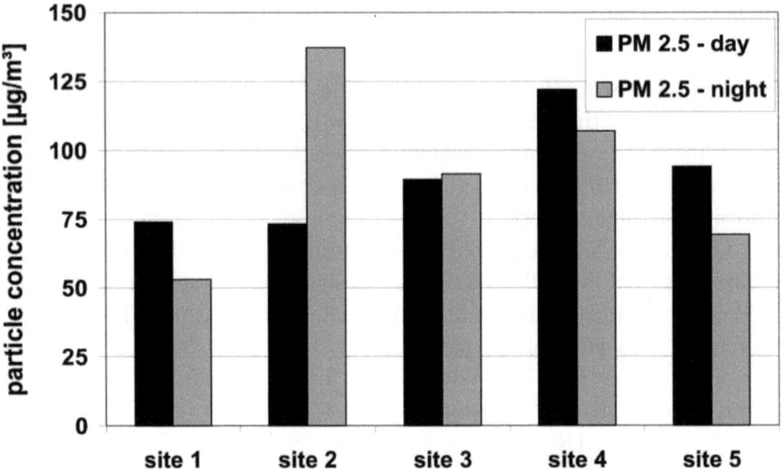

Fig. 2 Average $PM_{2.5}$ concentrations for all five sampling sites in 2006

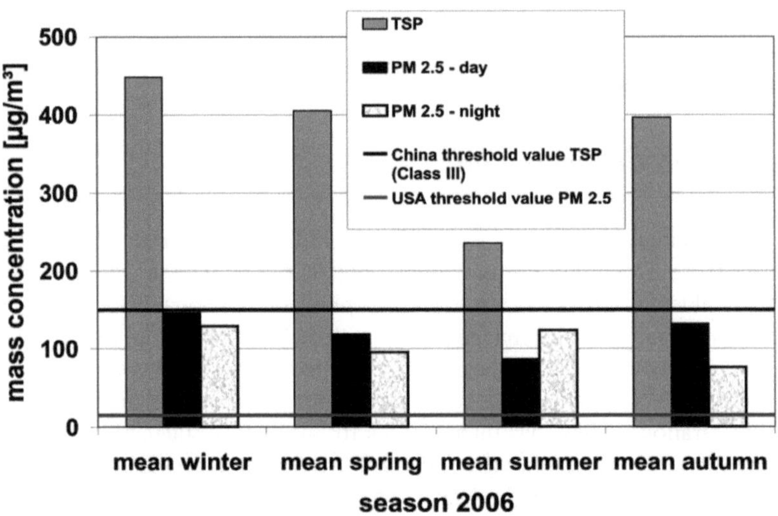

Fig. 3 Average TSP and $PM_{2.5}$ concentrations for each season (2006) at sampling site 4 in Beijing and the corresponding threshold values

Figure 3 illustrates the average seasonal concentrations for TSP and $PM_{2.5}$ concentrations at sampling site 4 whereas the average seasonal $PM_{2.5}$ concentrations for all five sites can be seen in Fig. 4. Site 1, which is located northwest of the city centre (Fig. 1), displayed the lowest $PM_{2.5}$ concentration throughout the whole year.

Site 2–4, which are central sites situated within the fifth Ring Road, proved to have higher particle mass concentrations. The seasonal course is illustrated in Fig. 5 with the results of TSP and $PM_{2.5}$ concentrations from site 4. The high spring

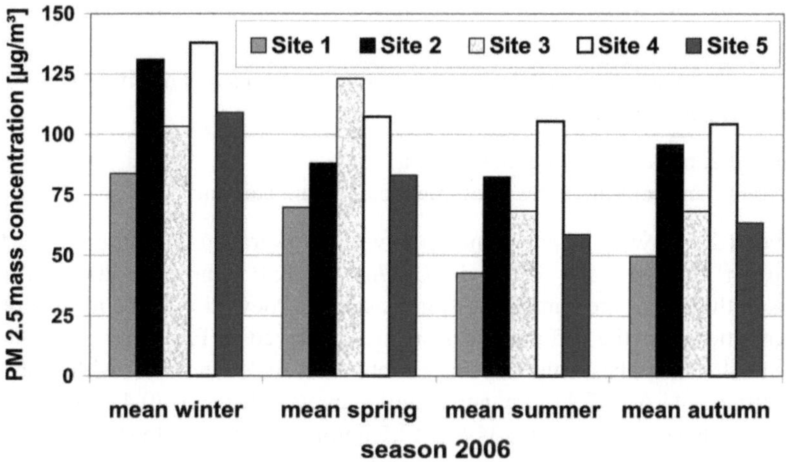

Fig. 4 Average $PM_{2.5}$ concentrations (mean day and night) for each season (2006) at all five sampling sites in Beijing

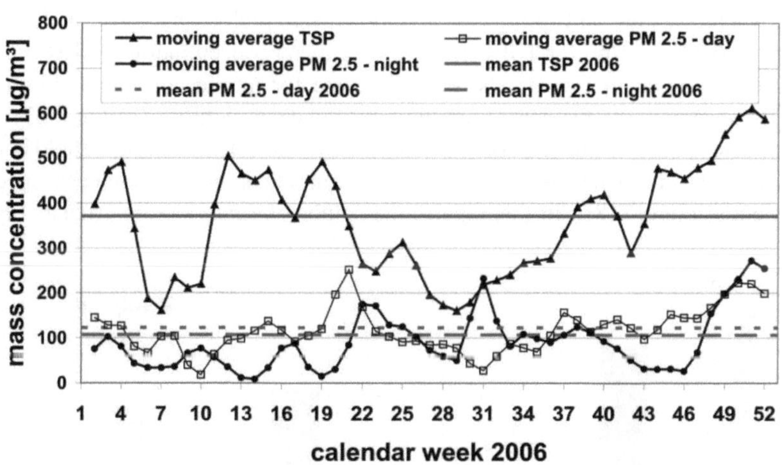

Fig. 5 $PM_{2.5}$ and TSP concentrations at site 4 for the course of the year 2006

concentrations occur especially in the coarse fraction (TSP ≤100 µm). This supports the assumption that the high amount of particulate matter during this time of the year mainly originates from long-distance transport from arid areas northwest of Beijing. During so-called dust storms days these geogenic particles accumulate in the city [4–6]. The high particle concentration in wintertime, however, is reflected both in the coarse and fine fractions. During this period, heating activities, which include a high amount of coal combustion, contribute noticeably to the high aerosol pollution.

Water-Soluble Ions

With ion chromatography (IC) four water-soluble anions (i) and five water-soluble cations (ii) could be detected:

1. Sulphate, nitrate, chloride and fluoride
2. Sodium, ammonium, potassium, magnesium and calcium

Sulphate (SO_4^{2-}) was the prevalent anion with an average concentration of 16.2 ± 10.2 μg/m³ for $PM_{2.5}$ and 33.9 ± 14.7 μg/m³ for TSP samples in 2006 (all mean values in this paper are named plus/minus standard deviation). The mean sulphate concentrations for the different seasons are displayed in Fig. 6 for TSP samples from site 4. The highest sulphate concentrations occurred in winter with an average concentration 41.6 ± 17.4 μg/m³ and a maximum weekly value 70.4 μg/m³. Summer values are also quite high with 35.6 ± 12.5 μg/m³, on average. In summer, a strong secondary formation of sulphate, caused by the strong solar radiation and high humidity, is assumable. In winter, the burning of fossil fuels, and especially of coal, probably contributes considerably to the high sulphate concentrations.

Nitrate, the second most abundant water-soluble anion, showed a different seasonal trend. Lowest nitrate concentrations were reached in summer time (8.2 ± 4.5 μg/m³ for TSP), whereas winter and spring concentrations were high (winter: 16.2 ± 7.5 μg/m³, spring: 11.5 ± 5.9 μg/m³ for TSP). The seasonal mean concentrations of TSP samples (site 4) are plotted in Fig. 6. The annual course for nitrate concentrations in TSP samples was similar to the course of total mass concentration ($r = 0.72$, $n = 53$).

Sulphate mainly originates from stationary sources (industrial or household burning of fossil fuels/coal), whereas the main source for nitrate are mobile sources (especially traffic emissions) [7, 8]). In this regard, the sulphate/nitrate ratio is crucial (Fig. 7).

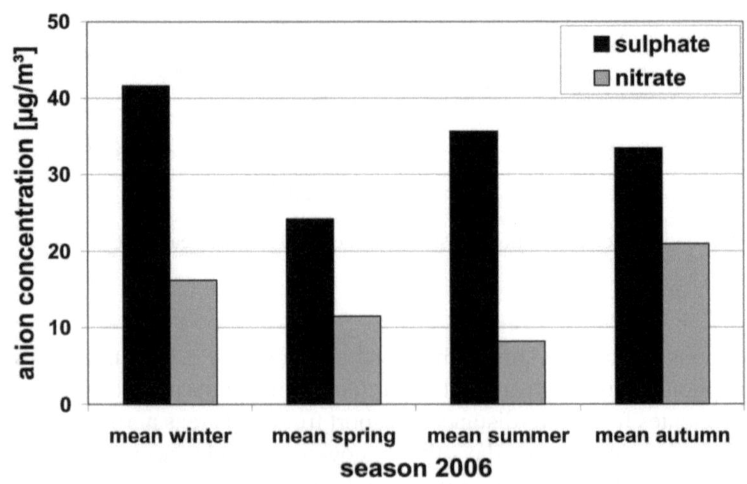

Fig. 6 Concentration of water-soluble sulphate and nitrate from TSP samples at site 4 for the different seasons in 2006

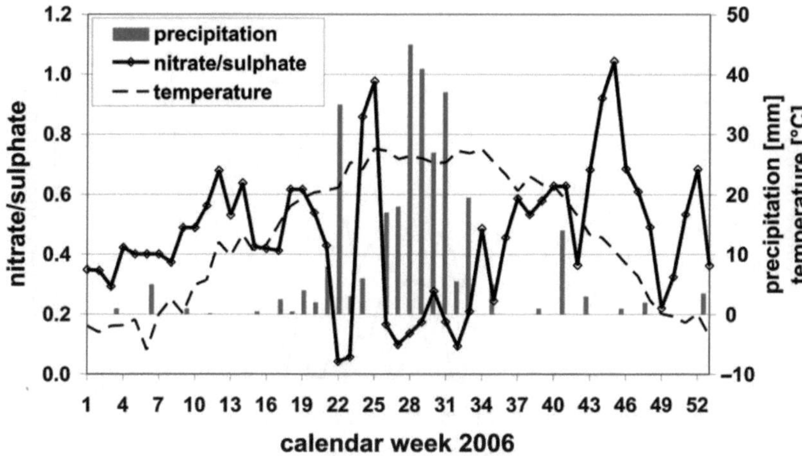

Fig. 7 Annual course for the ratio nitrate/sulphate from TSP samples at site 4 with additional meteorological data

The ratio is lowest in summer time when most rain occurs in Beijing. In this time, as mentioned above, the secondary formation of sulphate is high due to the high humidity and strong solar radiation. The temperature plays an indirect role to that effect and the long cold winters account for a long heating period in Beijing (usually from mid-November to mid-March [9]).

Fluoride and chloride concentrations were very low in $PM_{2.5}$ samples and therefore not detectable with IC (detections limits: 0.2 and 0.4 mg/L, respectively) in many samples. In TSP samples, the mean annual concentrations were 0.8 ± 0.6 μg/m^3 for fluoride and 3.3 ± 3.9 μg/m^3 for chloride.

The annual courses for selected cations in TSP samples are presented in Fig. 8. Calcium was the most abundant water-soluble cation with an annual mean concentration of about 13.4 ± 5.0 μg/m?. Usually, Ca can be seen as a geogenic element. In Beijing, however, Ca is also closely connected to construction activities. As a consequence of the fast growing city and lots of building activities for the Olympic Games in August 2008, there is a high density of construction sites in Beijing. Ca-rich particles in Beijing aerosols have been related to dust from construction sites [10]. Ca has also been used as an indicator element for construction dust in Beijing [9]. In contrast to the coarse particles, $PM_{2.5}$ samples showed lower Ca concentrations which partly could not be detected by IC. For fine particles, ammonium was the most important cation.

The percentage of measured water-soluble cations with respect to total dust concentrations of the according element varies between 35% and 72%. Total element concentrations were measured with ICP-MS after total digestion of one quarter of the filter.

Finally, the comparison between water-soluble anions and cations suggested a strong correlation between ammonium and sulphate ($r = 0.77$, $n = 53$) whereas

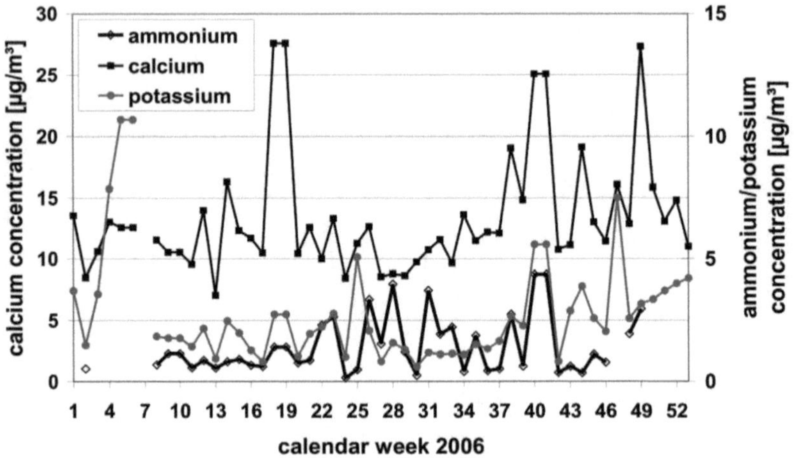

Fig. 8 Annual course for selected water-soluble cations in TSP samples from site 4 (missing values due to analytical problems)

ammonium and nitrate showed no correlation ($r = 0.10$, $n = 53$). Therefore, ammonium in Beijing's aerosols seems to be mainly present in the form of ammonium sulphate.

Conclusion

All over the year 2006, particulate matter concentrations in Beijing were high and thus atmospheric pollution was severe. $PM_{2.5}$ and TSP mass concentrations as well as the concentrations of the different ions analyzed revealed a different annual course. Seasonal differences are caused by meteorological influence, different air mass origin and different source contributions. Also spatial differences could be detected with higher particle concentrations in the city centre and lowest concentrations in the outskirts northwest of the city.

Water-soluble ions displayed an important part of particulate matter in dust samples from Beijing. Sulphate was the most abundant anion with highest concentrations in winter. Calcium was the dominant water-soluble cation in TSP samples. Further knowledge of the temporal and spatial aerosol distribution and the corresponding sources is important for the detailed assessment of the pollution of Beijing's atmosphere.

Acknowledgements This study was funded by the German Research Foundation as part of the DFG project "Investigation of relevant processes of aerosol pollution in Beijing – source identification, elemental composition and bio-availability with respect to different land use types". Ion chromatography for water-soluble cations was carried out at the German Weather Service (DWD) in Freiburg, Germany. In this respect the authors want to express their special thanks to M. Fricker for his support. All other analyses were performed at the Institute of Mineralogy and Geochemistry at the Universität Karlsruhe, Germany.

References

1. WHO (2001) Environment and People's Health in China
2. Raizenne M, Neas LM, Damokosh AI, Dockery DW, Spengler JD, Koutrakis P, Ware JH, Speizer FE (2006) Environmental Health Perspectives 105 (5) pp 492–499
3. Zhang J, Song H, Tong S, Li L, Liu B, Wang L (2000) Science of the Total Environment 262:63–71
4. Wang S, Yuan W, Shang K (2006) Atmospheric Environment 40:7975–7982
5. Wang Y, Zhuang G, Tang A, Zhang W, Sun Y, Wang Z, An Z (2007) Atmospheric Environment 41:1091–1106
6. Xie S, Yu T, Zhang Y, Zeng L, Qi L, Tang X (2005) Science of the Total Environment 345:153–164
7. Arimoto R, Duce RA, Savoie DL, Prospero JM, Talbot R, Cullen JD, Tomza U, Lewis NF, Rai, BJ (1996) Journal of Geophysical Research, 101:2011–2023
8. Wang Y, Zhuang G, Sun Y, An Z (2006) Atmospheric Environment 40:6579–6591
9. He K, Yang F, Ma Y, Zhang Q, Yao X, Chan CK, Cadle S, Chan T, Mulawa P (2001) Atmospheric Environment 35:4959–4970
10. Zhang D, Iwasaka Y (1999) Atmospheric Environment 33:3213–3223

Utilization of Meandering in a Receptor-Oriented Solution of the Line Source Dispersion Equation

Jari Härkönen, Janne Berger, Mari Kauhaniemi, Juha Nikmo, and Ari Karppinen

Introduction

Lateral long period fluctuation (meandering) of wind direction affects significantly on the results of dispersion models [1]. Anfossi et al. showed that meandering exists under all meteorological conditions [2]. Meandering seems to be independent on stability conditions, but is especially important in low wind speed conditions.

Line source dispersion modeling results are particularly sensitive to meandering because of the geometry of receptor – finite line source system. One may imagine that lateral fluctuation of wind direction during the averaging time changes effectively the real distance between the receptor point and line source as illustrated in Fig. 1. This influences substantially on the computed concentration at the receptor point, because the concentration is a strongly nonlinear function of the distance between point source S(x,0) and receptor R(x,y).

The aim of this work is to develop and test a specific meandering function applicable to analytical line source dispersion models in local scale when nearly parallel flows to the road are problematic.

Experimental Method

The measured NO_x data includes 1,038 cases from 1 January to 15 April 2002 near Oslo. The monitoring site with three monitors and the background monitor (B) is shown in Fig. 2.

J. Härkönen, M. Kauhaniemi, J. Nikmo and A. Karppinen
Finnish Meteorological Institute, Air Quality Research, Erik Palmenin Aukio 1, P.O. Box 503, FI-00101 Helsinki, Finland

J. Berger
Norwegian Institute for Air Research (NILU), P.O. Box 100, NO, 2027, Kjeller, Norway

S. Rauch et al. (eds.), *Highway and Urban Environment*, Alliance for Global Sustainability Bookseries 17,
DOI 10.1007/978-90-481-3043-6_11,

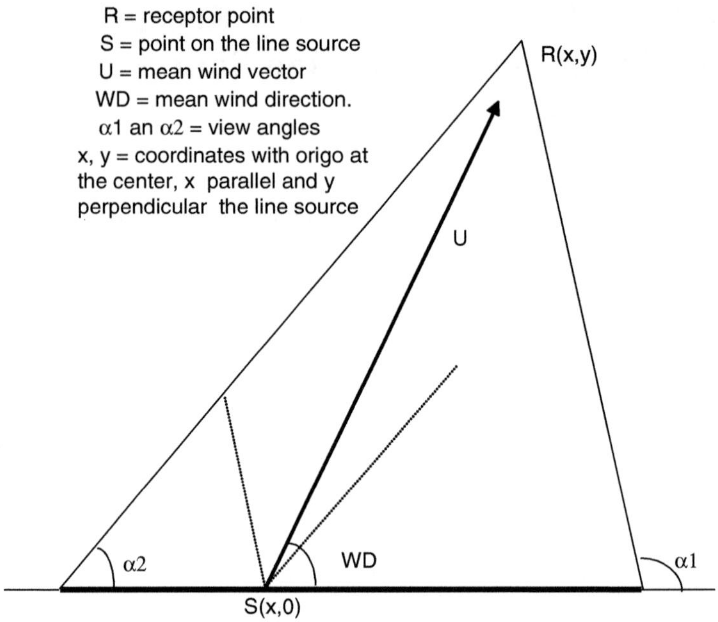

Fig. 1 Illustration of a finite line source – receptor point geometry

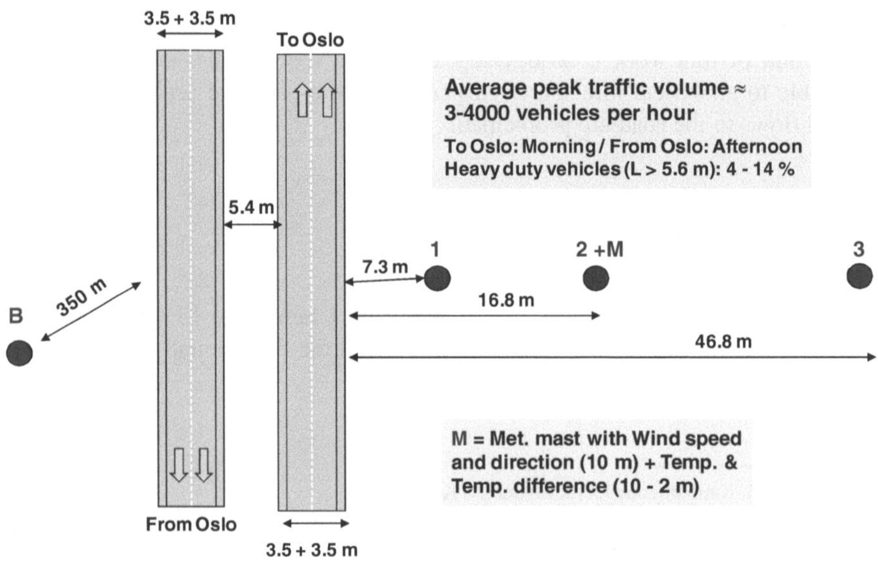

Fig. 2 The monitoring site Nordbysletta. The road is oriented approximately 60° due north

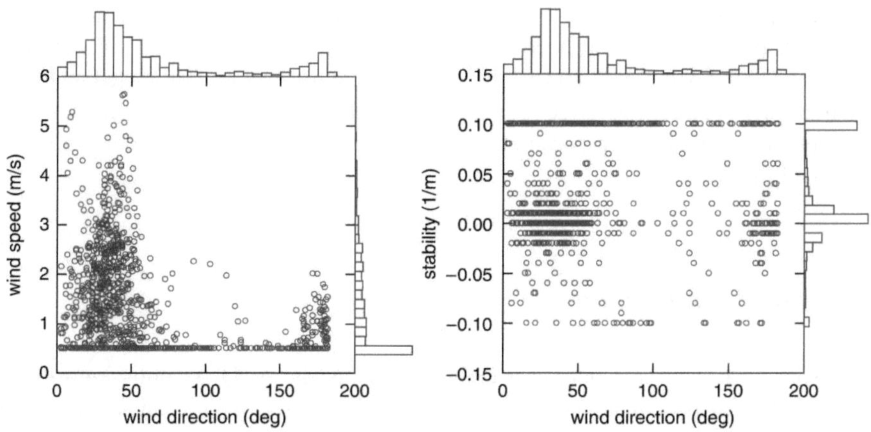

Fig. 3 Hourly wind speed and stability distributions against mean meteorological wind direction, with corresponding histograms at Nordbysletta monitoring site

Nordbysletta is located north-east of Oslo in a rural, relatively flat environment with average roughness length estimated to be 0.25 m. The road segment used in the campaign is ~850 m long with four lanes orientated ~60° due north (in the N-NE and S-SW sectors). Three stations measuring concentrations of NO_x and one background station were located at different distances from the road. The measurement height was 3.5 m. At station 2, there was also a meteorological mast measuring hourly wind speed data, wind direction data, temperature and vertical temperature difference between the heights of 10 and 2 m. In addition, hourly traffic counts and average speeds of light and heavy duty vehicles separately were available. The traffic flow is on average 36,000 vehicles/day, and the speed limit is 90 km/h [3].

Distributions of wind speed and stability are presented in Fig. 3. The mean meteorological wind direction is parallel to the road with values 0° and 180°. Minimum wind speed at 10 m height is fixed to 0.5 m/s. The histograms show that there are a lot of low wind speed cases distributed evenly over the directions. Near neutral stability prevails, but also the peak of very stable conditions is observed. The meteorological conditions, with many cases of flows nearly parallel to the road, are rather demanding for the purposes of dispersion modeling.

Computational Methods

The Gaussian fine line source model CAR-FMI used in this work is described, for example in [4]. The model is based on the general solution of [5]. The current version allows the use of separate emission data. The lowest wind speed in the plume is restricted to 0.3 m/s.

The objective of this work is to test a new method, where the influence of meandering on the dispersion could be accounted for. The new function described below is called by CAR-FMI after reading the meteorological time series. Computed receptor-oriented wind direction, effective distance, and probability are assigned to the computation of dispersion parameters and concentration. Only NO_x concentration is treated in this study.

Emission Computation

The Norwegian emission dataset is calculated using the AirQUIS emission module [6], which consists of an area source module, a point source module, and a line source module. The latter module needs information such as annual daily traffic, speed limit, road data and classification of vehicles. The emission is calculated as traffic based (function of fuel type, vehicle classes, driving speed, etc.) or as fuel based (with regard to CO_2, SO_2 the emission is calculated based on mass balance, fuel consumption and the content of carbon, sulphur, and lead in the fuel). In addition, factors describing the age of the vehicle, road gradient, the share of NO_2 in the total NO_x emissions are included. The emission factors are based on COPERT III [7].

Receptor-Oriented Mean Wind Direction

In the treatment of lateral fluctuation of wind direction, temporal and hourly mean values are presented with sub- and superscripts, correspondingly. Normal distribution is assumed for temporal wind direction wd with mean value WD and standard deviation σ_m. Chosen critical value for c, c(crit) = 3.21, corresponds to p-value < 0.001 in two tailed distribution. Standard deviation σ_m of meandering is grouped as a function of wind speed [8] and is applied to all stability regimes, because stability seems to have a minor influence on crosswise oscillations [2]. The range for temporal wind direction is accordingly

$$wd = WD \pm c\sigma_m \tag{1}$$

The range of the view angles ($\alpha 1$, $\alpha 2$), as illustrated in Fig. 1, is transformed to (HI, LO) determining the allowed wind sector, where $HI = \alpha 1 + c\,\sigma_m$ and $LO = \alpha 2 - c\,\sigma_m$. From the mean wind direction WD we calculate the probability P for the flow to be in the (HI, LO) sector according to Eq. 2. Standardized values ($x = (x - WD)/\,\sigma_m$) and cumulative normal distribution $\Phi(x)$ is applied to determine the probability P.

$$P = \Phi[HI - WD)/\sigma_m] - \Phi\,[(LO - WD)/\sigma_m] \tag{2}$$

The probability P is assumed to be independent from short term fluctuations being outside the integrand and becomes directly proportional to concentration. The expected wind direction EWD in the receptor-oriented (HI, LO) sector is defined according to Eqs. 3a and 3b. Consequently, the deviation from the meteorological mean wind direction WD is zero for winds perpendicular to the road, but the deviation increases as WD turns parallel to the road.

$$F = \Phi\{0.5[HI + LO)/2 - WD]/\sigma_m\} \quad (3a)$$

$$EWD = WD + \sigma_m \, \Phi^{-1}(F) \quad (3b)$$

As illustrated in Fig. 1, the distance between receptor R and the point source S on the line source changes with meteorological mean wind direction WD. Mean wind direction WD is now replaced by EDW. The effective distance DIS (Eq. 4), where y is the perpendicular distance of R from the line source, determines the distance of the receptor from the road in the expected mean wind direction EWD. As EWD becomes never 0° or 180°, the distance DIS has always a finite value allowing always an analytical solution of the line source dispersion model.

$$DIS = |y / \sin (EWD)| \quad (4)$$

The probability (Eq. 2) is directly used to scale the emissions. Expected wind direction (Eq. 3b) and effective distance (Eq. 4) are applied to the computation of vertical and lateral dispersion parameters σ_z and σ_y. The modification of the line source dispersion equation is restricted to the dilution term $Q/[U*\sin(WD)*\sigma_z]$, which is now presented as $P*Q/[U*\sin(EWD)*\sigma_z]$, where Q is emission strength and U is the mean wind speed.

Results and Discussion

The distributions of the parameters defined above against meteorological mean wind direction WD are graphically illustrated at the nearest monitoring site ST1 using the meteorological time series of Nordbysletta. In the second phase we compared the observed measurements with predictions (backgrounds subtracted) by CAR-FMI when meandering was accounted as defined in Eqs. 1–4.

Dependence of Probability, Expected Wind Direction and Effective Distance on Meteorological Mean Wind Direction

The mean flow is oriented perpendicular to the road when the meteorological mean wind direction WD is 90°. Probability of the plume reaching the receptor decreases to 0.5 when mean wind turns parallel to the road and is roughly 1.0 with nearly perpendicular flows, as seen in Fig. 4a. If WD is nearly parallel to the road, the probability may be slightly below 0.5 in case of finite line sources, which indicates that the allowed wind sector (HI, LO) is smaller than half of the temporal wind direction range defined in Eq. 1.

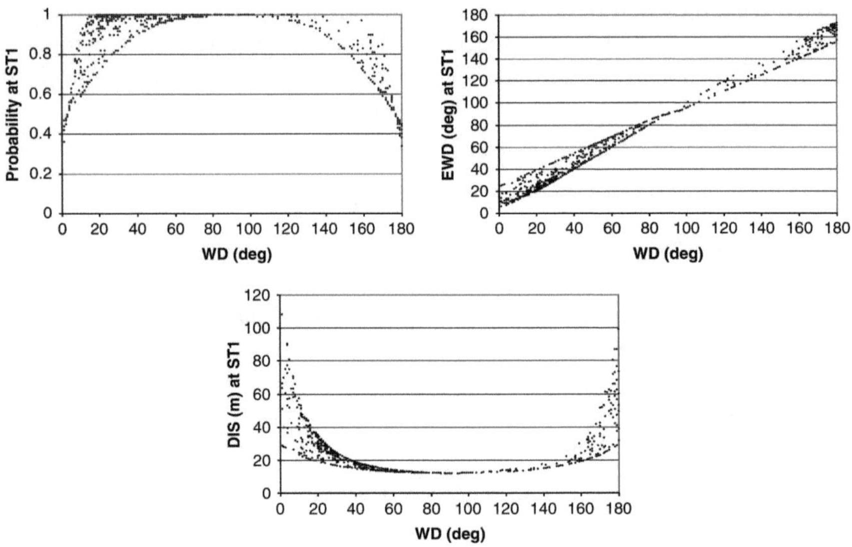

Fig. 4 Probability P, expected mean wind direction EWD (degrees) and effective distance DIS (m) against meteorological mean wind direction WD (degrees) at the station ST1 locating 7.3 m from the road side.

Correspondingly, the expected wind direction (EWD) is equal to the meteorological mean wind direction (WD) for perpendicular winds, while in other cases EWD turns away from the road direction (Fig. 4b). As a consequence, the effective distance (DIS) has a finite value also in the case of mean wind direction WD being parallel to the road as seen at the station ST1 in Fig. 4c. Variations of the computed parameters with a fixed meteorological mean wind direction associate with the changes of wind speed.

The statistical agreement between the measured and modeled concentrations is best at the nearest monitoring site (ST1), but decreases with the distance of monitoring from the road. Similarly the slope decreases with distance, indicating increasing overestimation at more distant sites. According to our analysis linear regression seems not to be sensitive to wind direction, but depends strongly on wind speed.

The influence of wind speed on the predictions was studied by computing relative bias RB defined as $(\overline{pred} - \overline{obs}) / \overline{obs}$ in four groups of wind speed. The results are presented in Table 1, where overestimation is limited to wind speeds <2 m/s, while stronger wind speeds associate with underestimation at all monitors. Remarkably high overestimation is found at the most remote monitor ST3 during very weak winds (<1 m/s at 10 m height). Surprisingly, the best performance of the model is observed just by the road side (ST1). The overestimated low wind speeds and underestimated higher winds are responsible to the moderate constants k_0 and slopes clearly below 1 in Table 2.

The results suggest that the mixing by the traffic flow cannot be described merely by an initial constant in vertical and lateral dispersion parameters as is the

Table 1 Linear regression coefficients k and k_0 and squared correlation R^2 between observed and predicted concentrations (µg/m³) are computed for all stations. Distance is measured from the road side and N is number of cases

Monitor	Distance (m)	N	k	k_0	R^2
ST1	7.3	1,038	0.784	31.5	0.765
ST2	16.8	1,038	0.651	28.8	0.732
ST3	46.8	1,038	0.540	23.4	0.661

Table 2 Relative bias RB is computed for observed and predicted concentrations (µg/m³) in four wind speed groups <1 m/s, 1–2 m/s, 2–3 m/s and >3 m/s. N is the number of cases

Group (m/s)	N	ST1	ST2	ST3
<1	269	0.012	0.227	0.457
1–2	254	0.081	0.214	0.223
2–3	165	−0.172	−0.160	−0.173
>3	350	−0.121	−0.116	−0.151
All	1,038	−0.018	0.105	0.201

current practice in CAR-FMI. On the other hand, the lowest wind speed regime is clearly outside the application area of analytical solutions of dispersion equations as commented by several authors.

Conclusion

We propose a procedure for line source dispersion modeling in low wind speed conditions, when meandering has a strong effect on the dispersion. The comparisons between predicted and observed NO_x concentrations show that the suggested meandering procedure applied to an analytical line source model:

- Improves predictions in situations of nearly parallel winds to the road, which are caused by lateral fluctuation of wind direction
- Wind speed limit for over- or underestimation regimes is about 2 m/s at 10 m height
- Allows temporal diffusion to the upwind side of the road by lateral fluctuation, which is important at receptor points locating within the first tens of meters from the road

References

1. Oettl D, Almbauer RA, Sturm PJ (2001) A new method to estimate diffusion in stable, low wind conditions. J. Appl. Meteorol. 40:259–268
2. Anfossi D, Oettl D, Degrazia G, Goulart A (2005) An analysis of sonic anemometer observations in low wind speed conditions. Boundary Layer Meteorol. 114:179–203

3. Hagen LO, Larssen S, Walker SE (2003) Forurensning som funksjon av avstand fra vei. Målinger på RV159 Nordby-sletta v/Skårer vinteren 2001–2002, og sammenligning med VLUFT. (NILU OR 22/2003)
4. Härkönen J, Valkonen E, Kukkonen J, Rantakrans E, Jalkanen L, Lahtinen K (1995) An operational dispersion model for predicting pollution from a road. Int. J. Environ. Pollut. 5:602–610
5. Luhar AK, Patil RS (1989) A general finite line source model for vehicular pollution prediction. Atmos. Environ. 23:555–562
6. Denby B (2004) AirQUIS Models Module – User's Guide. Kjeller (NILU TR 5/2004)
7. Ekström M, Sjödin Å, Andreasson K (2004) Evaluation of the COPPERT III emission model with on-road optical remote sensing measurements. Atmos. Environ. 38:6631–6641.
8. Leahey DM, Hansen MC, Schroeder MB (1994) Variations of wind fluctuations observed at 10 m over flat terrain under stable atmospheric conditions. J. Appl. Meteorol. 33:712–720.

Measurements of Heavy Metals in Dry and Wet Deposition in São Paulo City

Ana Paula G. Fontenele, Adalgiza Fornaro, and Jairo J. Pedrotti

Introduction

In Metropolitan Area of São Paulo (MASP), the air contamination is caused mainly by gases and particulate matters released by the large fleet of light and heavy vehicles, but also by industrial processes [1]. Besides environmental consequences, the particulate matters, (PM), have been linked to health damage because they contain a wide range of toxic metals and organic contaminants. Among the (PM) constituents, heavy metals have been investigated due to the elevated toxicity to living organisms. Some metals, like copper and zinc, are essential nutrients but they can cause harmful effects depending on their concentration levels. Cadmium is considered very toxic, classified as potential carcinogenic. Lead is also recognized by the World Health Organization (WHO) as one of the most dangerous chemical element for the human health. In MASP, epidemiological studies show that increases in hospital admissions and mortality have been associated with air pollution episodes [2]. Aerosols can be removed from the atmosphere by two main pathways: dry and wet deposition, depending on their physical characteristics such as particle size, vapor and solubility pressure in water, and chemical composition [3]. In this study the concentrations of zinc, copper, cadmium and lead in fine ($PM_{2.5}$) and coarse ($PM_{2.5-10}$) particle material in São Paulo rainwater were determined. The aim is to evaluate the extension of the contamination of the São Paulo city atmosphere as well as to identify the main deposition route of heavy metals in this urban area.

A.P.G. Fontenele and J.J. Pedrotti (✉)
Departamento de Química – Universidade Presbiteriana Mackenzie,
Rua da Consolação, 896 – São Paulo, SP
e-mail: jpedrotti@mackenzie.com.br

A.P.G. Fontenele and A. Fornaro
Departamento de Ciências Atmosféricas, Instituto de Astronomia, Geofísica e Ciências Atmosféricas (IAG/USP), Universidade de São Paulo, Rua do Matão, 1226. 05508-090, São Paulo, SP
e-mail: fornaro@model.iag.usp.br

S. Rauch et al. (eds.), *Highway and Urban Environment*,
Alliance for Global Sustainability Bookseries 17,
DOI 10.1007/978-90-481-3043-6_12, © Springer Science+Business Media B.V. 2010

Site Sampling

São Paulo City, with a territorial area over 1,528 km^2, is located in the southeastern region of São Paulo state, Brazil, ~45 km from the coast and 780 m above sea level. With almost 11 million inhabitants, it is the largest part of the Metropolitan Area of the São Paulo (MASP) with more than 17 million inhabitants. The main sources of pollution in the city are vehicular and industrial emissions. Diesel, hydrated ethanol, natural gas and gasohol (a mix of 22–25% in ethanol + gasoline) are the most common fuels used in more than 7 million heavy and light vehicles. The vehicular fleet was responsible for 40% of the 67,000 t PM_{10} emitted in the MASP in 2003. From this, 10% is released by industrial processes, 25% by secondary aerosol formation and the remaining by resuspended road dust [1]. The site sampling was located in the center of São Paulo City (23°32′ SE 46°39′ W) surrounded by large avenues with intense flux of vehicles.

Experimental

Rainwater samples were collected on the top of a building (15 m above ground) within the campus of Mackenzie University, by an automatic wet-only rainwater collector G.K. Walter model 211. After collection, the rainwater samples were filtered through a 0.22 μm cellulose acetate membrane and then acidified with HNO_3 (pH = 2) and stored in 50 mL high-density polyethylene flasks at 4°C until the analysis. The aerosol particles were sampled by a portable sampler called "Mini-Vol" (Airmetrics, USA) during the winter season; July 7 to September 5, 2003. The particle sampler separates the aerosol particles as coarse (2.5–10 μm equivalent aerodynamic diameter, EAD) and fine (<2.5 μm EAD) using Nuclepore filters with 8 and 0.4 μm pore size, respectively. The particle mass loading on the filters was determined gravimetrically. The soluble metals were extracted under mechanical shaking in 25 mL ultra-pure water for 1 h, followed of acidification with 0.1% HNO_3.

Reagents and Solutions

All reagents were of analytical grade. The ultrapure nitric acid used for conditioning the samples was prepared by sub-boiling distillation using Marconi system (Piracicaba, Brazil). The solutions were prepared by using ultrapure water obtained from a Nanopure® system (resistivity >18 MΩ cm). Copper, lead, cadmium and zinc stock solutions (0.010 mol/L) were prepared by dissolution of its nitrate salts in 10 mmol/L HNO_3 solutions. The solutions with lower metal ion concentration were diluted daily from the stock solution in 10 mmol/L HNO_3, which was also used as carrier electrolyte for copper, lead and cadmium. A 10 mmol/L acetate buffer solution (pH 4.5) was used as electrolyte for zinc determinations. Mercury was chemically purified and bi-distilled under vacuum.

Analytical Instrumentation

Pulse differential anodic striping analyses, DPASV were carried out by μAutolab Type II Potentiostat (Eco Chemie B.V. – Utrecht, Netherlands). A GPES 4.5 software was used for electrochemical instrumentation control, acquisition and data treatment. The flow injection analysis (FIA) system consisted of carrier reservoir, an Ismatec peristaltic pump model REGLO to propel the electrolyte carrier at 220 μL/min flow rate, a manually operated injector based on three solenoid valves which was provided with a loop of 200 μL, and microdegasser system to remove the interference of the dissolved oxygen in flowing solutions on Hg drop electrode, and the electrochemical detector. A 20-cm long flexible fused silica capillary (i.d. = 0.53 mm) was used to carry the degassed solution to the flow detector. The batch electrochemical cell from VA STAND 663 (Metrohm), provided with multimode mercury electrode, Ag/AgCl reference electrode and glassy carbon auxiliary electrode, was converted for flow measurements by using a PTFE adaptor in "L" shape fixed on the glass capillary and with its tip positioned ~0.5 mm apart from the mercury drop surface [4]. A siphon prevents the cell from flooding. It was connected to the outlet tubing which, together with the inlet channel from the FIA system, leaves the cell through a hole in the top of cell lid.

Procedure

After assuring a constant level of the electrolyte solution in the flow cell, the sample solution was injected into the FIA system. The deposition potential for cadmium, lead and copper on Hg drop electrode was applied (45 s after the sample injection) for 3 min at –0.8 V. The pre-concentration of zinc amalgam was carried out at –1.1 V also for 3 min. After this step, a 10 s rest period was observed, and then a 10 mV/s differential pulse anodic stripping voltammogram scan was started. The heavy metals quantification was made by standard addition method using three spikes of each element. The analytical detection limits were 0.50 μg/L for copper, 0.40 μg/L for lead, 0.010 μg/L for cadmium and 0.95 μg/L for zinc. The analytical precision was typically about ±5%. All measurements were carried out at 25 ± 1°C room temperature.

Results and Discussion

Wet Deposition

The amount of rainfall recorded from January to December 2003 was 720 mm. The dissolved metal concentrations were evaluated for 47 events of rain which represents a precipitation of 616 mm. The rainy period occurred between October

Table 1 Mean concentrations (standard deviation) and volume weight mean (VWM) values of the trace metals and hydrogen ion in 60 wet-only precipitation samples in downtown São Paulo (Feb–Dec 2003) and in other urban areas

	São Paulo[a]		Athens [6]	Guaíba [7]	São Paulo [8]	Candiota [9]
	Mean (sd)	VWM	Mean (sd)			
	μg/L		μg/L			
Cd	0.17 (0.19)	0.17	0.20 (0.14)	–		–
Pb	1.7 (1.2)	1.5	0.88 (1.0)	–	1.5(1.0)	–
Cu	3.6 (3.4)	2.8	15.4 (14.5)	6.8 (8.5)		2.00 (1.51)
Zn	9.8 (9.9)	9.9	33.5 (40.8)	29.6 (18.8)		51.5 (22.8)
pH	5.6	5.4	6.46	5.7	5.0	5.4

[a] Present study.

and March (spring and summer). During these months, the precipitation totalized 632 mm, which was equivalent to 88% of the total rainfall registered. The pH values in the rainwater ranged from 4.80 to 7.18, with a VWM of 5.4, indicating a slightly acidic precipitation. The lower pH values were observed in summer and spring seasons, corresponding to the rainy period. In São Paulo city, ammonia is the main neutralizing agent of the wet precipitation [5]. The dissolved metal concentrations showed great variability between the individual samples, as indicated by the extent of the standard deviations (Table 1). A large fraction of the measurements, especially in the case of Cd, was below the analytical detection limit of the FIA-DPASV technique. No significant correlation was found between the dissolved metal concentrations and the rainwater pH, or the rainfall quantity and rate. Higher metal concentrations were obtained from rainwater collected after long dry periods, usually over 15 days. This was evidenced in some rain events between April and August. (Autumn and winter seasons – dry periods). On August 7, 2003, a 370 mL rain sample was collected after 27 days without precipitation. The mean concentrations determined for copper, lead, cadmium and zinc were 8.5, 4.3, 0.24 and 30, respectively. In comparison with the mean concentration (Table 1), the Cu, Pb, Cd and Zn concentration found in rainwater sample collected on August 7 had enrichment factors of 136%, 152%, 41% and 206%, respectively, suggesting a significant accumulation of pollutants in the atmosphere during the period without precipitation. The comparisons of mean dissolved metal concentrations with literature data is presented in Table 1.

Cadmium concentration found in the present study was comparable with that found in Athens [6]. Lead concentration was not significantly different from a previous study carried out at the same site in São Paulo during January to May 2000 [8], and was higher than that observed in the capital of Greece. The mean concentration of copper observed in São Paulo Metropolis was usually lower than those recorded in Athens and Guaíba [7], and higher than in Candiota [9]. Values for dissolved zinc concentrations in this study were lower than those obtained in Athens and Guaíba, and much lower than in Candiota, a region where a large thermoelectric power plant, using mineral coal as fuel, is established.

Dry Deposition

The average mass concentrations of fine and coarse particle material were 31 ± 13 and 33 ± 22 μg/m, respectively [10]. The dry deposition was evaluated for copper, lead and cadmium. Tables 2 and 3 show the mean concentrations (standard deviation), minimum and maximum values of copper, lead and cadmium in fine (d < 2.5 μm) and coarse mode aerosol (2.5 < d < 10 μm) for diurnal and nocturnal periods. The results show high variability, suggesting influence of emissions and meteorological conditions. During the aerosol sampling period, the meteorological conditions in São Paulo showed 31 days with unfavorable conditions for pollutant dispersion and 59 days with thermal inversion (height < 400 m). The mean metal concentrations were higher in aerosols collected during night compared to samples collected during the day. This might be explained by the evolution of a stable nocturnal boundary layer and with intense vehicular traffic at the beginning of evenings.

Lead was the predominant metal, followed by copper and cadmium in the fine particle fraction (Table 2). On the other hand, in the coarse fraction, copper was the metal found in higher concentration (Table 3). The relative contribution of Cd, Pb and Cu in fine and coarse particulate matter is presented in Fig. 1, suggesting that metals are emitted from different sources. Lead contamination in the atmosphere can be attributed to industrial emissions, mainly non-ferrous smelters; while copper can be associated with emissions from motor vehicles using ethanol and gasohol (mixture of 22–25% in ethanol + gasoline) as fuels. In urban areas, like São Paulo, cadmium sources can be associated with both worn out tires and the white paint used for marking lines on pavement (asphalt) on streets and avenues [11]. The significant contribution of lead to $PM_{2.5}$ may be an indicative of the toxicity of inhalable particle matter, which has been correlated with pulmonary diseases.

Table 2 Average concentrations (standard deviation), and minimum and maximum values of the metals in fine particle (d < 2.5 μm) samples

	Diurnal			Nocturnal		
	Average (sd)	Min.	Max.	Average (sd)	Min.	Max.
	ng/m			ng/m		
Cd	0.61 (0.81)	0.13	2.4	1.5 (2.7)	0.17	8.2
Pb	13 (6.6)	5.7	27	25 (33)	4.5	111
Cu	4.7 (4.1)	1.7	14	12 (19)	1.9	63

Table 3 Average concentrations (standard deviation), and minimum and maximum values of the metals in coarse particle samples (2.5 < d < 10 μm)

	Diurnal			Nocturnal		
	Average (sd)	Min.	Max.	Average (sd)	Min.	Max.
	ng/m			ng/m		
Cd	0.49 (0.77)	0.17	2.4	0.60 (0.83)	0.17	2.3
Pb	8.6 (7.9)	2.1	27	10 (14)	2.6	48
Cu	11 (6.8)	2.3	26	18 (17)	8.2	62

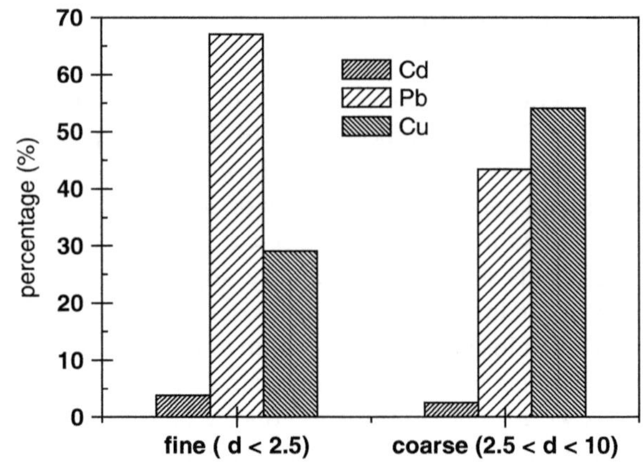

Fig. 1 Differences of the Cd, Pb and Cu distribution in fine and coarse particles of MASP

Table 4 Avarege atmospheric levels of Cu, Cd and Pb compared with other results reported in the literature

	Fine aerosol			Coarse aerosol		
	ng/m					
	Cd	Pb	Cu	Cd	Pb	Cu
Tubarão (winter) [13]	–	5.7	1.2	–	3.9	1.6
Rio de Janeiro [14]				0.4	16	22
Volta Redonda (2004–2005) [15]				4	140	263
MASP (1983–1985) [16]		94	11		51	19
MASP (1997) [12]	–	42	19	–	38	44
MASP (2003)*	2.1	38	17	1.1	19	29

Similar results of metal concentration distributions in fine and coarse dry deposition were observed in other studies carried out in the MASP during the winter 1997. The aim was to characterize the chemical composition of aerosols in São Paulo city by using proton induced x-ray emission (PIXE) technique [12].

Table 4 compares the results obtained in the present study with those previously obtained in the same area but also from other places.

Lead and copper concentrations in fine particles were higher in the present study than those found in Tubarão [13] and comparable with those obtained in MASP in 1997. For Cd, Pb and Cu concentrations in coarse particles, our results were also comparable with those obtained in Rio de Janeiro [14] but they were much lower than those found in Volta Redonda [15], an area of high industrial activity. In MASP, the lead aerosol contamination in fine and coarse modal has been declining since 1993 when the addition of tetraethyl lead to gasoline was prohibited in Brazil. In two decades, the contamination by lead has been reduced by 60% in fine, and more than 62% in coarse particles (Table 4).

Comparison Between Wet and Dry Deposition Fluxes

In order to evaluate the transport route for removal of trace metals from the atmosphere, wet and dry deposition rates were estimated. The wet deposition was calculated by using Eq. 1, according with what was previously described.

$$W_d = \frac{m_d}{(A) \times (d)} \tag{1}$$

where W_d is the trace metal deposition flux (μg m^{-2} day^{-1}), m_d is the collected trace metal mass at the sampling day, A is the deposition sampler collection area and d is the number of days with rain event. The dry deposition flux was estimated [3] by using Eq. 2:

$$D_d = C_i \cdot v_d \tag{2}$$

where D_d is the dry deposition flux (μg m^{-2} day^{-1}), C_i is the metal concentration (μg/m^3) for each day of sampling and v_d is the deposition velocity of particles (cm/s) [3]. Considering that the deposition velocity of particles depends on their size, a v_d ranging from 0.5 to 2 cm/s was used for coarse particles, while a 0.5 cm/s deposition velocity was adopted for fine particles. The deposition velocity of aerosols reflects a combination of experimental uncertainties and patterns which are obtained by various models such as the curves of the deposition velocity versus the diameter of the particle [3]. The wet and dry deposition fluxes are presented in Table 5. The order of the wet deposition fluxes for trace metals was Cd < Pb < Cu < < Zn. The wet deposition fluxes were 1.4–1.9 times larger than the dry deposition. Therefore, the precipitation plays an important role in trace metal deposition in this large urban area. For dry deposition, the coarse particulate matter showed higher deposition rates for cadmium and copper. A comparison of our dry deposition results for coarse aerosols with data from other urban areas reported in the literature shows that Cd, Pb and Cu levels in São Paulo are higher than those measured in Tokyo [17] but lower than those in Chicago, IL and South Haven, MI [18]. Furthermore, they were found to be extremely low compared to those recorded in Izmir, a highly industrialized area in Turkey [19].

Table 5 Wet and dry deposition flux during 2003 in São Paulo

	Deposition		
	Wet	Dry (fine)	Dry (coarse)
		μg m^{-2} day^{-1}	
Zn	118	–	–
Cd	1.8	0.04	0.02–0.95
Pb	15	0.82	0.42–17
Cu	34	0.35	0.6–24

Conclusions

The pH measurements indicate that the rainwater in São Paulo is slightly acidic, showing 5.4 pH_{VWM} values. The heavy metal concentrations in wet and dry deposition were comparable with others in large cities around the world. In wet deposition, the metal concentrations follow the order: Cd < Pb < Cu < Zinc. For fine and coarse dry deposition, higher metal concentrations were observed during nocturnal periods, associated with the evolution of the stable nocturnal boundary layer and with the intense vehicular traffic in early evenings. Lead was the predominant metal in fine particles while copper was found in higher concentrations in coarse particles. In conclusion, the main transport route to remove heavy metals in São Paulo was the wet deposition.

Acknowledgements The authors are grateful to Fundação de Amparo a Pesquisa do Estado de São Paulo – FAPESP, Mackpesquisa for financial support and CNPq for scholarship (A.P.G. Fontenele).

References

1. CETESB – Relatório da Qualidade do Ar no Estado de São Paulo – 2003, Secretaria do Meio Ambiente, Série Relatórios – ISSSN 0103–4103, São Paulo, 2004 (www.cetesb.sp.gov.br)
2. Freitas C, Bremner SA, Gouveia N, Pereira LAA, Saldiva PHN (2004) Hospital admissions and mortality: association with air pollution in São Paulo, Brazil, 1993 to 1997, Rev. Saúde Pública, 38:751–757
3. Seinfeld JH, Pandis SN (2006) Atmospheric Chemistry and Physics: From Air Pollution to Climate Change, 2nd ed., Wiley: New York
4. Pedrotti JJ, Gutz IGR (2003) Ultra simple adaptor to convert batch cells with mercury drop electrodes in voltammetric detectors for flow analysis. Talanta, 60:695–705
5. Santos MA, Illanes CF, Fornaro A, Pedrotti JJ (2007) Acid rain in downtown São Paulo city, Brazil, Water Air Soil Pollut: Focus, 7:85–92
6. Kanellopoulou EA (2001) Determination of heavy metals in wet deposition of Athens, Global Nest: Int. J. 1:45–50
7. Migliavacca D, Teixeira EC, Wiegand F, Machado ACM, Sanchez J (2005) Atmospheric precipitation and chemical composition of an urban site, Guaíba hydrographic basin, Brazil, Atm. Environ. 39:1829–1844
8. Richter EM, Pedrotti JJ, Angnes L (2003) Square-wave quantification of lead in rainwater with disposable gold electrodes without removal of dissolved oxygen, Electroanalysis, 15:23–24
9. Migliavacca D, Teixeira EC, Pires M, Fachel J (2004) Study of chemical elements in atmospheric precipitation in South Brazil, Atm. Environ. 38:1641–1656
10. Albuquerque TT, Andrade MF (2004) Proceedings of the XIII Brazilian Meteorological Congress. Fortaleza, Ceará, Brazil
11. Ozaki H, Watanabe I, Kuno K (2004) Investigation of the heavy metals sources in relation to automobiles, Water Air Soil Pollut. 157:209–223
12. Castanho ADA, Artaxo A (2001) Wintertime and summertime São Paulo aerosol source apportionment study, Atm. Environ. 35:4889–4902
13. Godoy MLDP, Godoy JM, Artaxo P (2005) Aerosol source apportionment around a large coal fired power plant – Thermoelectric Complex Jorge Lacerda, Santa Catarina, Brazil, Atm. Environ. 39:5307–5324

14. Toledo VE, Almeida Jr PB, Quiterio SL, Arbilla G, Moreira A, Escaleira V, Moreira JC (2008) Evaluation of levels, sources and distribution of toxic elements in PM10 in a suburban industrial region, Rio de Janeiro, Brazil, Environ. Monit. Assess, 139:49–59
15. Gioda A, Sales JA, Cavlacanti PMS, Maia MF, Maia LFPG, Aquino-Neto FR (2004) Evaluation of air quality in Volta Redonda, the main metallurgical industrial city in Brazil, J. Braz. Chem. Soc. 15:856–864
16. Orsini CQ, Tabacniks MH, Artaxo P, Andrade MF, Kerr AS (1986) Characteristics of fine and coarse particles of natural and urban aerosols of Brazil, Atm. Environ. 20:2259–2269
17. Sakata M, Marumoto K (2004) Dry deposition fluxes and deposition velocities of trace metals in the Tokyo Metropolitan Area measured with a water surface sampler, Environ Sci Technol. 38:2190–2197
18. Yi SM, Shahin U, Sivadechathep J, Sofuoglu SC, Holsen TM (2001) Overall elemental dry deposition velocities measured around Lake Michigan, Atm. Environ. 35:1133–1140
19. Muezzinoglu A, Cizmecioglu SC (2006) Deposition of heavy metals in a Mediterranean climate area, Atm. Res. 81:1–16

[illegible]

Exposure of Cyclists to Ultra Fine Particles

Luc Int Panis, Nico Bleux, Rudi Torfs, Vinit Mishra, Bas de Geus, Romain Meeusen, Grégory Vandenbulcke, and Isabelle Thomas

Introduction

Health research has sufficiently demonstrated that regular physical activity is beneficial for health and lowers medical risks associated with a sedentary lifestyle. Commuter cycling conveniently combines both the level of activity and the frequency necessary to improve fitness [1]. A popular review of evidence can be found in [2]. On the other hand, many people worry over the possible negative health effects of commuter cycling: the accident and injury risks and the exposure to traffic related air pollution. All of these possible negative effects, as well as their geographical distribution, are studied simultaneously in the Belgian SHAPES research project on cycling and health. In the SHAPES project we collect data on the exposure of cyclists to particulate air pollution. A novel measuring strategy was developed using small battery driven optical sensors which allows us to collect data on levels of Particulate Matter (PM) while cycling in real traffic. Traffic is generally blamed for high PM levels in urban regions.

Initial work focused on simultaneous measurements of PM10, PM2.5, PM1 and Ultra Fine Particulate (UFP) matter. A combination of both weight based PM measurements and particle numbers proves to be very useful because both are related to distinct sources which we can identify from simultaneous video recordings [3].

In this paper we report UFP levels measured while cycling on a cycling track and the effect of the distance of the cycling track to the road on the exposure of cyclists.

L.I. Panis, N. Bleux, R. Torfs and V. Mishra
VITO, Boeretang 200, 2400, Mol, Belgium

B. de Geus and R. Meeusen
VUB, Brussels, Belgium

G. Vandenbulcke and I. Thomas
UCL, Louvain-la-Neuve, Belgium

S. Rauch et al. (eds.), *Highway and Urban Environment*, Alliance for Global Sustainability Bookseries 17,
DOI 10.1007/978-90-481-3043-6_13, © Springer Science+Business Media B.V. 2010

Experimental Method

In the present study a bicycle equipped with a number of different instruments (collectively called AeroFlex II) was used to measure aerosol concentrations in the ambient environment and to evaluate personal exposure of cyclists. It is used here to capture the time-activity exposure patterns of individuals in an urban transport microenvironment. It consists of four different instruments: a GRIMM 1.108 Dust monitor, a TSI P-TRAK, a commercial GPS and a video camera. The GRIMM 1.108 spectrometer is a portable environment dust monitor which can simultaneously measure PM1.0, PM2.5, PM10 and TSP. It has two optical sensors which provide near real-time particle number concentration measurements at a maximum logging rate of 1 min. The size range covered by the instrument is 0.3–20 μm over 15 channels. The design features of the GRIMM 1.108 provide ease of portability: a rechargeable Li-Fe battery (providing 6 h of runtime) and 4 MB of internal data storage [3]. This makes the GRIMM 1.108 readily deployable in the field and useful for mobile measurements in the SHAPES project.

Ultrafine particle counts at 1-s resolution were made, using P-TRAK Ultra fine Particle Counters (TSI Model 8525), for particles in the size range 0.02–1 μm. The P-TRAK is a hand-held field instrument based on the condensation particle counting technique using isopropyl alcohol. It has a relatively robust performance while in motion, rapid warm-up, battery-powered, and it has an operation time which is longer than more sophisticated instruments. The P-TRAK has the ability to detect high concentrations (maximum detectable limit: 500,000 cm^{-3}). Moreover, it has a high data-logging resolution and a fast response time [3].

Initial pilot tests revealed that the P-TRAK was rather sensitive to shocks leading to temporary gaps in the data collected. This was circumvented by designing a shock absorbing suspension and by manually bypassing the electronic shock detection mechanism. This final set-up was used for all later measurements.

All measurements discussed in this paper were executed while cycling in real traffic in and around the city of Mol, Belgium. A test trajectory was chosen that includes a cycling track parallel to a major regional road N18 (two lanes, ~13,000 vehicles/day including buses and trucks, 70 km/h speed limit). The cycling lane is between 1.8 and 2 m wide and is mostly separated from the road to the left by a 1.8 m wide parking lane except at a signalized intersection (see Fig. 1). There is also some traffic on parking places around the shops situated to the right of the cycling track.

During the test there was a light wind blowing from NW at a 30° angle to the road (which is oriented N-S). Meteorological data was not measured at the site shown in Fig. 1, but at the nearby VITO campus.

A first set of validation tests was conducted to test the reliability of the P-TRAK instruments, as well as the set-up on the bicycle. Two P-TRAK instruments were carried in a pannier and supplied with sampled air through two tubes of variable length fitted to an extendible post so that any sampling height and direction could be chosen (see Fig. 2).

It was first tested whether the two available P-TRAK instruments can provide identical results when sampling tubes are placed at the same location (Fig. 2a).

Fig. 1 Case study site along the N18 in Mol, Belgium

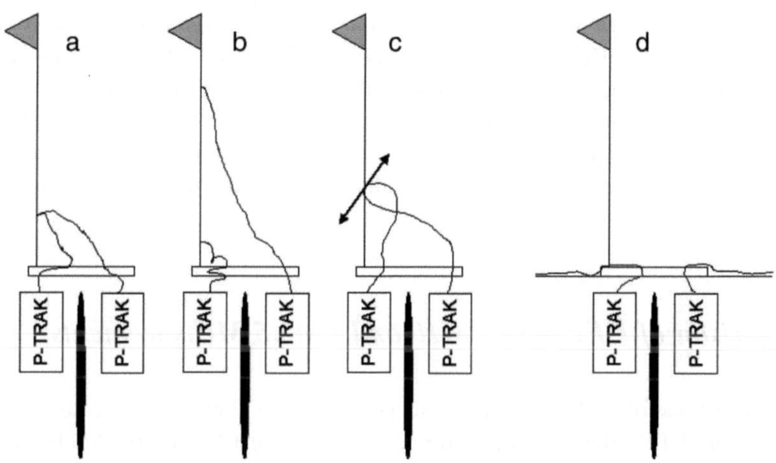

Fig. 2 Experimental set-up for validation of the methodology. (**a**) Sampling at 1 m above the ground for both instruments. (**b**) Sampling at 0.8 and 1.6 m above the ground. (**c**) Sampling in different directions at the same height. (**d**) Set-up to determine the effect of distance on UFP exposure on the cycling track. The horizontal distance between both sampling tubes was 2 m in the pilot test and 1 m during the final test

Subsequently the effect of sampling height (Fig. 2b) and sampling direction (Fig. 2c) were tested. Subsequently the bicycle was set-up to measure UFP concentrations on a cycling track in the city of Mol. Simultaneous measurements were taken at a horizontal distance of 2 m while cycling (Fig. 2d). Measurements were taken during trips in both directions on three different days in November 2007 (N = 6). The main purpose was to determine the set-up and number of replicates necessary to resolve the expected difference.

After analysis of the initial results it was decided to reduce the horizontal distance between sampling points to just 1 m to improve maneuverability and safety while measuring. A new set of ten trips in both directions were made on the 3rd and 4th of April 2008 under similar meteorological conditions. Part of the trajectory is shown in Fig. 1. The total trajectory is exactly 4 km long and was cycled in about 12 min with an average speed of 19.6 km/h (SD 1.4 km/h). Measurements were compared by subjecting average particle numbers per trajectory for each device to a Wilcoxon matched pairs signed ranks test.

Results

The ratio and 95% CI between measurements taken during the pilot test by both instruments were 1.00 (0.94–1.05) for case a, 0.98 (0.86–1.09) for case b and 1.01 (0.94–1.09) for case c respectively.

We concluded that the set-up was reliable and reproducible although a small correction was sometimes needed to off-set systematic differences between instruments. There is no effect of either sampling height or sampling direction on the results. The results of the measurements will therefore adequately reflect the concentration in the breathing zone of the cyclist and his/her external exposure to UFP. The facts that results are independent of sampling direction confirm the results in [4]. Here it is also demonstrated that results are not influenced by cycling speeds below 21 km/h.

Comparison of UFP with PM10 and PM2.5 Measurements

Results from one of the preliminary tests shown in Fig. 3 clearly indicate that peaks of PM10 and UFP often occur at different sites on a test trajectory. Analysis of the video footage revealed that PM10 peaks were often related to construction activities while UFP was more closely related to traffic (although UFP peaks attributed to construction equipment also can be seen at construction site 1).

Exposure of UFP to Cyclists on a Cycling Track

The pilot test revealed an unexpected, but important difference over a horizontal distance of 2 m. Measurements taken at the side of the road were on average 10% higher than those taken simultaneously at the other side of the bicycle (Ratio 1.096; stdev 0.044). Average particle numbers during the 3 days of the pilot test were around 30,000 cm^{-3}. Non-parametric tests revealed a difference that was barely significant (Critical value of 0 for $p = 0.05$ at $N = 6$ in the Wilcoxon Matched pair signed ranks test) and only after correction for an unusually large systematic difference

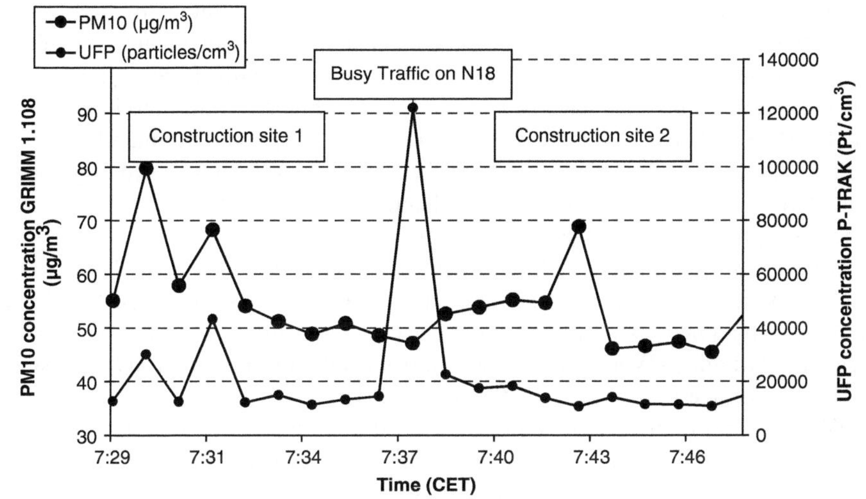

Fig. 3 Patterns of PM10 and UFP are related to different sources

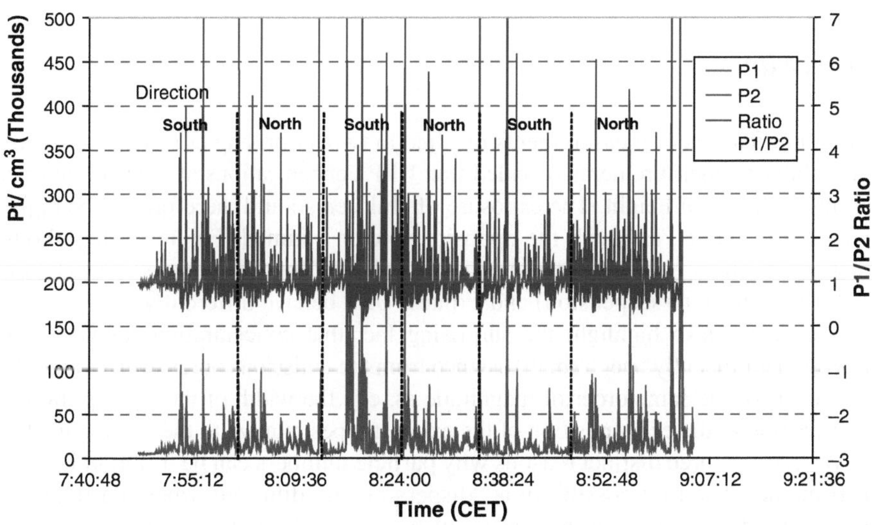

Fig. 4 Number of particles on the cycling track (P1 closest to the road, P2 1 m further away from the road). Six trips on April 2, 2008

(~4%) between both instruments. It was therefore decided to repeat the test with N = 10 which should allow us to determine the significance of the difference at the p = 0.01 level.

The preliminary results were confirmed in the second test although average particle numbers (~20,000 cm^{-3}) were significantly lower than during the pilot test (*t*-test $p < 0.001$). Some results are shown in Fig. 4.

The measurements are characterized by an important number of high peaks coinciding with motorized traffic on the road and alternating with much lower values. Values around 10,000 particles per cm^3 are most common while frequent peaks over 100,000 particles per cm^3 have been recorded. The highest values are in excess of the P-TRAK range (500,000 cm^{-3}). P1 (closest to the road) frequently records higher values than P2 does (only 1 m further away from the road). Although occasionally P2 values are marginally higher than P1, values of P1 are often three to five times as high during peak UFP events (right axis in Fig. 4).

On average, 1,150 (stdev: 722) more particles cm^{-3} were measured closest to the road. Although there is a lot of variation and data is obviously not distributed normally, non-parametric tests reveal that the average P1 values for each trajectory are significantly higher (Wilcoxon matched pairs signed ranks test: $p < 0.005$ one-sided test). The average P1/P2 ratio is 1.06 (stdev 0.04).

We could not demonstrate an effect of wind direction on the measured concentrations. Measured particles numbers are not significantly different and independent of whether the wind is blowing emissions towards the cycling track (trips in northern direction) or to the opposite side of the road (away from the cyclist, trips in southern direction).

Discussion

These limited sets of measurements resulted in two surprising conclusions. Firstly, our validation measurements revealed that UFP concentrations at the cycling track are independent of height. The exposure of children should therefore not be higher than that of adults, although the opposite is often claimed. Future studies should however confirm if heavier particles such as TSP (Total Suspended particles) or PM10 (e.g. from re-suspension) and exhaust UFP behave differently.

Our results also highlight the surprising fact that concentrations of UFP may decrease significantly over a small distance from the emission source even though this distance is of the same order of magnitude as, e.g. the width of the cycling track or the variation in the exact position of the exhaust pipes relative to the cycling track.

We can see three distinct reasons why particle numbers can be different for both instruments. The first possibility is "dispersion and dilution" where particles are transported by wind and at increasing distances are diluted in a larger volume of air. This process certainly occurs at larger scales but at this scale level we do not see a consistent difference between both measurements that would be in line with this explanation. Possibly this pattern may emerge during measurements at higher wind speeds.

The second possibility is that particles are not yet uniformly mixed and erratic turbulence (caused by traffic itself) causes the exposure of one of the instruments (but not the other) to the emission plume. This would explain why differences are occasionally very large and differences in the other direction (although smaller) can also occur. The fact that there appears to be no significant effect of cycling direction (i.e. the side of road) on the number of particles measured, may imply that the process is not primarily driven by wind dispersion but by turbulence.

The third possibility is that due to high local concentrations and turbulence during peaks, particles collide and coagulate thus reducing the number of particles but not the mass. This assumption will be tested in a future experiment.

Simplified dispersion models can be used to demonstrate that under stable meteorological conditions, as was the case during our study, dispersion over short ranges is only determined by distance and the initial mixing volume which depends on the size and wake of the vehicles. This is entirely consistent with our measurements.

Further studies are required to corroborate the tentative conclusions of this experiment. In this paper we have focused on mobile measurements in real traffic because these are most relevant for exposure of UFP to cyclists, and also of interest to policy makers. Unfortunately the GPS readings are not accurate enough to determine the exact distance between the emission and the P-TRAK sampling tube. To test which explanation effect best accounts for the measurements, we will use fixed monitors at different distances from the road, measuring both UFP and PM mass. In addition, this will allow us to confirm the results for different wind speeds and directions and determine the rate of decrease at different distances. This decay function will then be used to construct a CFD model in ENVI-met (www.envi-met.com) that will allow us to make more general predictions as well as study scenarios that may modify exposure of UFP on the cycling track (such as separation distance, provision of physical separation by parked cars, sound screens or green hedges) [5, 6].

True exposure to cyclists should be defined as a combination of concentrations and inhalation. Therefore (in other parallel experiments) simultaneous measurements are taken of physical load (breathing rate, oxygen consumption and power output). These combined measurements will allow us to determine the exact quantity of particulate matter that is inhaled. This quantity is important to determine accurately because in the end it determines whether cyclists are exposed to a higher internal dose of PM than car drivers are. No other project has ever designed such simultaneous measurements and most authors have either used rough estimates to account for the higher breathing rate of cyclists [4] or used modeled PM levels [7]. Given that the ultimate goal of the SHAPES project is to provide conclusions and recommendations for all Belgian regions, it is important to take the effects of slopes on breathing rate into account. All test trajectories therefore include sections with slopes.

Conclusion

A P-TRAK device can conveniently be used for making UFP measurements while cycling. UFP measurements prove to be a very useful supplement to PM10 and PM2.5 measurements as they are more closely related to emissions of motorized traffic.

Measurements show that at least under some conditions UFP fall rapidly with increasing distances from the emission source. This reveals important new possibilities to decrease the exposure of people to PM in the light of the new European directive that calls for a 20% reduction of the exposure of the population. Modifications to the design criteria for cycling infrastructure and motorized vehicles as well as changes to the organization of traffic and the highway code may all contribute to this objective.

Future research should confirm these results for other particle sizes (PM10) at different locations and under different meteorological conditions to elucidate the mechanism explaining our observations.

In the next phase of the SHAPES project the amount of UFP and PM10 that is actually inhaled by cyclists will be compared to the exposures to car drivers.

Acknowledgement The work reported in this paper was partly financed by the Belgian science policy under the Science for Sustainable Development program and the VITO UFP strategic research project.

References

1. de Geus B, De Smet S, Nijs J, Meeusen R (2007) Determining the intensity and energy expenditure during commuter cycling. Br. J. Sports Med. 41:8–12
2. Cavill N, Davis A (2007) Cycling and health, what's the evidence? Cycling England, London, 51 pp
3. Berghmans P, Bleux N, Int Panis L, Mishra VK, Torfs R, Van Poppel M (2008) Exposure assessment of a cyclist to PM10 and Ultrafine particles. Science of the total environment doi:10.1016/j.scitotenv.2008.10.041
4. den Breejen E (2006) Fietsers en verkeersuitstoot. Verkenning van de blootstelling van fietsers aan fijn en ultrafijn stof. Wetenschapswinkel Biologie. Utrecht: Universiteit Utrecht, P-UB-2006–02, 86 pp. (in Dutch)
5. Vankerkom J, De Maerschalck B, Janssen S, Tonneijck F (2007). Update en gevoeligheidsanalyse van de vegetatiemodule in het ENVI-met model. VITO, 48 pp. (in Dutch)
6. Bruse M, Fleer H (1998) Simulating surface–plant–air interactions inside urban environments with a three dimensional numerical model. Environ Model Software 13:373–384
7. Hertel O, Hvidberg M, Ketzel M, Storm L, Stausgaard L (2007) A proper choice of route significantly reduces air pollution exposure – a study on bicycle and bus trips in urban streets. Sci Total Environ 389:58–70

Air Pollution Exposure in an Unventilated Tunnel in Leeds, UK

Anil Namdeo and James Tippins

Introduction

Traffic emission is known as the major source of air pollution in urban areas, which contribute substantially to ambient levels of particulate matter and a range of gaseous pollutants. The continual traffic growth has raised concerns over the impact of traffic emissions on human health and urban environmental quality. Roadside concentrations of ultra-fine particles, NO and other pollutants can be up to 20 magnitudes greater than at residential locations. Such spatial variations are a factor of traffic volume, driver behavior, fleet composition, meteorological and the surrounding built environment. Canyons and tunnels, along with other road features are known potential pollution hotspots [1–4]. In canyons and tunnels, exposures to pollutants are often elevated in comparison to elsewhere so that individuals may gain a significant contribution to daily total exposure [2].

Particles have been documented to increase mortality, morbidity and decreased lung function [5]. Industry and vehicular traffic are major anthropogenic sources of particles. Most of the studies have linked health effects to mass concentration of particles under 10 μm (PM_{10}) in size. However, there is considerable uncertainty about which physical and/or chemical characteristics of PM are most important of determinants of health effects. Recent studies have shown that ultrafine particles (with $d < 0.1$ μm) are more toxic than larger particles [6, 7]. Ultra fine particles (UFP) are important because of their larger surface area and the ability to deposit in high alveolar region and the ability to induce inflammation [8–12]. Particle number

A. Namdeo
Institute for Transport Studies, University of Leeds, Leeds, LS2 9JT, UK
Transport Operations Research Group, Newcastle University,
Newcastle upon Tyne NE1 7RU, UK

J. Tippins
JMP Consulting, Blackfriars House, Parsonage, Manchester, M3 2JA, UK

S. Rauch et al. (eds.), *Highway and Urban Environment*,
Alliance for Global Sustainability Bookseries 17,
DOI 10.1007/978-90-481-3043-6_14, © Springer Science+Business Media B.V. 2010

concentrations (PNC) are dominated by UFPs [2, 13] suggesting that this metric could be more indicative of potential health impacts of particles. Because of this reason many researchers have paid increasing attention to UFP in the atmosphere, especially close to road traffic which contributes significantly to particle number concentrations [9, 13, 14].

Exposure of drivers and passengers in vehicles and in restricted environments (e.g. tunnels and canyons) is of particular concern because of elevated levels of particle concentrations in these situations compared to elsewhere [15, 16].

This study thus aimed to investigate the exposure of drivers and passengers travelling through a tunnel on a busy urban motorway in Leeds, UK. This study also aimed to investigate if the particle levels were higher inside the tunnel than the outside. Spatial and temporal variations in PNCs in the tunnel and car cabin were determined to quantify personal exposures. This study also helped to test if mobile measurements can be used to investigate temporal and spatial exposure to ultrafine particles. Currently, there are no limit or guideline values for particles. This study could help in understanding personal exposure of commuters and thus help in developing legal standards for UFPs.

Experimental Method

Study Area: The Westgate tunnel (Fig. 1) in Leeds, UK is one of the longest unventilated tunnels in Europe, and relies solely on natural ventilation generated by wind movements and the piston effect of moving vehicles forcing the movement of air. The eastbound bore is 392 m in length and the westbound bore slightly shorter at 380 m due to the curvature of the tunnel's alignment (Fig. 2); however both bores

Fig. 1 Westgate tunnel, Leeds, UK

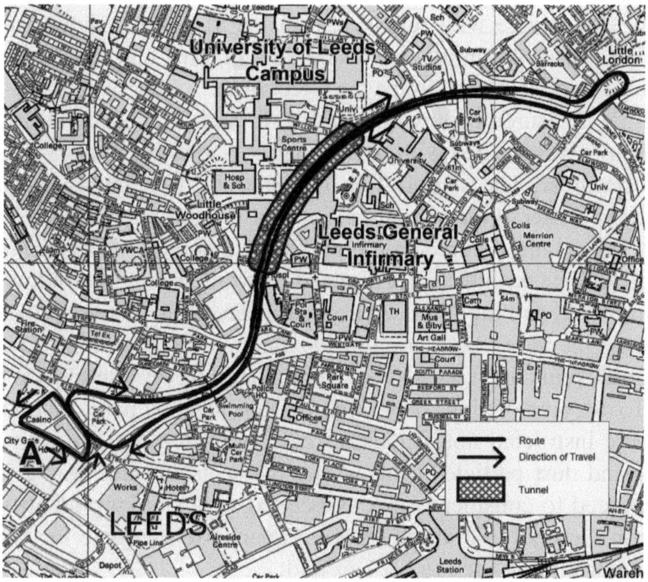

Fig. 2 Sampling route (© Crown Copyright/database right 2007. An Ordnance Survey/EDINA supplied service)

are greater than the advised maximum 240 m in length for unventilated tunnels [ASHRAE 1999 cited in 17]. The tunnel is also on a slight gradient with the westbound traffic flowing downhill. The tunnel forms part of the Leeds Inner Ring Road, and was constructed within a vertical cutting to limit its intrusion into the surrounding environment.

Monitoring Setup: TSI Model 3007 Condensation Particle Counter (CPC) was chosen in this study for ultrafine particle measurements. This model has successfully been used in a previous study to measure UFPs on a mobile vehicle platform [18]. This is a portable unit and is capable of detecting particles in the range of 10 nm to >1 μm [19].

Prior to each session the alcohol cartridge was replaced, and the CPC allowed to complete a full 600-s warm up, before carrying out a 15 s zero test using the factory supplied zero filter. The internal clock was updated to GPS time, which was available from car's CAN (Computer Aided Network) tablet interface. The logging interval for all sampling was set to one second to ensure the highest resolution of data was available for analysis. A 2003 Ford Mondeo hatchback (2.0 TDCi LX) was used as the platform for the monitoring equipment. The car was fitted with a data acquisition PC with wireless tablet interface, which logged data from the onboard Differential GPS receiver and CAN at 10 Hz.

Two sampling locations were chosen to monitor PNCs; the roof rack of the car for external and the back of the driver's headrest for internal measurements. Two sampling lines were used to connect to two CPCs which were located in the boot space of the vehicle with rubber foam used to provide shock dampening as experience by [20]. Both sample lines were 2.66 m long, and made from an appropriate anti-static material to minimize the effect of the tubing walls [20]. The sample line for internal measurements was threaded through the left corner of the parcel shelf and the inlet was located within the breathing zone of the front passenger, and fixed to the headrest. The sample line for external measurements was threaded through the right corner of the parcel shelf and passed through the partially open rear driver-side window and secured to the roof rack assembly using zip ties, 155 cm from the road surface with the inlet directed forward. The rear window was sealed using adhesive tape, and the lines secured with care being taken to avoid kinking.

The standard set up for ventilation inside the monitoring vehicle was as follows; windows closed, vents open, fans set to 'recirculate' (level 2 setting) and air conditioning off. To discount obvious non-combustion sources of particles both driver and passenger were instructed not to make excessive or abnormal movements, which could re-suspend dust particles from within the seats. As this was a fleet vehicle there was no need to consider issues associated with smoking identified by [21]. After each run the boot and all doors were opened to ventilate the car and bring the internal environment back to background.

The route (Fig. 2) chosen for this study was 4.2 km long and was selected to facilitate efficient passage through both the east and westbound bores of the Westgate tunnel whilst also providing a suitable location to stop and pre-condition the cabin before each run. This was a small lay-by (marked as A in Fig. 2) behind a car park. All measurements from within the tunnel were taken from the right-hand lane as it had been identified that pollution concentrations varied with distance across the carriageway [22]. As well as the tunnel microenvironment, the route encountered other urban highway features such as; open carriageway, covered carriageway (bridges), slip roads and two busy gyratories.

Results and Discussion

Sampling was conducted on Tuesday 26th and Wednesday 27th June 2007, with 07:30–09:30 and 12:30–14:30 representing the peak and inter-peak periods respectively. The duration of each sampling period was limited to 2 h as this was a conservative estimate of the lifespan of the battery powering the onboard computer hardware. A total of 21 trips were completed, with 11 carried out during the peak period and ten during the inter-peak period. All particle number counts (PNCs) were measured in particles cm^{-3}.

External concentrations: Figure 3 shows external PNCs as recorded by the CPC measuring air at the roof rack of the car. PNCs are presented in four groups; the first group of data indicates averaged concentrations for the whole route, while the second set indicated concentrations for the non-tunnel part of the route; and lastly the third and fourth sets present concentrations in the East and West bores of the tunnel. Concentrations in each set are separated for peak and non-peak periods.

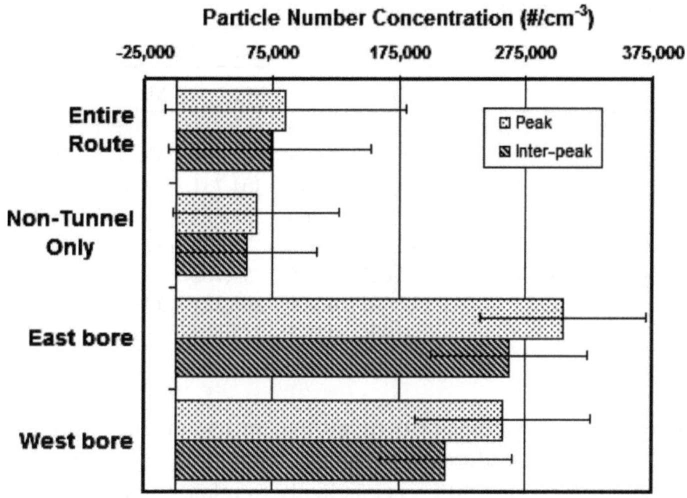

Fig. 3 External peak and inter-peak PNC concentrations

As displayed in Fig. 3, irrespective of the location, average PNCs are higher during the peak period, with the greatest actual variation between sampling periods evident within the two tunnel bores, showing decreases of between 42,000 and 45,000 particles cm^{-3}. This clearly shows the influence of road traffic on UFP levels. Levels within the eastbound bore are 4.86 and 4.76 magnitudes greater that the rest of the non-tunnel section of the route and greater by a slightly smaller factor of 4.11 and 3.86 for the westbound bore. This demonstrates that absence of any ventilation and the restricted topography has resulted in more than fourfold increase in UFP concentration. A comparison between the two bores shows that PNC levels within the eastbound bore are 1.2 magnitudes greater than those of the westbound bore for the peak and inter-peak periods, with an average difference of 48,400 particles cm^{-3}. Interrogating the CAN data showed that the average speed of the monitoring vehicle through both bores was recorded as being similar; however, during sampling, the driver commented that whilst traversing the westbound bore it was not necessary to use the accelerator to maintain the vehicle's momentum. Interrogation of the CAN data showed that the accelerator pedal is further depressed whilst ascending the eastbound bore and 'hardly depressed' whilst in the westbound bore. It is therefore probable that an increased load on the engine and associated increase in emissions is the contributing factor. During the inter-peak, traffic flows are the same in each bore and this explanation therefore still holds.

The highest maximum particle concentration level was within the eastbound bore during the peak period, measuring 397,604 particles cm^{-3} (Table 1). This can be directly compared to [18], who also used a TSI Model 3007 CPC, and recorded a maximum value of 800,000 particles cm^{-3}; however this was sampled along highways and attributed to high polluting diesel vehicles, not tunnels. Within the Westgate tunnel, highest mean concentrations are seen during the peak period, measuring

Table 1 Internal and external UFP concentrations

		Internal PNC (#/cm^{-3})			External PNC (#/cm^{-3})			Ratio – external/ internal
Description	Session	Mean	Max	SD	Mean	Max	SD	
Entire Route	Peak	14,108	41,624	5,650	85,711	397,604	94,366	6.1
Non-tunnel only	Peak	14,130	41,624	5,738	62,664	391,874	65,616	4.4
East bore	Peak	10,931	19,987	2,843	304,731	397,604	64,715	27.9
West bore	Peak	11,434	28,874	2,366	257,625	388,860	68,885	22.5
Entire route	Inter-peak	14,036	47,627	4,954	73,539	394,952	79,755	5.2
Non-tunnel only	Inter-peak	13,933	47,627	4,855	55,102	383,408	54,857	4.0
East bore	Inter-peak	17,302	17,777	4,319	262,221	394,952	61,043	15.2
West bore	Inter-peak	15,205	30,996	4,997	212,507	318,803	52,152	14.0

304,731 particles cm^{-3}, whereas [9] found that measurements within the Grande Mare tunnel were higher during the afternoon period, averaging in excess of 400,000 particles cm^{-3}. This difference is likely to relate to the different lengths of the two tunnels, different diurnal profiles and fleet composition. To put these values into perspective, [23] sampled PNCs within residential flats near a busy road and measured indoor concentrations of approximately 10,000 particles cm^{-3}. Assuming this observation is typical of indoor environments, particle concentrations within the Westgate tunnel are therefore up to 30 magnitudes greater than an indoor environment, where sources of particles are controlled and limited to infiltration through the infrastructure of the building.

Spatial variation in external concentrations: Averaged external PNC profiles along the route for the peak and inter-peak periods are presented in Figs. 4 and 5. Although similar shaped profiles can be seen for the two sampling periods, concentrations are higher within both tunnel bores during the peak period. Steep increases are seen prior to entering both tunnel bores, starting on average 124 m from the tunnel entrances, with the greatest relative increase seen at the westbound tunnel where concentrations rise by a factor of 3.43 and 5.35 for peak and inter-peak periods respectively, representing an increase of 1,361 and 1,892 particles cm^{-3} m^{-1} travelled for these periods. For the eastbound bore, concentrations increase by a factor of 2.35 and 2.60 for the two periods - representing an average increase of 1,264 particles cm^{-3} m^{-1} travelled. Away from the tunnel, other PNC events are evident in both peak and inter-peak periods at similar distances (and therefore geographical locations) along the route.

Internal concentrations: Figure 4 shows the variation in internal exposure levels for both the peak and inter-peak periods (plotted on secondary axis). Averaged over the entire route, internal concentrations are 6.1 and 5.2 magnitudes less than the corresponding external concentrations for the peak and non-peak period respectively (Table 1). This shows that keeping windows closed and recirculating air in the vehicle period considerably reduce in-cabin exposure of particles along the entire route. Within the tunnel microenvironment, during the peak period internal concentrations are up to 22 to 27 magnitudes less than external levels. This ratio is less during the non-peak period (14–15.2 for West and East bores respectively).

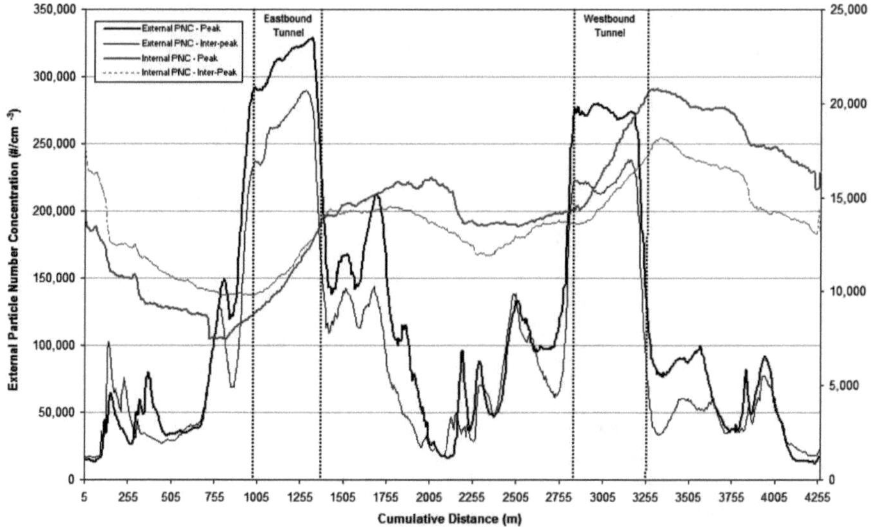

Fig. 4 External and internal peak and inter-peak PNCs

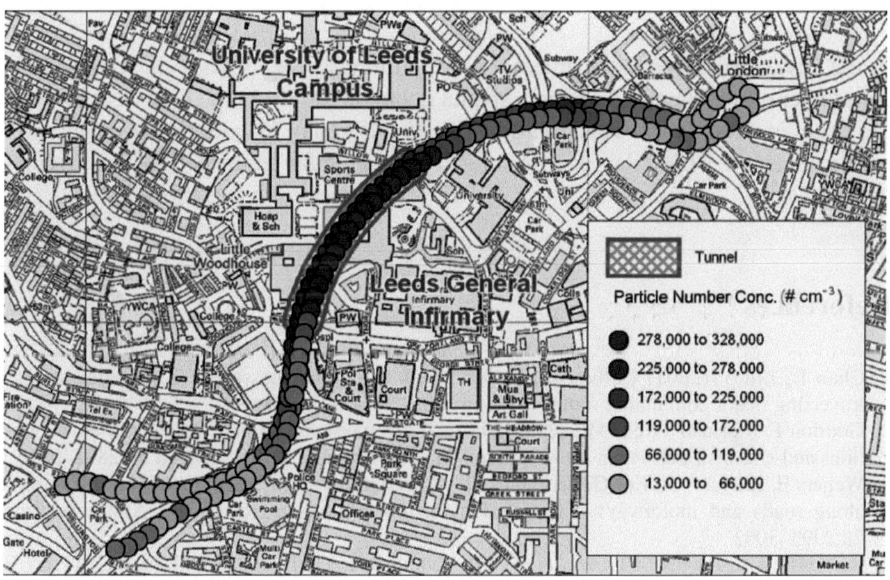

Fig. 5 External inter-peak PNCs

Figures 4 and 5 show that internal PNCs do not exactly track external PNCs temporally and spatially. A delayed penetration of particles into the vehicle cabin is evident when spatial profiles (Figs. 5 and 6) of external and internal PNCs are compared. This effect is markedly evident from internal profiles in Fig 4.

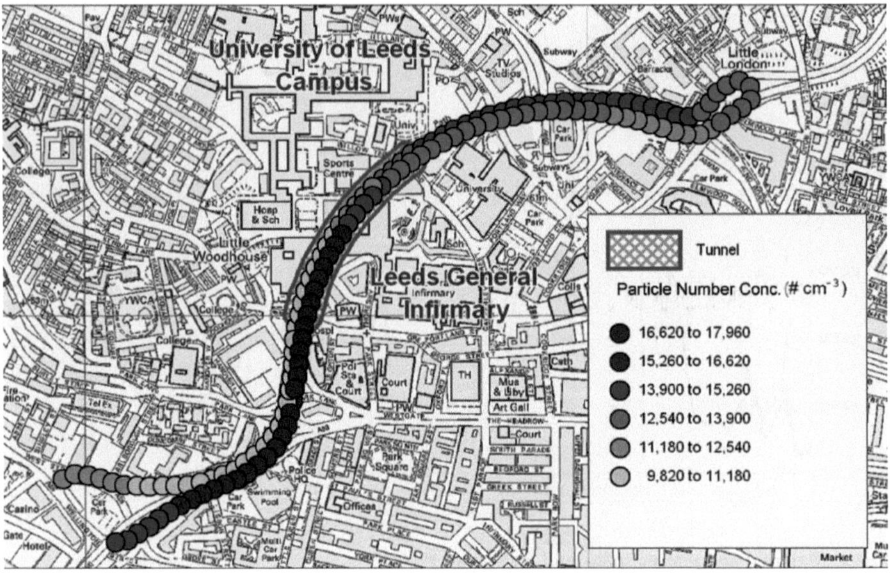

Fig. 6 Internal inter-peak PNCs

In the case of the West bore, internal concentrations peaked when the vehicle was about to exit the tunnel indicating a spatial delay of 190 m. In the case of East bore, the delay is 400–500 m. This could be due to a time lag in external air penetrating inside the cabin. This delay was prolonged due to the fan in the car set to recirculation mode.

References

1. Chan L, Liu Y (2001) Carbon monoxide levels in popular passenger commuting modes traversing major commuting routes in Hong Kong. Atmos Environ 35:2637–2646
2. Gouriou F, Morin JP, Weill ME (2004) On-road measurements of particle number concentrations and size distributions in urban and tunnel environments. Atmos Environ 38:2831–2840
3. Weijers E, Khlystov A, Kos G, Erisman J (2004) Variability of particular matter concentrations along roads and motorways determined by a moving measurement unit. Atmos Environ 28:2993–3002
4. Kristensson A, Johansson C, Westerholm R, Swietlicki E, Gidhagen L, Wideqvist U, Vesley V (2004) Real-world traffic emission factors of gases and particles measured in a road tunnel in Stockholm, Sweden. Atmos Environ 38:657–673
5. Pope CA, Dockery DW (2006) Health effects of fine particulate air pollution: lines that connect. J Air Waste Manage Assoc 56:709–742
6. Donaldson K, Brown D, Clouter A, Duffin R, MacNee W, Renwick L, Tran L, Stone V, (2002) The pulmonary toxicology of ultrafine particles, J Aerosol Med. 15:213–220
7. Li N, Sioutas C, Cho A, Schmitz D, Misra C, Sempf J, Wang M, Oberley T, Froines J, Nel A (2003) Ultrafine particulate pollutants induce oxidative stress and mitochondrial damage, Environ. Health Perspect. 111(4):455–460

8. Wichmann HE, Spix C, Tuch T, Wölke G, Peters A, Heinrich J, Kreyling W, Heyder J (2000) Daily Mortality and Fine and Ultrafine Particles in Erfurt, Germany, Part I: Role of Particle Number and Particle Mass. Research Report number 98. Health Effects Institute, Cambridge, MA
9. Hughes LS, Cass GR, Gone J, Ames M, Olmez I (1998) Physical and chemical characterization of atmospheric ultrafine particles in the Los Angeles area. Environ. Sci. Technol. 32:1153–1161
10. Oberdörster G (2000) Toxicology of ultrafine particles: in vivo studies. Philos Trans R Soc Lond A 358:2719–40
11. Oberdörster G (2001) Pulmonary effects of inhaled ultrafine particles. Int Arch Occ Env Health 74:1–8
12. Pekkanen J (2006) Exposure to ultrafine particles and respiratory health effects. Toxicol Lett 164:S33–S33
13. Woo KS, Chen DR, Pui DYH, McMurry PH (2001) Measurement of Atlanta aerosol size distribution: observations of ultrafine particle events. Aerosol Sci Tech 34:75–87
14. Kaur S, Clark RDR, Walsh PT, Arnold, SJ, Colvile RN, Nieuwenhuijsen M (2006) Exposure visualisation of ultrafine particle counts in a transport microenvironment. Atmos Environ 40: 386–398
15. Brunekreef B, Janssen NA, Hartog J, Harssema H, Knape M, Vliet P-v (1997) Air pollution from truck traffic and lung function in children living near motorways. Epidemiology 8:298–303
16. Diapouli E, Chaloulakou A, Spyrellis N (2007) Levels of ultrafine particles in different microenvironments – implications to children exposure. Sci Total Environ 388:128–136
17. Chow WK, Chan MY (2003) Field measurement on transient carbon monoxide levels in vehicular tunnels. Building Environ 38:227–236
18. Westerdahl D, Fruin S, Sax T, Fine P, Sioutas C (2005) Mobile platform measurements of ultrafine particles and associated pollutant concentrations on freeways and residential streets in Los Angeles. Atmos Environ 39:3597–3610
19. TSI (2001) Model 3007 Condensation Particle Counter: Operation and Service Manual. TSI Incorporated (USA).
20. Bukowiecki N, Dommen J, Prévôt A, Richter R, Weingartner E, Baltensperger U (2002) A mobile pollutant measurement laboratory – measuring gas phase and aerosol ambient concentrations with high spatial and temporal resolution. Atmos Environ 36:5569–5579
21. Abi Esber L, El-Fadel M, Nuwayhid I, Saliba N (2007) The effect of different ventilation modes on in-vehicle carbon monoxide exposure. Atmos Environ 4:3644–3657
22. Adams H, Niewenhuijsen M, Colvile R, McMullen M, Khandelwal P (2001) Fine particle ($PM_{2.5}$) personal exposure levels in transport microenvironments, London, UK. Sci Total Environ 279:29–44
23. Zhu Y, Hinds W, Krudysz M, Kuhn T, Froines J, Sioutas C (2005) Penetration of freeway ultrafine particles into indoor environments. Aerosol Sci 36:303–322

Assessment of Aerosol Concentration Sampled at Five Sites in Beijing from 2005 Till 2007

Stefan Norra, Nina Schleicher, Doris Stüben, Fahe Chai, Yzhen Chen, and Shulan Wang

Introduction

Beijing, the capital of China, comprises about 15 million inhabitants. This megacity is highly affected by atmospheric pollution from various sources, such as combustion processes in industry, households and traffic, further industrial emissions, abrasion processes in traffic, and construction activities. These anthropogenic sources are superimposed by dust storms episodically occurring in spring and transporting particles from western deserts towards Beijing.

Typical characteristics of Beijing influencing the air pollution are a high population density and high energy use intensity (Beijing covers 1,040 km^2; on 6% of this area 50% of the population, 80% buildings and 80% energy consumption are concentrated), an intensive coal consumption (about 70% of the energy is provided by coal) and the number of vehicle increased very rapidly in recent years. Soil borne dust contributes to a high amount to the TSP: Statistically, about more than 5,000 construction sites are underway at the same time. It is predicted that more and more construction sites will be under way during the preparation period of the Olympic Games in 2008, so it makes the TSP high due to the effects of demolishing and removal of buildings, of uncovered bulk material piles, of material loss during the transportation, and of uncovered earth. Industrial air pollutants affect local districts, such as Shi Jingshan District with power plants, cement plants, iron and steel plants. Fa Tou is another district with intense air pollution caused by chemical factories, where odor and harmful gas pollution are serious.

Aerosols are considered to harm human health in various ways, especially via the respiration tract and the cardio-vascular system [1]. The impact of aerosols on human health is currently assessed mainly by the total mass concentration, although

S. Norra, N. Schleicher, and D. Stüben
Institute of Mineralogy and Geochemistry, Universität Karlsruhe (TH), Germany

F. Chai, Y. Chen, and S. Wang
China Research Academy of Environmental Sciences, Beijing, China

S. Rauch et al. (eds.), *Highway and Urban Environment*,
Alliance for Global Sustainability Bookseries 17,
DOI 10.1007/978-90-481-3043-6_15,

special regards are spent to soot and sulphate containing particles [2–6]. However, also metals are considered to contribute to the negative health impacts of particles [7–13]. Epidemiological studies often refer to the general mass concentrations of particles because of the complexity of the specific toxicities of relevant substances in atmospheric particles, which are for large parts still not known [6, 14]. Since the total $PM_{2.5}$ mass concentration and the concentrations of metals in $PM_{2.5}$ are important to assess the air quality of megacities, this study aims to investigate exemplarily their spatial distributions over two years in Beijing.

Methods

Samples of $PM_{2.5}$ were taken at five stations established along a 50 km long transect running through Beijing from North-West to South-East. The specific locations are presented in [15]. $PM_{2.5}$ was taken during day and night time, from 7 a.m. to 7 p.m. and 7 p.m. to 7 a.m., respectively. Each sampling interval lasted for 1 week. For sampling, mini-volume samplers were used pumping 200 L air per hour. Samples were collected on quartz fiber filters of 50 mm in diameter. Gravimetric analysis was performed after at least 48 h equilibration with a microbalance (Sartorius MP2) for five times.

Chemical Analyses

For the analysis of trace metals, one quarter of the filters were digested in Teflon vessels with concentrated HNO_3 (Merck, subboiled), concentrated HF (Merck, p.a) and concentrated $HClO_4$ (Merck, p.a). Determination of elements was performed with HR-ICP-MS (Axiom). Quality control was performed by additional analysis of reference material SRM 1648 (urban particulate matter) acquired from NIST (National Institute of Standards, USA). Results of standard material are within ±10% of the certified values for element concentrations.

Results and Discussion

Average $PM_{2.5}$ mass concentrations at all five stations are presented for each meteorological season in Fig. 1. It is obvious that the lowest $PM_{2.5}$ mass concentration occurred during summer time, although this concentration is still far above the newly introduced $PM_{2.5}$ threshold value of 15 μg/m^3 in the US. Nevertheless, with respect to the Olympic Games taking place in Beijing in August 2008, it is at least good information. Up to more than 120 mg/m^3 $PM_{2.5}$ mass concentration was observed during other seasons, especially during winter time.

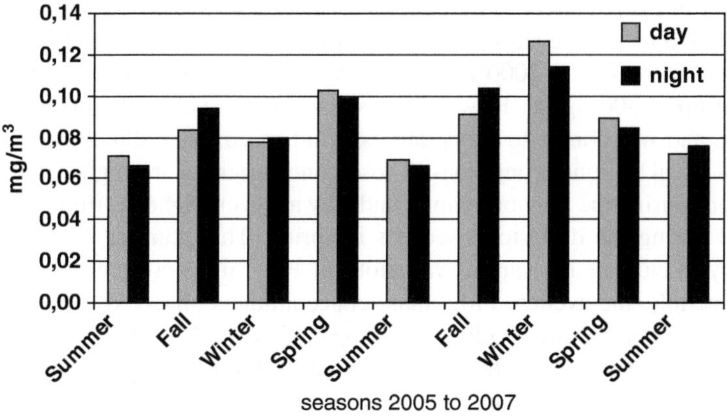

Fig. 1 Average $PM_{2.5}$ mass concentrations in different seasons from 2005 till 2007

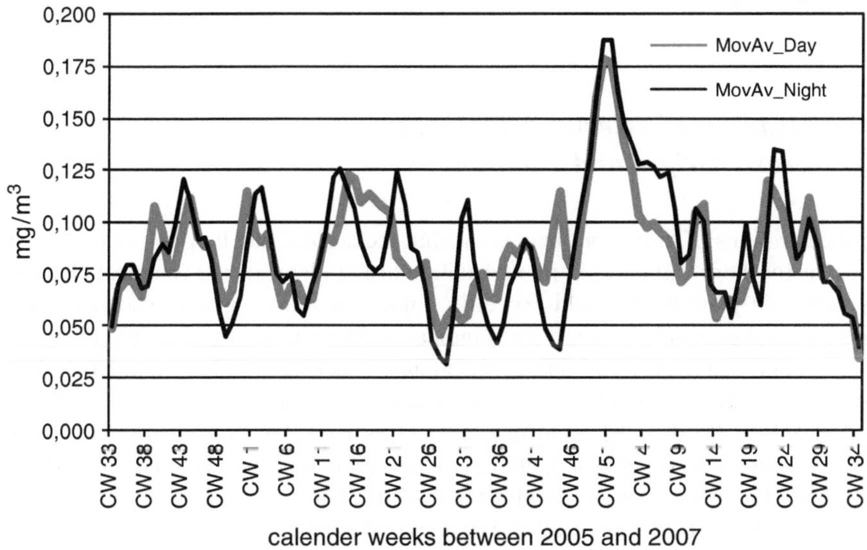

Fig. 2 Average mass concentrations of PM2.5 sampled during daytime and nighttime between 2005 and 2007

In winter time, emissions from combustion processes for heating and industry accumulate and produce high concentrations of $PM_{2.5}$, especially in cases of low wind speed and southern and eastern wind directions [16]. Figure 2 presents the mean $PM_{2.5}$ concentrations of all five stations calculated as moving average.

This was calculated according to the formula: $ac_{wn} = (c_{wm} + 2c_{wn} + c_{wo})/4$; ac_{wn}: moving average of mean aerosol concentration of all five sites of week n, c_{wm}: mean aerosol concentration of all five sites of week *m*, which is the week before week $_{n}$,

c_{wn}: mean aerosol concentration of all five sites of week $_n$, c_{wo}: mean aerosol concentration of all five sites of week *o*, which is the week following week *n*.

Especially the winter 2006/2007 showed very high concentrations with maxima values of up to 300 μg/m^3. However, reasons causing such high concentrations are complex, and will comprise temperatures, wind speed, wind directions and general meteorological regional conditions. Those concentrations are higher than the concentrations during the previous winter and they are even higher as the concentrations occurred during the dust storm seasons in spring. Thus, further investigations are necessary to eluidate the various variations of $PM_{2.5}$ mass concentrations along the year. Generally, the average $PM_{2.5}$ mass concentrations observed in Beijing exceed those investigated in various cities of Europe [17] or in New York [18] but are still in the range of $PM_{2.5}$ concentrations measured in Beijing in the years 1999 to 2003 [19–21]. Thus, although many measures to reduce the aerosol concentration took place during the last years, $PM_{2.5}$ mass concentration could not be reduced to concentrations similar to that found in cities of Europe and the US. Most probably, the economic development with all its consequences of enormous construction activities or growing number of vehicles counteracts aerosol reducing measures.

The Meaning of Selected Metal Loads for Aerosol Origin and Pollution

Concentrations of indicator elements within the aerosol can help to understand the reasons for those variations in more detail. Indicator elements are chemical elements, which mainly are associated to specific sources or source groups, such as anthropogenic and geogenic sources. Such indicator elements are for example Pb and Ti. Lead mainly indicates anthropogenic sources, such as combustion of petroleum based fuel, where Pb is naturally abundant, and industrial metal processing activities. Nevertheless, Pb also can occur in geogenic particles, but in most cases to a lesser degree, compared to anthropogenic particles. Titanium indicates mainly geogenic sources, although also used in some colors. Since humans nowadays almost use any chemical element for some technical material, elements purely indicating geogenic sources are hardly to be found [22, 23].

However, in Beijing, Pb can be used as a good indicator for anthropogenic and Ti for geogenic sources as shown in Fig. 3. The ratio of Pb/Ti indicates periods dominated by anthropogenic or geogenic sources of particles. A high TSP mass concentration during autumn 2005 only shows a relative low ratio of Pb/Ti, indicating a predominance of geogenic particles whereas during winter 2005/2006 a high ratio indicates the peak of the heating time. Again, in spring 2006, during the period of dust storms, the ratio drops to its minimum within the period of observation because of the predominating geogenic origin of the particles. For the rest of 2006, Pb/Ti is more or less constantly increasing, indicating predominant anthropogenic emissions as the cause for increasing TSP mass concentrations.

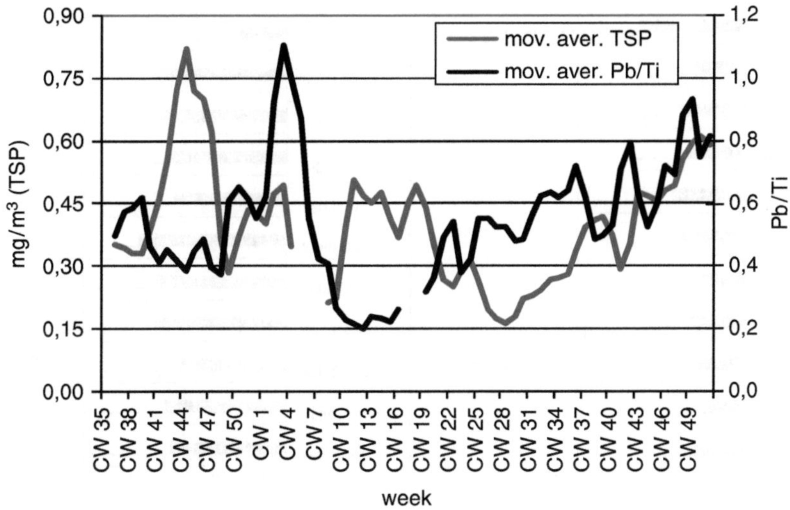

Fig. 3 Moving average of TSP mass concentration and the ration of Pb and Ti concentration between mid 2005 and end of 2006 at site 4 (SE Beijing)

Besides Pb, also Zn indicates the ubiquitous anthropogenic aerosol pollution in urban systems, since it is released during various combustion and abrasion processes. For the $PM_{2.5}$ fraction, Pb and Zn mass concentrations calculated as μg/m^3 stay below 0.5 μg/m^3, but are considerable higher than what is found in New York and Baltimore [18, 24–26]. Up to more than twofold of Pb and Zn concentrations (μg/m^3) in $PM_{2.5}$ were observed in the TSP fraction. However, Pb and Zn concentrations calculated as μg/g were lower in TSP than in $PM_{2.5}$.

Thus, the particle size fraction above 2.5 μm considerable contributes to the overall particle mass, but dilutes the metal concentrations within the particle mass. Nevertheless, within the TSP fraction, Pb and Zn concentrations still reach on average 1,000 and 2,500 μg/g, respectively. Lead and Zn concentrations in $PM_{2.5}$ exceed up to more as twice as much those in TSP. It is noteworthy to highlight the situation at site three located just near to the central Tienamen Square. Here the Pb and Zn mass concentrations expressed in μg/m^3 are less than at the sites northwest and southeast, but highest if expressed in μg/g. Those high concentrations might contribute to the health harming potential of aerosols although no standards are set for harmful metal concentrations in μg/g, yet (Fig. 4).

Conclusions

This study demonstrated that $PM_{2.5}$ concentrations are varying in the Beijing area. Lowest aerosol concentrations occurred during summer, whereas spring show high aerosol concentrations predominantly caused by geogenic particles, and winter is a

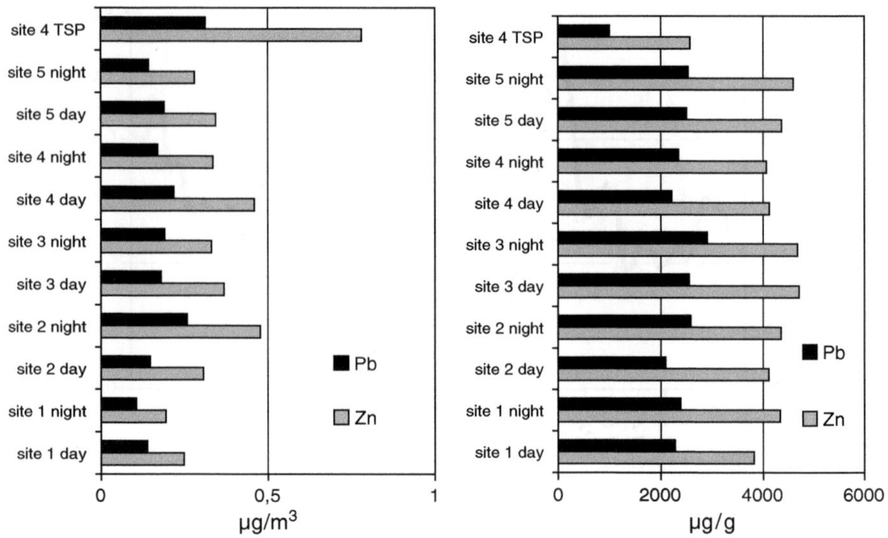

Fig. 4 Medians of Pb and Zn concentrations relative to air volume (μg/m³) and to aerosol mass (μg/g) at the five $PM_{2.5}$ sampling stations and the TSP sampling station for the period from September 2005 till end of 2006

time of high aerosol concentrations caused by anthropogenic emission processes. $PM_{2.5}$ concentrations during night are in average equal to those during day. Indicator elements help to understand the different roles of geogenic and anthropogenic sources contributing to the aerosol pollution in Beijing. Futhermore, beside the $PM_{2.5}$ mass concentrations, health studies should consider harmful metal concentrations, but not only calculated as μg/m³ but also as μg/g. This kind of relation provides important additional information for the assessment of the particular chemistry of single particles.

Acknowledgements Thanks to all colleagues of the China Research Academy of Environmental Sciences, who have collected the samples for such a long period. Additionally the authors thank the technical staff of the Institute of Mineralogy and Geochemistry (Norbert Bauer, Bernd Grunvinck, Inna Haag, Cornelia Haug, Pirimze Lakashvilli, Claudia Mössner, Peter Schaupp, Heinz Vögele) for numerous support with respect to sampling technology and chemical analyses. Finally, the German Research Foundations is gratefully mentioned for funding this project.

References

1. Kappos A, Bruckmann P, Eikmann T, Englert N, Heinrich U, Höppe U, Koch E, Metz N, Rauchfuss K, Rombout P, Schabronath J, Schulz-Klemp V, Spallek MF, Wichmann H-E, Kreyling WG, Krause GHM (2003) Bewertung des aktuellen wissenschaftlichen Kenntnisstandes zur gesundheitlichen Wirkung von Partikeln in der Luft. Umweltmedizin in Forschung und Praxis 8:257–278

2. Dockery DW, Pope CA, XU XP, Spengler JD, Ware JH, Fay ME, Ferris BG, Speizer FE (1993) An association between air pollution and mortality in six U.S. cities. The New England Journal of Medicine 329:1753–1759
3. Oettinger R, Drumm K, Knorst M, Krinyak P, Smolarski R, Kienast K (1999) Production of reactive oxygen intermediates by human macrophages exposed to soot particles and asbestos fibers and increase in NF-kappa B p50/p105 mRNA. Lung 177:343–354
4. Gardiner K, Tongeren MV, Harrington M (2001) Respiratory health effects from exposure to carbon black: results of the phase 2 and 3 cross sectional studies in European carbon black manufacturing industry. Occupational Environmental Medicine 58:496–503
5. Rushton L (2001) Cancer and air pollution. In: Brimblecombe P, Maynard RI (Hrsg.): The Urban Atmosphere and Its Effects. Imperial College Press, London, pp. 129–161
6. Diabaté S, Günther R, Völkel K, Thiele D, Wottrich R (2004) Gesundheitseffekte durch inhalierbare Feinstäube aus technischen Verbrennungsanlagen: In vitro Untersuchungen zur Wirkung feiner und ultrafeiner Partikel auf kultivierte Lungenzellen. Forschungsbericht FZKA-BWPLUS
7. Chillrud SN, Epstein D, Ross JM, Sax SN, Pederson D, Spengler J, Kinney PL (2004) Elevated airborne exposures of teenagers to manganese, chromium, and iron from steel dust and New York City's subway system. Environmental Science and Technology 38:732–737
8. Dreher KL, Jaskot RH, Lehmann JR, Richards JH, McGee JK, Ghio AJ, Costa DL (1997) Soluble transition metals mediate residual oil fly ash induced acute lung injury. Journal of Toxicology and Environmental Health 50:258–305
9. Kodavanti UP, Jaskot RH, Costa DL, Dreher KL (1997) Pulmonary proinflammatory gene induction following acute exposure to residual oil fly ash: roles of particle-associated metals. Inhalation Toxicology 9:679–701
10. Shi X, Mao Y, Knapton AD, Ding M, Rojanasakul Y, Gannett PM, Dalal N, Liu K (1994) Reaction of Cr(VI) with ascorbate and hydrogen peroxide generates hydroxyl radical and causes DNA damage: role of a Cr(IV)-mediated Fenton-like reaction. Carcinogenesis 15:2475–2478
11. Witholt R, Gwiadza RH, Smith DR (2000) The neurobehavioral effects of subchronic manganese exposure in the presence and absence of pre-parkinsonism. Neurotoxicology and Teratology 22:851–861
12. Zereini F, Skerstupp B, Rankenburg K, Dirksen F, Beyer J-M, Claus T, Urban H (2001) Anthropogenic emission of platinum-group elements into the environment. Journal of Soils and Sediments 1:44–49
13. Gwiazda RH, Lee D, Sheridan J, Smith DR (2002) Low cumulative manganese exposure affects striatal GABA but not Dopamine. Neurotoxicology 23:69–76
14. McClellan RO (2002) Setting ambient air standards for particulate matter. Toxicology 181–182:329–347
15. Schleicher N, Norra S, Fahe C, Chen Y, Wang S, Stüben D (2009) Seasonal trend of water-soluble ions at one TSP and five PM2.5 sampling sites in Beijing, China. In: Morrison GM; Rauch S (eds.): Highway and Urban Environment, Springer, Berlin
16. Norra S, Hundt B, Stüben D, Cen K, Liu C, Dietze V, Schultz E (2007) Size, morphological and chemical characterization of aerosols polluting the Beijing atmosphere in January/ February 2005. In: Morrison GM, Rauch S (eds.): Highway and Urban Environment, Springer, Berlin, pp. 167–180
17. Sillanpää M, Hillamo R, Saarikoski S, Frey A, Pennanen A, Makkonen U, Spolnik Z, Van Grieken R, Branis M, Brunekreef B, Chalbot M-C, Kuhlbusch T, Sunyer J, Kerminen V-M, Kulmala M, Salonen RO (2006): Chemical composition and mass closure of particulate matter at six urban sites in Europe. Atmospheric Environment 40:S212–S223
18. Qin Y, Kim E, Hopke PK (2006) The concentration and sources of PM2.5 in metropolitan New York City. Atmospheric Environment 40:S312–S332
19. He K, Yang F, Ma Y, Zhang Q, Yao X, Chan CK, Cadle S, Chan T, Mulawa P (2001) The characteristics of PM2.5 in Beijing, China. Atmospheric Environment 35:4959–4970
20. Wang Y, Zhuang G, Tang A, Yuan H, Sun Y, Chen S, Zheng A (2005) The ion chemistry and the source of PM2.5 aerosol in Beijing. Atmospheric Environment 39:3771–3784.

21. Zheng M, Salmon LG, Schauer JJ, Zeng L, Kiang CS, Zhang Y, Cass GR (2005) seasonal trends in PM2.5 source contributions in Beijing, China. Atmospheric Environment 39:3967–3976
22. Hashimoto Y, Sekine Y, Otoshi T (1992) Atmospheric aluminium from human activities', Atmospheric Environment 26B:295–300
23. Nath B, Norra S, Chatterjee D, Stüben D (2007) Fingerprinting of land use related chemical and pollution patterns of street sediments: an environmental forensic case study from Kolkata, India. Environmental Forensics Journal 8:313–328
24. Ito K, Xue N, Thurston G (2004) Spatial variation of PM2.5 chemical species and source-apportioned mass concentrations in New York City. Atmospheric Environment 38:5269–5282
25. Ondov JM, Buckley TJ, Hoe PK, Ogulei D, Parlange MB, Rogge WF, Squibb KS, Johnston MV, Wexler AS (2006) Baltimore supersite: highly time- and size-resolved concentrations of urban PM2.5 and its constituents for resolution of sources and immune responses. Atmospheric Environment 40:S224–S237
26. Qureshi S, Dutkiewicz VA, Khan AR, Swami K, Yang KX, Husain L, Schwab JJ, Demerjian KL (2006) Elemental composition of PM2.5 aerosols in Queens, New York: solubility and temporal trends. Atmospheric Environment 40:S238–S251

Emissions Pollutant from Diesel, Biodiesel and Natural Gas Refuse Collection Vehicles in Urban Areas

José M. López, Nuria Flores, Felipe Jiménez, and Francisco Aparicio

Introduction

The emission certification for heavy-duty applications is done by running engines in an engine dynamometer, not with the complete vehicle. The rationale for this is that a certain engine can be used for a number of different vehicle applications.

However, the interest to carry out dynamic emission testing with complete vehicles, either on a chassis dynamometer or on the road is increasing. According to the current directives, the measurement of emissions for the official approval of Euro IV industrial vehicle diesel engines must be performed in accordance with three types of test: ESC, ELR and ETC; the maximum acceptable emission values for these tests are shown in Table 1.

- ESC: European Steady Cycle
- ELR: European Load Response Test
- ETC: European Transient Cycle

Data Collection

Vehicles Description

The refuse collection trucks are lateral loaders. Due to the severe service operation high engine rating are required. The trucks used have difference power output, but the same curb weight. Some of the specifications for the instrumented trucks are presented in Table 2.

The CNG engine was turbocharged in stoichiometric ratio with multipoint port fuel injection and equipped with a three-way catalyst.

J.M. López, N. Flores, F. Jiménez and F. Aparicio
Polytechnic University of Madrid (UPM), University Institute for Automobile Research

S. Rauch et al. (eds.), *Highway and Urban Environment*,
Alliance for Global Sustainability Bookseries 17,
DOI 10.1007/978-90-481-3043-6_16, © Springer Science+Business Media B.V. 2010

Table 1 Maximum acceptable emission figures for ESC, ELR and ETC tests on Euro IV industrial vehicles

		European heavy-duty emission limit values	
EURO IV		ESC and ELR	ETC
CO	g/kW h	1.50	4.00
HC	g/kW h	0.46	–
NMHC	g/kW h	–	0.55
CH_4[a]	g/kW h	–	1.10
NO_x	g/kW h	3.50	3.50
PM	g/kW h	0.02	0.03
Smokes	m^{-1}	0.50	–

ESC, European Steady Cycle; ELR, European Load Response Test; ETC, European Transient Cycle.
[a] Only for CNG engines.

Table 2 Instrumented Trucks Specifications

	Diesel truck	CNG truck
	Diesel engine (ID)	Otto engine
Engine type	Turbocharged with intercooler	Turbocharged with intercooler
Displacement	7,790 cm^3	7,790 cm^3
Cylinders	Six, in line	Six, in line
Bore/stroke	115/125 mm	115/125 mm
Compression ratio	16:1	11.5:1
Engine power rating	228 kW (310 CV) at 1,950–2,400 rpm	200 kW (272 CV) at 1,100–1,800 rpm
Engine torque rating	1,300 N m (133 kg m) at 1,200–1,675 rpm	1,100 N m (112 kg m) at 1,100–1,800 rpm
Curb weight	26,000 kg	26,000 kg

Measurement Equipment

The "on-board" measurement equipment is a Horiba OBS 2200 device (Fig. 1). It can acquire data under real-driving conditions of vehicle use. Such data are instantaneous, and accumulated CO, CO_2, THC and NO_x emissions, fuel consumption, and the speed at any one instant are recorded. The set of gas analyzers provides real mass emission results (g/s) and instantaneous emissions (expressed as percent or ppm, depending on the gases involved). The analyzers are:

- Multiple CO, CO_2 and H_2O analyzer based on the Heated Non-Dispersive Infrared Detection (HNDIR).
- THC analyzer based on the Heated Flame Ionization Detection (HFID).
- NO_x analyzer based on the Heated Chemiluminescent Detection (HCLD).

The equipment also incorporates an exhaust gas flow meter that includes a calibrated pipe with its Pitot tubes, the sample gas outlet and the temperature sensor.

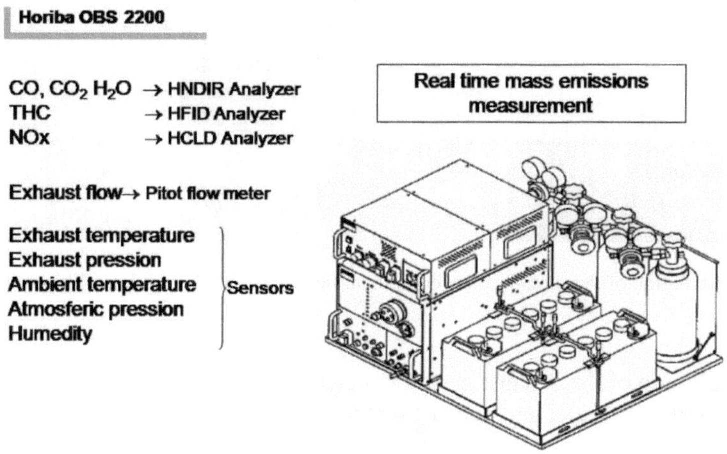

Fig. 1 Emissions measurement equipment Horiba OBS 2200

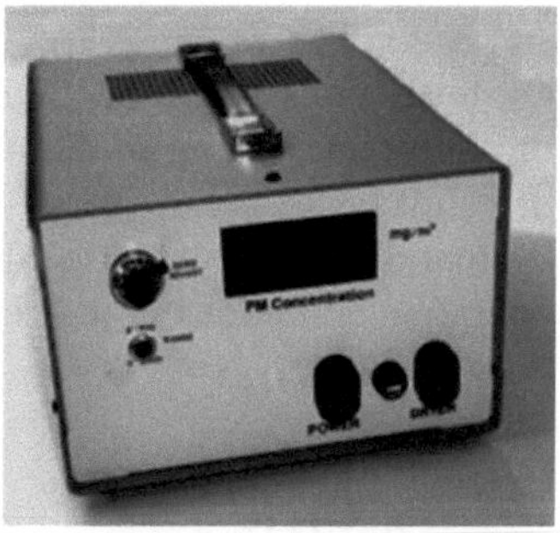

Fig. 2 MAHA particle measurement equipment

By using the carbon balance method, fuel consumption can be very accurately determined from the mass emissions of CO, CO_2 and THC. Also, given that the exhaust gas and fuel flow rate is known (calculated by carbon balance), the lambda factor can be accurately calculated.

The equipment also includes sensors for measuring humidity and ambient temperature and a GPS system for measuring the speed and position at every instant. It also has a laptop with specific software for system control, data collection and display and for generating test result reports.

Also integrated is an EIU (External Input Unit) with free analog ports to connect other equipment such as smoke opacity meters, thermocouples, etc. These analog inputs have allowed a MAHA particle measurement device to be connected (Fig. 2). This device provides the amounts of particles in mg/m^3 measured by laser technology.

The installation of the instrumentation equipment inside a vehicle is shown in the following images. Figure 3 shows the installation of the on-board emission measurement device – Horiba OBS 2200-, and the MAHA particle measurement equipment.

Fig. 3 Installation of the measurement equipment in a refuse truck with lateral loading

The installation of the exhaust gas flow meter is also shown in Fig. 3, with its Pitot tubes, sample gas outlet and exhaust temperature sensor.

Analysis Methodology

On-board emissions measurements were conducted in real-world driving conditions on a specific driving route in the city of Madrid. Although chassis dynamometer drive cycles exists for heavy duty vehicles [1, 2], on-road test represent the field evaluation of the exhaust after-treatment system development, under real-world operating conditions in the vehicle [3–5].

Driving Cycle

The first step in the data analysis was the separation of in-use vehicle data into a number of distinct operational segments or categories specific to the refuse application. Figure 4 shows a sample of vehicle speed data based on which the operational categories were defined [6]. Route segment include:

- Approach
- Collection
- Transfer to Dump
- Dump
- Return

Data sequences for each segment were analyzed to determine their emission contributions. The characteristics of all the segments significantly differ. For example, the Approach, Transfer to Dump, and Return segments involve no refuse compactions (i.e., no use of body hydraulics or power-take-off operation). Collection segment is a combination of kinematic and hydraulic operation. The Dump category does not involve significant vehicle motion, basically power-take-off operation.

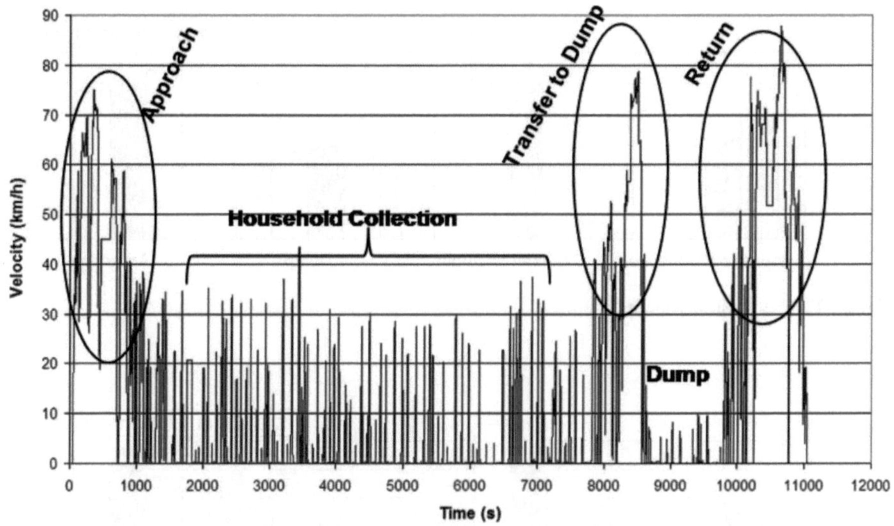

Fig. 4 Vehicle sample speed collection route data

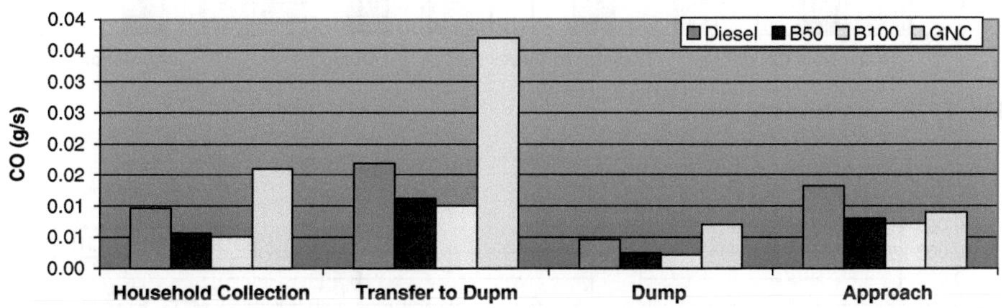

Fig. 5 CO emissions results

Results

The comparative study is carried out with regard to CO, HC, NO_x, PM and CO_2 emissions and fuel consumption for each segment of the driving cycle, in respect of two types of engine with three different fuels [7]. Figures 5–10 show the comparative results obtained in a real-world driving cycle emissions measurement.

The results for CO are given in Fig. 5. CO emissions have decreased with increasing amount of methyl ester in the blend. The most notable decrease was demonstrated during the Dump process. The CNG vehicle has higher CO emissions than the diesels.

Figure 6 presents tailpipe THC emissions. THC emissions have decreased with increasing amount of methyl ester in the blend. The differences among the selected segments are not relevant. The CNG vehicle had the same or even lower HC emissions than diesels except Transfer to Dump.

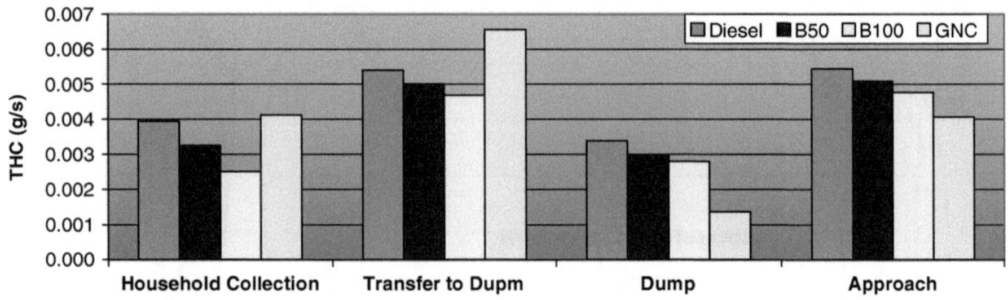

Fig. 6 THC emissions results

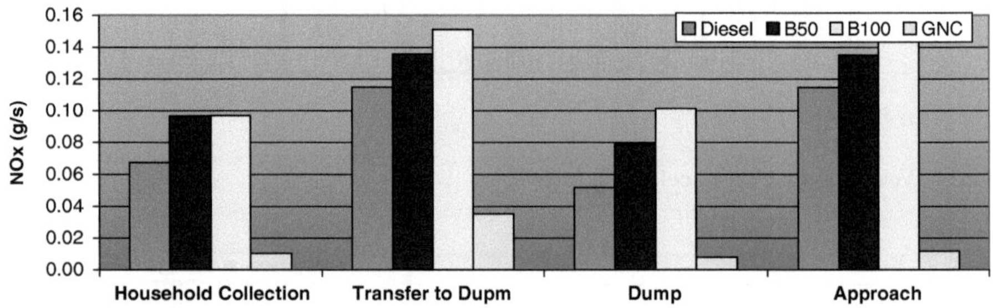

Fig. 7 NO_x emissions results

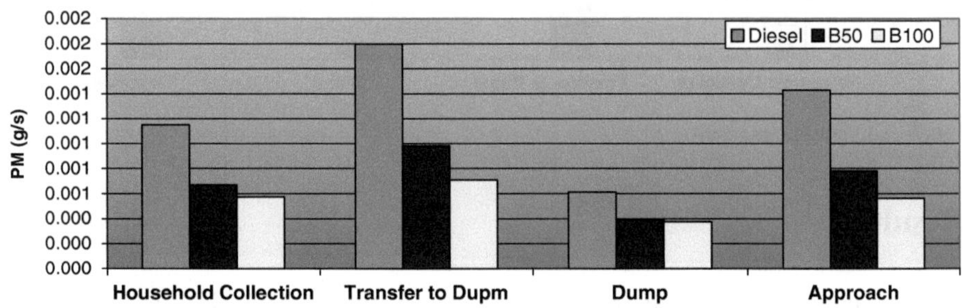

Fig. 8 PM emissions results

Figure 7 shows that with increasing amount of biodiesel in the blend, NO_x emissions present a noticeable increase. The CNG vehicle has lower NO_x emissions than the diesels.

The results for PM are given in Fig. 8. PM emissions have decreased with increasing amount of methyl ester in the blend. The most notable decrease was demonstrated during the Dump process. The CNG vehicle did not emit PM emissions. Fossil fuel combustion is the main source of global CO_2 levels. Figure 9 shows the CO_2 emissions in g/s of the vehicle when operated under the driving cycle. It can be observed

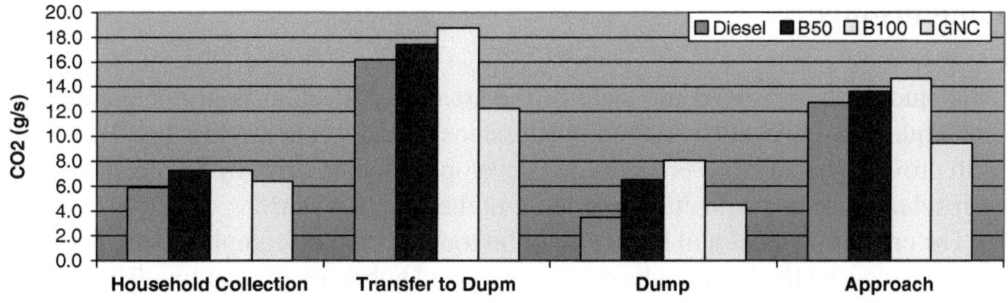

Fig. 9 CO_2 emissions results

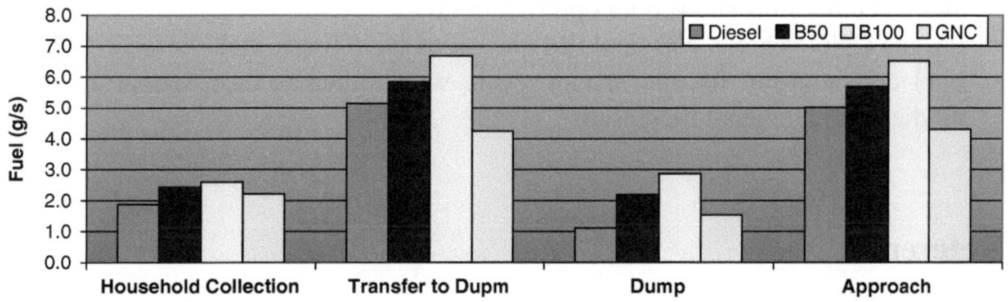

Fig. 10 Fuel consumption results

Table 3 Results for complete driving cycle

			B50	B100	CNG
CO	g/km	2.528	1.521	1.369	5.044
HC	g/km	1.009	0.857	0.699	1.151
NO_x	g/km	18.185	24.783	25.538	3.807
PM	g/km	0.295	0.169	0.140	–
CO_2	g/km	1,859.067	2,105.778	2,164.077	2,017.089
Fuel	g/km	590.535	731.727	801.550	698.318

that CO_2 was increased with increasing amount of methyl ester in the blend. The CNG vehicle results in lower CO_2 emissions than the diesels during Transfer to Dump and Approach and slightly the same during the rest segments.

Fuel consumption presented an increase for biodiesel fuel blends in all categories. These results are calculated using a carbon balance and measured emission data. The heating value of the methyl ester and the blends was lower than the reference diesel fuel, while the density was higher. Since the fuel is delivered to the cylinders on a volumetric basis, a higher mass flow is pumped to the cylinders with the biodiesel blends for the same fuel volume. Table 3 shows the results for a complete driving cycle.

Conclusions

This study has presented the state-of-the-art data collection instrumentation and techniques used to evaluate pollution emissions in real-world driving. It will be used with proven data analysis procedure to develop a detailed drive cycle representative or residential refuse collection operation in the city of Madrid.

The emission model will be a very useful tool to compare complete fleet, different fuels and new technologies: advanced engines and hybrid configurations. The experimental results may lead to the following conclusions:

- For the diesel vehicle, NO_x and CO_2 emissions increased with the addition of biodiesel when the vehicle was operated over the specific driving test. CO, HC and PM emissions showed an opposite trend.
- For the CNG vehicle, NO_x and PM emissions were lower than diesels.
- Fuel consumption for the driving cycle was higher as the concentration of biodiesel in the blend increased.

References

1. Nylund N, Erkkilä K, Lappi M, Ikonen M (2004) Transit Bus Emission Study: Comparison of Emissions from Diesel and Natural Gas Buses: Research Report VTT 2004
2. Walkowicz K, Proc K, Wayne S, Nine R, Campbell K, Wiedermeier G (2003) Chassis Dynamometer Emission Measurements from Refuse Trucks Using Dual-Fuel™ Natural Gas Engines: SAE Paper No. 2003-01-3366
3. Frey H, Rouphail N, Zhai H, Farias T, Gonçalves G (2007) Comparing real-world fuel consumption for diesel-and hydrogen-fueled transit buses and implication for emissions: Transportation Research Part D 12. 281–291
4. Pelkmans L, Debal P (2006) Comparison of on-road emission with emissions measured on chassis dynamometer test cycles: Transportation Research Part D 11. 233–241
5. Wang W, Gautam M, Sun X, Bata R, Clark N, Palmer M, Lyons D (1993) Emissions Comparisons of Twenty-Six Heavy Duty Vehicles Operated on Conventional and Alternative Fuels: SAE Paper No. 932952
6. Ivani Z (2007) Data Collection and Development of New York City Refuse Truck Duty Cycle: SAE Paper No. 2007-01-4118
7. Tzirakis E, Karavulakis G, Zannikos F, Stournas S (2007) Impact of Diesel/Biodiesel Blends on Emissions from a Diesel Vehicle Operated in Real Driving Conditions: SAE Paper No. 2007-01-0076

Part III
Trace Elements in the Environment

Platinum-Group Element Distribution in Hathersage, a Small Town Near Sheffield, UK: Evidence for a Single Source from Catalytic Converters

Hazel M. Prichard, Jean Sampson, and Matt J. Jackson

Introduction

More than 50% of platinum (Pt) and palladium (Pd) and more than 80% of rhodium (Rh) is now used in the manufacture of catalytic converters fitted to vehicles [1]. There have been several reviews of the accumulation of Pt, Pd and Rh in the urban environment as a result of emissions from catalytic converters fitted to vehicle exhaust systems [2, 3]. However there have been relatively few investigations of the abundances of these precious metals in small towns. The concentrations of Pt, Pd and Rh in road dust are thought to be dependant on traffic flow, vehicle speed [4, 5] and weather conditions [6]. These precious metals are dispersed from roads into natural and artificial drainage [7–10] and have been found concentrated in sewage and sewage ash [11, 12]. Within large cities it is difficult to be certain whether platinum-group elements (PGE) in urban wastes are derived from diffuse sources from catalytic converters fitted to vehicles, or from people with, for example, Pd-bearing dental fillings, or point sources such as hospitals or industry [13, 14].

The aim of this study was to examine the values of precious metals in the town of Hathersage, 7 km west of the western edge of the city of Sheffield. Hathersage lacks any industry and its inhabitants mainly work in Sheffield. It has one major road (the A6187) traversing the centre of the town and it is situated on the north side of the Hope Valley. It is drained by a small (2–3 m wide) stream (with minor tributaries) that flows southwards to join the river Noe which flows from west to east along the main valley. In this small town the study of precious metals is much less complicated than in large cities, which can be sub-divided into residential, commercial and industrial. The town is almost exclusively residential with a few shops along the main road. There are no major point sources for PGE and very little

H.M. Prichard (✉), J. Sampson, and M.J. Jackson
School of Earth and Ocean Sciences, Cardiff University, Main Building,
Park Place, Cardiff, CF10 3AT
e-mail: Prichard@cf.ac.uk

S. Rauch et al. (eds.), *Highway and Urban Environment*,
Alliance for Global Sustainability Bookseries 17,
DOI 10.1007/978-90-481-3043-6_17, © Springer Science+Business Media B.V. 2010

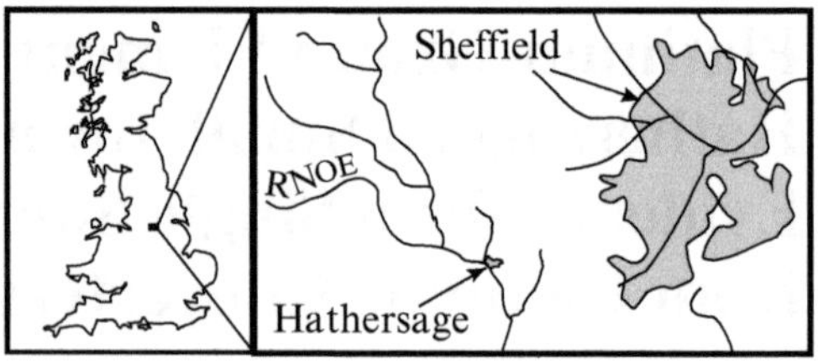

Fig. 1 Location of Hathershage in the United Kingdom

potential for location of point sources of PGE higher up in the drainage catchment for Hathersage which is small and occupied by moor land (Figs. 1 and 2). Therefore this study identifies the distribution of PGE derived predominantly from catalytic converters in this small town with a population of nearly 2,000 inhabitants (1,867 people) [15]. In addition, the analyses of samples for gold (Au) allows a comparison of the distribution of Au with Pt, Pd and Rh and analyses of osmium (Os), iridium (Ir) and ruthenium (Ru) provide a complete record of the PGE content in these samples. Any values above 2 ppb for the PGE are above background values of natural abundances, which are extremely low away from the major geological sources of PGE in South Africa and northern Siberia [16].

Sampling and Methods

Eleven sites on roads (A–K) were chosen for collection of paired road dust and gully waste samples. Sites A, B, C, D and E are on low traffic roads in residential areas up slope of the main road. Sites F and I are located on the main road with F in the centre of Hathersage (by the George Hotel) and I on the eastern edge of Hathersage. Site J is outside the school and north east of the main road and sites G, H and K are on minor roads down slope and to the south of the main road (Fig. 2).

Samples were collected in summer and winter with a set of road dust samples collected on 12 December 2005 in dry conditions, and paired sets of road dust and gulley samples collected on 11 and 12 July 2006 respectively, also in dry conditions and again in the winter on 11 December 2006 in rainy conditions (Table 1). Traffic flow per hour during the day, not during rush hour periods, was counted at all sites in winter 2006 and summer 2006 and averaged values for each site are given in Table 1. Four samples of stream sediment were collected from the main small stream that drains Hathersage, one upstream and two just downstream of the main road and one 150 m further downstream. Additionally one sample was collected from the much larger (approximately 15 m wide) River Noe (Figs. 1 and 2).

The samples were dried and sieved to less than 1 mm, crushed and analyzed for all six PGE and Au using Ni sulphide fire assay and an ICP-MS finish.

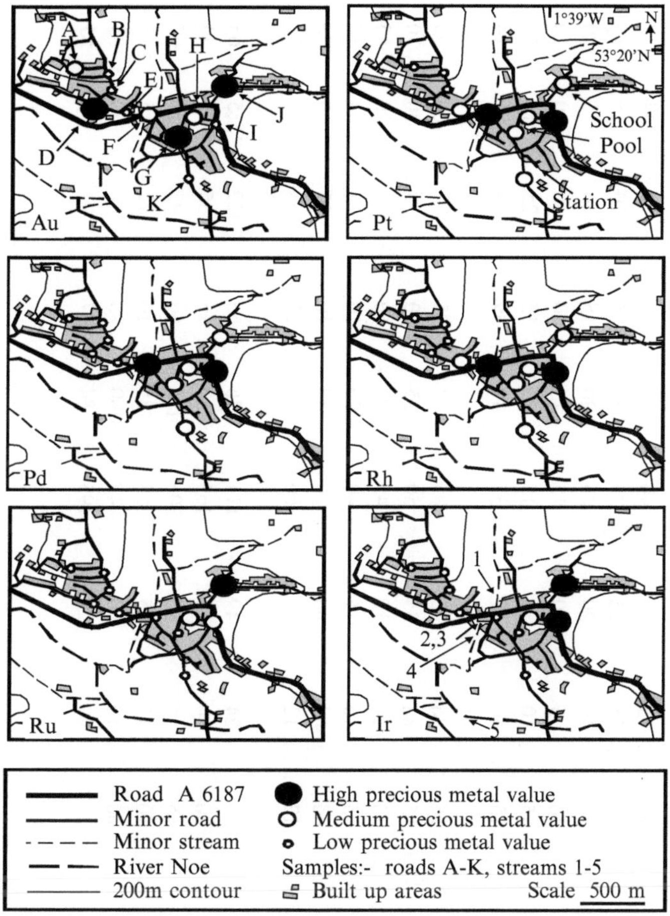

Fig. 2 Precious metal distribution in Hathersage. *Circles* correspond to average concentrations of Au, Pt, Pd, Rh and Ru for the three road dust samples taken in winter 2005, summer and winter 2006. Ir values are for winter 2005. Au values are low 0–20 ppb, medium 21–40 ppb and high 41–55 ppb. Pt values are low 0–30 ppb, medium 31–60 ppb and high 61–67 ppb. Pd values are low 0–40 ppb, medium 41–90 ppb and high 91–195 ppb. Rh values are low 0–9 ppb, medium 10–20 ppb and high 21–26 ppb. Ru values are low 0–5 ppb, medium 6–10 ppb and high 11–15 ppb. Ir values are low 0 ppb, medium 1–9 ppb and high 10–22 ppb. Zero (0) is taken as values below detection limit

The samples collected in winter 2005 were analyzed at Genalysis laboratories in Western Australia and the subsequent analyses of samples collected in 2006 were analyzed by Ultratrace, also in Western Australia. Standards AMIS-0002 and SARM65 were analyzed with the samples and measured values are consistent with the certified values. Internal standards, blanks and repeats of samples were also run and gave confidence in the sample values.

Table 1 PGE data

Sample site		A	B	C	D	E	F	G	H	I	J	K
Traffic flow/hour		7	37	37	78	182	535	212	58	310	147	205
Au												
Winter 2005	Road dust	34	15	14	15	11	40	17	77	12	144	7
Summer 2006	Road dust	25	5	10	75	25	25	60	25	20	15	15
Winter 2006	Road dust	5	14	20	10	20	45	20	15	10	5	20
Summer 2006	Gully	10	5	10	5	10	15	15	20	10	10	10
Winter 2006	Gully	10	5	5	10	5	15	20	15	25	10	10
Average for	Road dust	21	11	15	50	19	37	49	39	14	55	14
Pt												
Winter 2005	Road dust	15	22	26	26	86	85	72	95	97	100	38
Summer 2006	Road dust	nd	nd	5	nd	35	50	35	10	35	30	30
Winter 2006	Road dust	nd	28	10	10	10	45	25	60	70	10	45
Summer 2006	Gully	nd	10	5	10	15	40	20	10	25	nd	15
Winter 2006	Gully	nd	nd	nd	nd	nd	10	40	10	35	20	20
Average for	Road dust	5	17	12	12	44	60	44	55	67	47	38
Pd												
Winter 2005	Road dust	17	32	29	24	49	114	76	66	122	125	65
Summer 2006	Road dust	nd	5	50	20	4	165	50	15	55	35	40
Winter 2006	Road dust	5	15	15	15	25	305	125	60	95	15	60
Summer 2006	Gully	nd	30	10	15	5	55	35	10	25	5	15
Winter 2006	Gully	nd	nd	nd	10	nd	45	90	5	50	20	35
Average for	Road dust	7	17	31	20	26	195	84	47	91	58	55

Rh												
Winter 2005	Road dust	3	6	6	6	22	27	19	14	33	25	12
Summer 2006	Road dust	nd	nd	nd	nd	10	20	10	nd	10	5	10
Winter 2006	Road dust	nd	nd	nd	nd	5	30	15	15	20	nd	10
Summer 2006	Gully	nd	nd	nd	nd	nd	10	5	nd	5	nd	nd
Winter 2006	Gully	nd	nd	nd	nd	nd	5	10	nd	10	nd	5
Average for	Road dust	1	2	2	2	12	26	15	10	21	10	11
Ru												
Winter 2005	Road dust	nd	nd	nd	6	nd	nd	4	nd	30	25	nd
Summer 2006	Road dust	nd	2	1	1	2	3	1	1	1	20	3
Winter 2006	Road dust	nd	nd	nd	nd	nd	nd	nd	20	nd	nd	nd
Summer 2006	Gully	nd	nd	2	6	1	2	2	2	1	nd	nd
Winter 2006	Gully	nd	2	1	2	4	nd	7	nd	8	nd	2
Average for	Road dust	nd	1	nd	2	1	1	2	7	10	15	1
Ir												
Winter 2005	Road dust	nd	nd	nd	2	nd	nd	nd	4	22	14	nd
Summer 2006	Road dust	nd	1	nd	nd	nd	nd	nd	nd	nd	nd	nd
Winter 2006	Road dust	nd	nd	nd	nd	nd	nd	nd	nd	nd	nd	nd
Summer 2006	Gully	nd	1	nd	nd	nd	nd	nd	nd	nd	nd	nd
Winter 2006	Gully	nd	nd	nd	1	nd	nd	nd	nd	nd	nd	nd
Average for	Road dust	nd	nd	nd	nd	nd	nd	nd	nd	nd	nd	nd

Values in ppb, nd = not detected. Pt, Pd and Rh show extremely similar distributions, see Fig. 2.

Results

The highest values of Pt, Pd and Rh are found in samples taken from the main A6187 road (sites F and I) with maximum values of 97 ppb Pt, 305 ppb Pd and 33 ppb Rh. The lowest values are located in low traffic flow residential areas on minor roads up slope of the main road in the northwest of Hathersage (sample sites A–E). Intermediate values of Pt, Pd and Rh are located in residential areas, with higher traffic flow, both down slope of the main road near community buildings including a swimming pool and railway station (sites G, H and K) and north east of the main road in front of the school (site J).

The distributions of Au, Ru and Ir are different from that of Pt, Pd and Rh (Fig. 2). Highest Au values occur outside the school with a maximum value of 144 ppb, in residential areas at sites D, G and J, and not on the main road at sites F or I (Table 1). Medium Au values occur in residential areas at sites A and H. Highest Ru values occur outside the school entrance, on one of the main road site I and in a residential area at site H. Ir concentrations are low, being detected only in the winter of 2005 at the same localities as the highest Ru values, on the main road at site I and outside the school at site J. The maximum Ru value is 22 ppb. Os was only detected at 2 ppb at one site.

Average values for the three sets of road dust data collected in the winter of 2005, summer of 2006 and winter of 2006 for Pt, Pd, Rh and Au are plotted versus traffic flow and show that Pd and Rh, and to a lesser extent Pt concentrations, increase with traffic flow but that this is not the case for Au (Fig. 3a–d).

Gully samples consistently contain lower PGE and Au values than the surface road dust at each site and gully samples with higher PGE and Au concentrations are

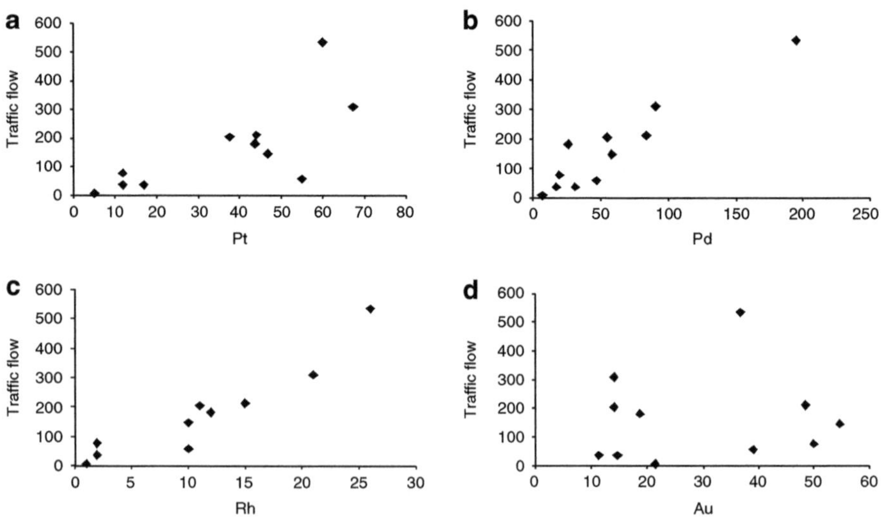

Fig. 3 (**a**)–(**d**) are graphs of Pt, Pd, Rh and Au, respectively, versus traffic flow

paired with road dust samples with higher precious metal values. The decrease in the Au, Pt, Pd and Rh concentrations in the gullies compared with the road dust is most marked at sites where the precious metal concentration in the road dust is highest. For example at site F, collected in summer 2006, the concentrations in road dust are reduced in gully waste from 25 to 15 ppb Au, from 50 to 40 ppb Pt, from 165 to 55 ppb Pd and from 20 to 10 ppb Rh.

The Pt/Pd ratio for all the samples is approximately 1.0 but varies from 0.2 to 8.8. For the paired summer 2006 road dust and gully sediment collected during dry conditions Pt/Pd ratios are generally similar (Fig. 4a, Table 1). In contrast the Pt/Pd ratios for the paired winter 2006 samples collected in rainy conditions show higher Pt/Pd ratios for the gullies than for the road dust. Pt/Rh ratios vary from 2 to 6 with an average of 3.7 and do not show any systematic variation between winter and summer samples or between road dust and gully samples (Fig. 4b, Table 1). Similarly there is no systematic difference between Pt/Au ratios in the summer or winter.

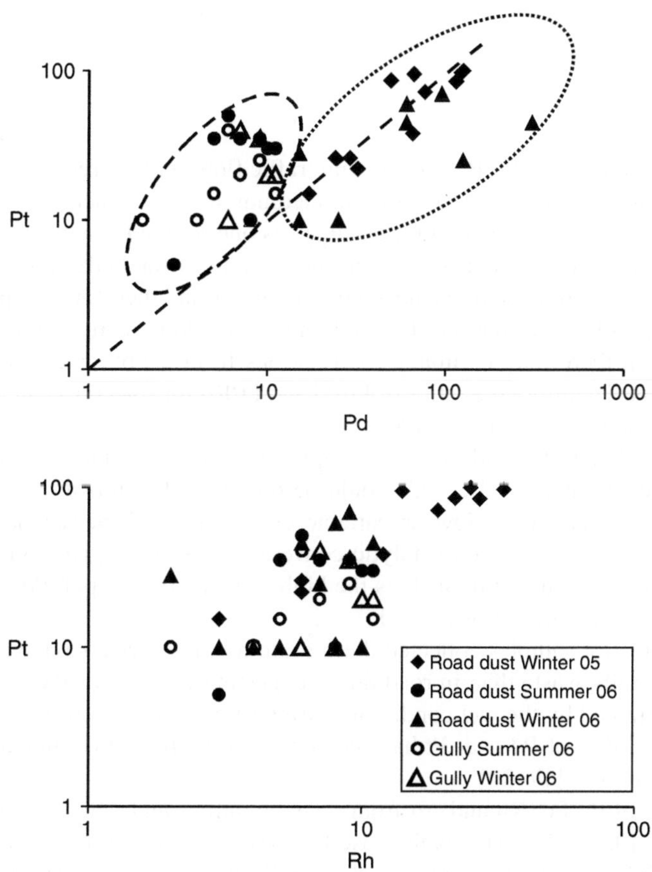

Fig. 4 Graphs of (**a**) Pt versus Pd and (**b**) Pt versus Rh. *Dashed* and *dotted circles* show separation of data sets

Table 2 Stream data

Sample	Au	Pt	Pd
1	10	nd	5
2	20	40	25
3	5	nd	nd
4	5	nd	10
5	15	nd	nd

Locations of samples 1–5 are shown in Fig. 2.

The maximum values in the stream samples occur in sediment collected just downstream of the bridge carrying the main road. They are lower than in road dust or gully samples, with maximum values of 40 ppb Pt, 25 ppb Pd, 20 ppb Au and undetected Rh, Ru, Os and Ir (Table 2).

Discussion

The increase of average Pt, Pd and Rh with traffic flow, and the remarkable similarity of the pattern of distributions of high, medium and low concentrations of each of these elements at sites A–K (Fig. 2), suggests one source from catalytic converters fitted to vehicles. Hathersage has no point sources for PGE emission as it is dominantly residential, lacking a hospital or industry, and therefore is a particularly clear example of a distribution of Pt, Pd and Rh predominantly from one source type, although there may be minor contributions from people with jewellery and dental fillings. Therefore the ratios of Pt/Pd and Pt/Rh for road dust can be taken as broadly characteristic of UK vehicle emission ratios for 2005–2006. Individual values of Pt, Pd and Rh and ratios in samples collected in winter or summer vary from the average at each site. This could be due to PGE removal by intermittent rainfall or variable traffic flow or variable emissions, with occasional loss of a larger particle of PGE from a catalytic converter. This could give unusually high values for an individual PGE such as the highest value of 305 ppb Pd collected in winter 2006 at high traffic flow site F.

Pt/Pd ratios for samples collected during rainfall in winter 2006 have higher Pt/Pd ratios in gully waste than in road dust, suggesting removal of Pd in the gullies, perhaps facilitated by the wet conditions during sample collection (Fig. 4). Similar increased mobility of Pd over Pt has been observed both in environmental [9] and geological samples [17–20].

Dispersal of Pt, Pd, Rh and Au from roads to gullies and then streams has been documented [9, 21–25]. This is the case for these samples with decreasing values for all six PGE and Au from road dust, to gully waste and into streams. The low PGE values in the River Noe are to be expected as the PGE concentrations will be diluted by the river sediment.

The presence of Au predominantly in residential areas of lower traffic flow, especially outside the school, suggests a different origin, perhaps from jewellery worn by pedestrians. The occurrence of highest Ru and Ir values outside the school also indicates a different source maybe due to loss from equipment. For example is Ru used in computer hard drives [1].

Conclusions

This study has shown that Pt, Pd and Rh are significantly above natural background levels in Hathersage and their concentrations all increase in road dust and gully waste with increasing traffic flow and in steam sediment samples with proximity to main roads. This indicates a source from vehicle emissions which is the only major source of these elements in this small town. There is no marked difference between precious metal concentrations in samples collected from sites in the winter or summer but a possible indication that Pd is more mobile and removed from gully waste in wet weather conditions. Gold, Ru and Ir are more abundant in residential areas especially outside the school and may be related to the presence of more people rather than vehicles.

Acknowledgements We thank the Royal Society who granted funding through the Senior Brian Mercer Award for 2004. We thank Genalysis and Ultratrace for analysis.

References

1. Jollie D (2008) Platinum Johnson Matthey, 56 pp
2. Zereini F, Alt F (1999) Anthropogenic PGE emissions. Berlin: SpringerVerlag; 639 pp
3. Zereini F, Alt F (2006) Palladium emissions in the environment: analytical methods, environmental assessment and health effects. Berlin: Springer Verlag; 485 pp
4. Higney E, Olive V, MacKenzie AB, Pulford ID (2002) Isotope dilution ICP-MS analysis of platinum in road dusts from west central Scotland. Appl Geochem 17:1123–1129
5. Parry SJ, Jarvis KE (2006) Temporal and spatial variation of Pd in the roadside environment. In: Zereini F, Alt F (eds.) Palladium Emissions in the Environment: Analytical Methods, Environmental Assessment and Health Effects. Berlin: Springer, pp. 419–432
6. Whiteley JD (2005) Seasonal variability of platinum, palladium and rhodium (PGE) levels in road dusts and roadside soils, Perth, Western Australia. Water Air Soil Pollut 160:77–93
7. Rauch S, Hemond HF, Peucker-Ehrenbrink B, Barbante C, Owari M, Wass U, Morrison G (2006) Regional and global transport of platinum-group elements from automobile catalysts. In: Zereini F, Alt F (eds.) Palladium Emissions in the Environment: Analytical Methods, Environmental Assessment and Health Effects. Berlin: Springer, pp. 295–305
8. Turner A, Crussell M, Millward GE, Cobelo-Garcia A, Fisher AS (2006) Adsorption kinetics of platinum-group elements in river water. Environ Sci Technol 40:1524–1531
9. Jackson MT, Sampson J, Prichard HM (2007) Pt and Pd variations through the urban environment: evidence from 11 sample types from Sheffield, UK. Sci Total Environ 385:117–131
10. Prichard HM, Jackson MT, Sampson J (2008) Dispersal and accumulation of Pt, Pd and Rh derived from a roundabout in Sheffield (UK): from stream to tidal estuary. Sci Total Environ 03.037; 401:90–99

11. Laschka D, Nachtwey M (1997) Platinum in municipal sewage treatment plants. Chemosphere 34:1803–1812
12. Eriksson J (2001) Concentrations of 61 trace elements in sewage sludge, farmyard manure, mineral fertiliser, precipitation and in oil and crops. Swedish Environmental Protection Agency, Stockholm URL: http://www.naturvardsverket.se/bokhandeln/pdf/620–62468.pdf.
13. Schwesig D, Rubel A, Furtmann K (2006) Palladium in waste waters and surface waters of North Rhine-Westfalia. In: Zereini F and Alt F (eds.) Palladium Emissions in the Environment: Analytical Methods, Environmental Assessment and Health Effects. Berlin: Springer, pp. 307–324
14. Stüben D, Kupper T (2006) Anthropogenic emission of Pd and traffic related PGEs – results based on monitoring with sewage sludge. In: Zereini F and Alt F (eds.) Palladium Emissions in the Environment: Analytical Methods, Environmental Assessment and Health Effects. Berlin: Springer, pp. 325–342
15. Government Housing Corporation population figures (2005) http://www.housingcorp.gov.uk/upload/pdf/Rural_Gazetteer_tables.pdf
16. Terashima S, Mita N, Nakao S, Ishihara S (2002) Platinum and palladium abundances in marine sediments and their geochemical behaviour in marine environments. Bull Geol Sur Japan 53:725–747
17. Wood SA, Vlassopoulos D (1990) The dispersion of Pt, Pd and Au in surficial media about two PGE-Cu-Ni prospects in Quebec. Can Miner 28:649–663
18. Bowles JFW, Gize AP, Vaughan DJ, Norris SJ (1994) Development of platinum-group minerals in laterites – initial comparison of organic and inorganic controls. Trans Inst Min Metal Sect B 103:B53–B56
19. Prichard HM, Lord RA (1994) Evidence for differential mobility of PGE in the secondary environment in the Shetland ophiolite complex. Trans Inst Min Metal Sect B 103:B79–B86
20. Oberthur T, Weiser TW, Gast L (2003) Geochemistry and mineralogy of PGE at Hartley platinum mine, Zimbabwe. Min Dep 38:344–355
21. Wei C, Morrison GM (1994) Platinum analysis and speciation in urban gullypots. Anal Chim Acta 284:587–592
22. Jarvis KE, Parry SJ, Piper JM (2001) Temporal and spatial studies of autocatalyst-derived platinum, rhodium and palladium and selected vehicle derived trace elements in the environment. Environ Sci Technol 35:1031–1036
23. De Vos E, Edwards SJ, McDonald I, Wray DS, Carey PJ (2002) A baseline survey of the distribution and origin of platinum-group elements in contemporary fluvial sediments of the Kentish Stour, England. Appl Geochem 17:1115–1121
24. Rauch S, Hemond HF (2003) Sediment-based evidence of platinum concentration changes in an urban lake near Boston, Massachusetts. Environ Sci Technol 37:3283–3288
25. Ek KH, Morrison GM, Rauch S (2004) Environmental routes for PGE to biological materials – a review. Sci Total Environ 34–35:21–38

Platinum and Lead in South African Road Dust

Sebastien Rauch and Olalekan S. Fatoki

Introduction

Automobile traffic is an important source of pollutants into the environment. Elevated Pb concentrations have been attributed to the use of tetraethyllead as an anti-knocking additive in gasoline. The recognition of human health effects, especially in children, has resulted in a progressive worldwide phase out of leaded gasoline [1]. Leaded gasoline was banned in South Africa in 2006. Until the ban, Pb was added to gasoline at a concentration of 0.33 g/l. Combined with a gasoline consumption of 7,000 Mt/year, this equates to a Pb consumption of 2,300 t/year, making South Africa the second largest consumer of Pb gasoline additives in Africa [2]. Lead concentrations are now expected to decrease in the South African environment owing to the ban of leaded gasoline.

Catalytic converters are placed in the exhaust system of cars to reduce the amount of gaseous pollutants (CO, hydrocarbons and NO_x) emitted as by-products of fuel combustion in the engine. Because these catalysts are poisoned by the presence of Pb, the ban of leaded gasoline and the subsequent availability of unleaded gasoline support the introduction of exhaust catalysts in South Africa. Platinum and other Pt group elements are the main active components in catalysts and some Pt is emitted into the environment [3]. As a consequence, Pt concentrations are increasing in the urban and roadside environment [4], and elevated Pt concentrations are now found in airborne particles, road dust, as well as in roadside soil, plants, and biota [5, 6]. Therefore, environmental Pt concentrations are expected to increase in South Africa as a result of automobile emissions. However, South Africa is the

S. Rauch (✉)
Department of Civil and Environmental Engineering, Water Environment Technology,
Sven Hultins gatan 8, 41296 Göteborg, Sweden
e-mail: Sebastien.Rauch@chalmers.se

O.S. Fatoki
Faculty of Applied Sciences, Cape Peninsula University of Technology, Cape Town,
South Africa

S. Rauch et al. (eds.), *Highway and Urban Environment*,
Alliance for Global Sustainability Bookseries 17,
DOI 10.1007/978-90-481-3043-6_18, © Springer Science+Business Media B.V. 2010

world's largest Pt producer [7] and mining activities might be an additional source of Pt. Mining occurs in the Bushveld Igneous Complex, a 65,000 km^2 mafic intrusion containing approximately 75% of the world's Pt resources [8]. Platinum emissions from the mines has until now not been quantified.

The study presented here aims at determining current Pb and Pt levels in road dust in South Africa and at assessing the relative importance of automobile and mining emissions. Road dust is an attractive medium for urban pollution studies and it is important to determine current Pb and Pt concentrations in road dust, as Pb concentrations are still high and expected to decrease in the near future, and Pt emissions by automobile traffic are still limited.

Methodology

Road dust samples were collected at several locations in Cape Town (2.9 million inhabitants), Pretoria (1.8 million), Port Elizabeth (1.0 million) and Rustenburg (0.4 million) in October–November 2007. Selected cities are shown in Fig. 1. Rustenburg was selected for its vicinity to the Bushveld Igneous Complex where Pt is mined, while other cities are important urban areas with high traffic volumes.

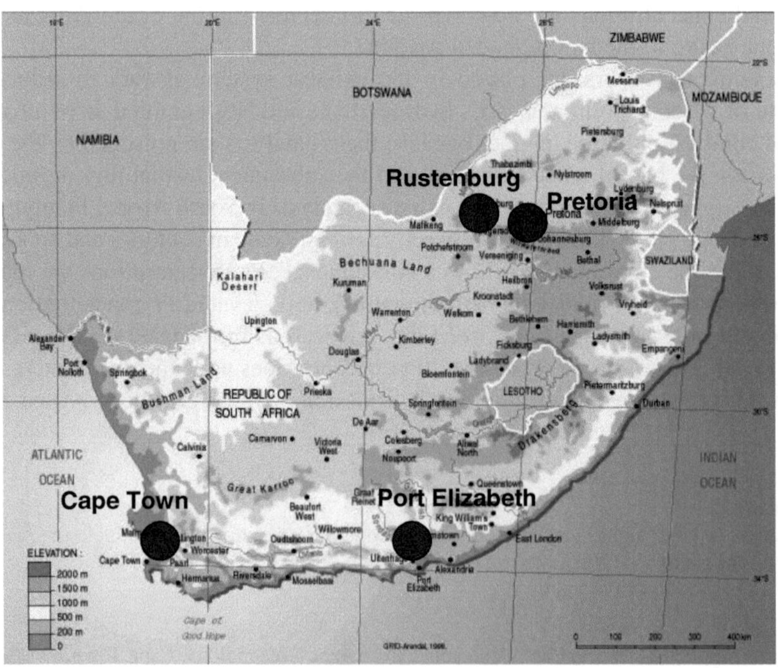

Fig. 1 Map of South Africa with cities selected for road dust sampling

Sampling was performed on asphalt covered roadside. Road dust was collected using a brush and placed in ziplocked PE bags for transport and storage.

Samples were dried overnight at 105°C. Dry samples were then sieved and the <125 μm fraction was retained for analysis. Samples were prepared by microwave-assisted acid digestion using closed Teflon vessels (Mars5, CEM, USA). Approximately 0.5 g of dry dust was placed in microwave digestion vessels. Digestion was performed in closed vessels after addition of 8 ml *Aqua regia* (6 ml HCl + 2 ml HNO_3) using 2-step temperature increase to 185°C and a maximum allowed pressure of 200 psi. It is important to note that the procedure does not provide a full digestion, but a strong leach usually suitable for the study of anthropogenic metals in road dust. The leachate was then slowly taken to dryness on a hot plate and redissolved in 5% HNO_3.

Prepared samples were analyzed by inductively coupled plasma-mass spectrometry (ICP-MS) using a quadrupole system (Elan 6000, Perkin Elmer, USA) and standard operating conditions. Calibration was performed by the analysis of multi-element standard solution, except for Pt for which a single element solution was used. Interferences on Pt analysis were corrected mathematically [9, 10] and reference material BCR-723 was analyzed to assess the accuracy of Pt concentrations.

Results and Discussion

Average Pb and Pt concentrations in road dust in South African cities are presented in Table 1.

Lead Concentrations

Average Pb concentrations were found to range from 345 μg/g in Rustenberg to 775 μg/g in Pretoria (Table 1), with a minimum concentration of 103 μg/g at Joubert Street in Rustenburg and a maximum concentration of 1,928 μg/g at Church Street in Pretoria (Fig. 2). For comparison, urban concentrations in Accra, Ghana were reported to be 365 ± 93 μg/g [11]. The higher concentrations in South Africa are likely the result of the larger consumption of Pb gasoline additives [2].

Large concentration variations were observed in individual cities; Pb concentrations in Pretoria range for instance from 186 μg/g at Schueman Street to 1,928 μg/g at Church Street (Fig. 2). Although automobile traffic is expected to be the main source of Pb, there is no direct correlation between Pb concentrations and traffic density. The absence of correlation may be the result of factors other than Pb emission from automobile traffic and subsequent deposition. Because leaded gasoline is now banned in South Africa, it is possible that removal efficiency play a more important role than Pb deposition.

Table 1 Pb and Pt concentrations in road dust in selected South Africa cities. Concentrations are presented as average ± standard deviation (standard deviation is presented in brackets when it exceeds average concentration)

City	Pb (μg/g)	Pt (ng/g)
Cape Town	706 ± 271	4 ± 2
Port Elizabeth	434 ± 228	6 ± 1
Pretoria	775 (999)	23 ± 13
Rustenberg	345 ± 172	223 ± 116

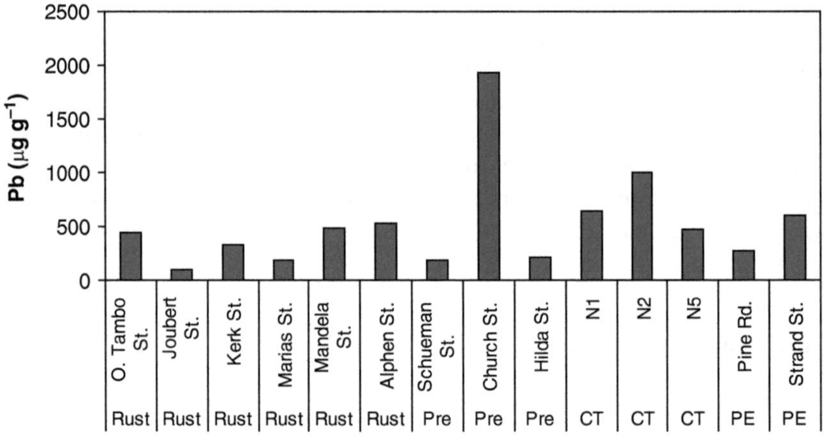

Fig. 2 Lead concentrations in road dust in Rustenburg (Rust), Pretoria (Pre), Cape Town (CT) and Port Elizabeth (PE)

Platinum Concentrations

Average Pt concentrations range from 4 ng/g in Cape Town to 223 ng/g in Rustenburg (Table 1) with a minimum concentration of 2 ng/g along highway N2 in Cape Town and a maximum concentration of 391 ng/g at Mandela Street in Rustenburg (Fig. 3). For comparison, average Pt concentrations in Ghana ranged from 1.5 ng/g at a background site to 55 ng/g near a heavy traffic road [11].

Concentrations in Rustenburg and Pretoria range from 108 to 391 ng/g and from 13 to 37 ng/g, respectively, and are clearly higher than in other South African cities where concentrations range from 2 to 7 ng/g. The relatively low concentrations in Cape Town and Port Elizabeth indicate that automobile traffic is a minor source of Pt in South Africa, possibly due to the currently limited number of vehicles equipped with exhaust catalysts. Therefore, the relatively high Pt concentrations found in Rustenburg and Pretoria are attributed to non-automobile sources. Since Rustenburg is located on the edge of the Bushveld Igneous Complex, near major Pt mines, mining activities are a likely source of Pt in road dust in Rustenburg.

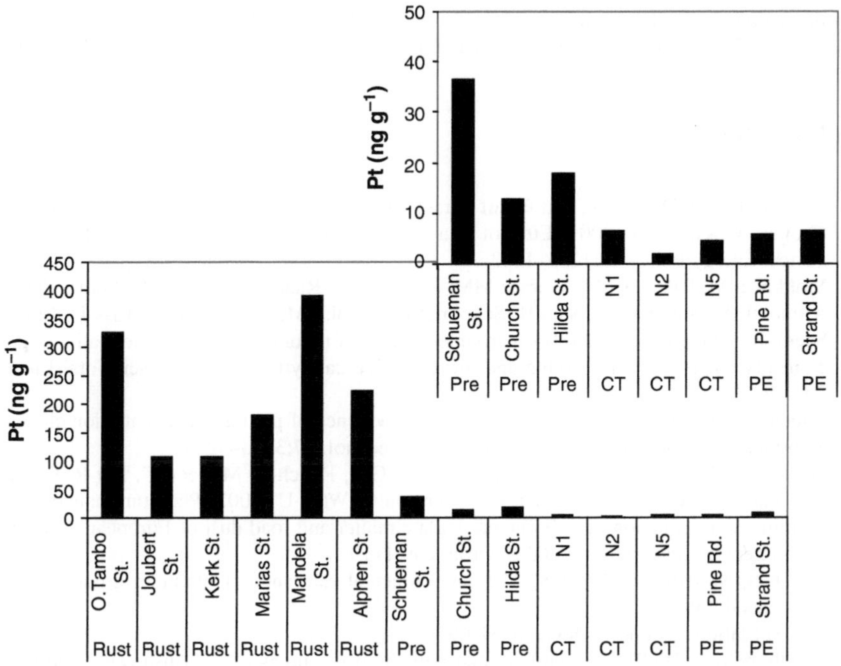

Fig. 3 Platinum concentrations in road dust in Rustenburg (Rust), Pretoria (Pre), Cape Town (CT) and Port Elizabeth (PE)

Pretoria is located approximately 150 km from Rustenburg and elevated Pt concentrations in Pretoria are attributed to atmospheric transport of Pt emitted from the mines.

Conclusions

Elevated Pb and Pt concentrations were found in South African road dust. The occurrence of Pb is attributed to automobile traffic owing to the use of leaded gasoline. Leaded gasoline has recently been banned in South Africa and environmental Pb concentrations are expected to decrease in the near future. In contrast, automobile emissions are a minor source of Pt due to the limited number of vehicles currently equipped with an exhaust catalyst. South Africa is the world's leading Pt producer and elevated Pt concentrations were found in road dust collected near Pt mines in Rustenburg.

Atmospheric dispersion of emitted Pt also results in contamination further from the mines and elevated Pt concentrations were also found in Pretoria. Further studies are needed to characterize emissions from Pt mines and associated risks of should be assessed. Automobile traffic may become an additional source of Pt in South Africa owing to the introduction of automobile exhaust catalysts.

Acknowledgements This study was performed with financial support from the Swedish International Development Agency (SIDA) and the South African National Research Foundation.

References

1. Seyferth D (2003) The rise and fall of tetraethyllead. 2. Organometallics 22:5154–5178
2. Thomas V, Kwong A (2001) Ethanol as a lead replacement: phasing out leaded gasoline in Africa. Energy Policy 29:1133–1143
3. Moldovan M, Palacios MA, Gomez MM, Morrison G, Rauch S, McLeod C, Ma R, Caroli S, Alimonti A, Petrucci F, Bocca B, Schramel P, Zischka M, Pettersson C, Wass U, Luna M, Saenz JC, Santamaria J (2002) Environmental risk of particulate and soluble platinum group elements released from gasoline and diesel engine catalytic converters. Sci. Total Environ. 296:199–208
4. Rauch S, Hemond HF (2003) Sediment-based evidence of platinum concentration change in an urban lake near Boston, MA. Environ. Sci. Technol. 37:3283–3288
5. Gomez B, Palacios MA, Gomez M, Morrison GM, Rauch S, McLeod C, Ma R, Caroli S, Alimonti A, Schramel P, Zischka M, Pettersson C, Wass U (2002) Platinum, palladium and rhodium contamination in airborne particulate matter and road dust of European cities. Risk assessment evaluation. Sci. Total Environ. 299:1–19
6. Rauch S, Morrison GM (2008) The environmental relevance of platinum group elements. Elements 4:259–263
7. (2007) Platinum. London: Johnson Matthey
8. Cawthorn RG (1999) The platinum and palladium resources of the Bushweld Complex. S. Afr. J. Sci. 95:481–489
9. Moldovan M, Rauch S, Gomez M, Palacios MA, Morrison GM (2001) Bioaccumulation of palladium, platinum and rhodium from urban particulates and sediments by the freshwater isopod *Asellus aquaticus*. Water Res. 35:4175–4183
10. Rauch S, Lu M, Morrison GM (2001) Heterogeneity of platinum group metals in airborne particles. Environ. Sci. Technol. 35:595–599.
11. Kylander ME, Rauch S, Morrison GM, Andam K (2003) Impact of Automobile Emissions on the Levels of Platinum and Lead in Accra, Ghana. J. Environ. Monit. 5:91–95

Part IV
Urban Water Contamination

Reduction of Stormwater Runoff Toxicity by Wet Detention Ponds

Tove Wium-Andersen, Asbjørn H. Nielsen, Thorkild Hvitved-Jacobsen, and Jes Vollertsen

Introduction

Stormwater runoff contains a wide range of contaminants [1]. Pollutants like PAH's, biocides, heavy metals and nutrients are often found in varying, but significant concentrations. Due to the content of various contaminants, stormwater runoff has been found to be toxic to the aquatic environment [2–4], with potential negative ecological impacts on receiving waters.

Toxic effects are estimated by exposing test organisms to the potential toxic sample or compound followed by a calculation of the inhibition of the test organism. The inhibition can be growth inhibition, mobility inhibition, inhibition of the reproduction and etceteras. Inhibition is expressed as a given effect concentration (EC) after a given exposure time. That is, EC50 is the concentration giving an inhibition equal to 50% after a certain exposure time and EC10 is the concentration giving an inhibition equal to 10% after a certain exposure time. Toxicity is not an absolute estimation but a relative measurement characteristic for the test organism applied, resulting in different EC50 values for the same compound. Therefore, it is important to consider which organisms that is relevant to use as test organisms for environmental samples such as stormwater runoff. The optimal solution is to use a battery of toxicity tests to reveal the ecological impacts of a sample [5].

To reduce the content of nutrients, organic matter and the different organic and inorganic micro pollutants, the stormwater runoff can be led through a wet detention pond where sedimentation and uptake by plants reduce the concentration of the contaminants. To improve sedimentation, a flocculent such as aluminum can be added to the water phase. It is well known that aluminum can reduce the content of phosphorous in the water phase in lakes [6], but whether aluminum addition also reduces the toxic effect of the water phase has not been identified.

T. Wium-Andersen, A.H. Nielsen, T. Hvitved-Jacobsen, and J. Vollertsen
Aalborg University, Section of Environmental Engineering, Sohngaardsholmsvej 57, 9000, Aalborg, Denmark

S. Rauch et al. (eds.), *Highway and Urban Environment*, Alliance for Global Sustainability Bookseries 17,
DOI 10.1007/978-90-481-3043-6_19,

The objective of this study is to investigate how addition of aluminum affects the toxicity of stormwater runoff. The toxic effect is estimated applying three different toxicity tests. The results of the study can be applied to determine whether addition of aluminum to wet detention ponds can reduce the toxic effect of stormwater by adding similar quantities of aluminum as typically used to remove phosphorous from the water phase.

Method

Collection and Storage of Samples

Stormwater runoff samples were collected from a wet detention pond located in the southern part of Odense, Denmark. The wet detention pond is constructed as a part of the EC LIFE-treasure project [7] and is placed in a green area next to a catchment with light industry and associated roads (Fig. 1). The catchment has an area of 27.4 ha of which 11.4 ha are impervious, resulting in an estimated runoff of 55,500 m^3/year. The stormwater runoff is pretreated in a grit chamber before entering the pond. Stormwater runoff for analysis was collected from the center of the pond. The samples were collected in glass bottles over a 2 month period and stored at 5°C until analysis.

Preparation of Samples Prior to Analysis

Four samples were collected over a 2 month period, two samples in February 2008 (sample I and II) and two at the beginning of April 2008 (sample III and IV). Due to a natural content of microalgae in the samples, the samples where filtrated

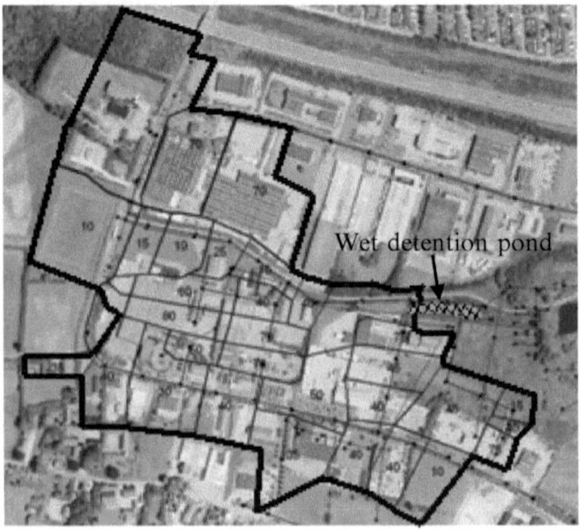

Fig. 1 Wet detention pond from where the samples were collected. The appurtenant catchment is outlined

through a 0.45 μm filter prior to toxicity tests. All four samples were analyzed for toxic effects using different toxicity tests. The last two samples were treated by aluminum addition to investigate its potential influence on the toxic effect.

The samples collected in April were divided into aliquots of 320 mL in 500 mL glass bottles. Aluminum sulfate ($Al_2(SO_4)_3$ $16H_2O$) was added to a final aluminum concentration of 0, 2, 5, 7 and 10 mg Al/L (sample III) and a concentration of 0, 10, 20, 30 and 40 mg Al/L (sample IV). The aliquots were placed on a shaker for 1 h to ensure thoroughly mixing and afterwards left to settle for 3 days at 5°C. Three days was chosen as this is the minimum hydraulic retention time for a typical Danish wet detention pond for stormwater treatment [8]. After the 3 days, the supernatant was extracted without disturbing the settled precipitate. The supernatant was tested for toxic effects by three different toxicity tests. The natural pH-value for sample III and IV prior to aluminum addition was 8.76 and 8.78, respectively. The products of the dissociation and hydration of the aluminum salt at these pH-values is mainly aluminum hydroxide ($Al(OH)_3$) and to a lesser extent aluminate ion ($Al(OH)_{4-}$). Aluminum hydroxide could be observed as visible flocks in the aliquots with an aluminum concentration equal to 5 mg Al/L and above.

Toxicity Tests

Toxicity tests were chosen on the basis the following criteria: The tests should be validated and comply with a standard; an existing protocol should be available; the tests should use organisms on different trophic levels; and the tests should be cost-time efficient. These criteria led to three different tests, using bacteria (Vibrio fischeri), algae (Selenastrum capricornutum) and crustaceans (Daphnia magna) as test organisms. Vibrio fisheri is a marine bacterium and will therefore not be found naturally in wet detention ponds. However, the test is commonly used and relative fast, giving the result after only 30 min and was on this basis included in the test battery. The last two tests use organisms which are indigenous to freshwater. All three tests are designed to measure acute toxic effects.

The test using Vibrio fisheri is based on the capacity of the bacteria to emit luminescence, meaning that the less luminescence the bacteria produce, the more stressed are the bacteria due to exposure to toxic substances. Specific volumes of the samples and different dilutions hereof were mixed with bacteria suspension (duplicate for each dilution). The luminescence was measured after 0, 5, 15 and 30 min of exposure to the sample on a M500 analyzer from SDI in the Microtox® Acute mode. Decrease in luminescence relative to a control gives an expression for the toxic effect and EC-values can be estimated. As Vibrio fischeri is a marine bacterium, the runoff samples were adjusted to a salinity of 2% prior to analysis. An advantage of this test is that for each dilution, a very large number of test individuals are applied, minimizing statistical fluctuations. The whole test was carried out according to the international standard DS/EN ISO 11348-1.

The algal test is based on growth inhibition of the algae due to toxic substances. A specific volume of an algae suspension is added to a dilution series of a sample. Triplicates were made for each dilution. The growth rate was hereafter estimated by measuring the optical density (OD) at 670 nm after an exposure time of 0, 24, 48 and 72 h. The growth rate of the algae is inhibited by toxic substances and a relative inhibition to a control can be calculated. In this study the toxicity test kit Algaltoxkit FTM, MicroBioTests Inc., was applied and the tests were carried out using 10 cm cuvettes as incubation chambers. An UVmini-1,240 spectrophotometer from Shimadzu was used to measure the OD670. Approximate 10,000 algae cells/mL were added to each cuvette at the beginning of each test, which due to the high number of individuals minimized the statistical fluctuations. The test is based on the international standard DS/EN ISO 8692.

The last test in this study applied the Crustacea Daphnia magna as test organism. The ability of the daphnia to swim is affected by toxic compounds and is therefore used as an expression of the toxic effect. The daphnia were exposed to dilutions of a sample and immobile daphnia were counted after 24 and 48 h. Twenty daphnia were exposed to each dilution of the sample, which was a much lower number of individuals than for the two other tests, resulting in larger statistical fluctuations. For each test, a control was made to ensure that the immobilization was caused by the toxic effect and not by any physical stress. In this study the toxicity test kit Daphtoxkit FTM, MicroBioTests Inc., was applied, based on the international standard DS/EN ISO 6341.

Results and Discussion

Data obtained from the four samples and the three toxicity tests were analyzed according to the respective standards. Dose-response curves were obtained and EC50 and EC10 values were estimated when possible. It was a general conclusion for the samples tested that the bacteria test was less sensitive than both the Daphtoxkit FTM and Algaltoxkit FTM. This is consistent with previous reported observations [2, 9]. The stormwater runoff had no observable effect on the luminescence readings (data not shown), indicating that the samples were non-toxic. The daphnia were only inhibited by undiluted aliquots, that is 100% stormwater runoff. Estimations of EC50 values were therefore not possible. The inhibition by 100% stormwater runoff samples was in the magnitude of 20–50% after 48 h of exposure, indicating a low toxic effect. The algal test was on the other hand much more sensitive to the toxic compounds in the runoff. Hence, calculation of EC50 values was possible, giving EC50 values in two groups: sample I and II had a low EC50 about 10% dilution (after 72 h of exposure) and sample III and IV had a relative high EC50 about 85% dilution (after 72 h of exposure). The significant difference in toxic effect may be due to a strong natural algal blossom in the wet detention pond, beginning by the end of March. This algal blossom may cause removal of toxic compounds from the water phase due to uptake by the algae and sorption at their surfaces.

When the samples were filtrated prior to analysis, all contaminants absorbed or adsorbed by the algae were removed from the sample. Even though there were large differences between the toxic effects of the samples, the samples were in general toxic to the test organisms and treatment by adding aluminum was relevant. The EC50 values obtained by the algal test (72 h exposure) and standard deviation for the different aliquots from the sample III and IV, are shown in Fig. 2.

EC50 values above 100% dilution were calculated by extrapolation of the dose-respond curve and not by concentration of the samples, for example by evaporation. As shown in Fig. 2 the toxic effect of the stormwater runoff decreased with increasing aluminum concentration up to 7 mg Al Al/L. Above 7 mg Al/L the toxic effect increased with increasing aluminum concentration. In the aliquots added 30 and 40 mg Al Al/L, the toxic effect was greater than for the pure sample. That is, aluminum reduced the toxic effect of the stormwater runoff up to a certain level. Above this level the aluminum itself became the source of the toxic effect. In Fig. 3, EC10

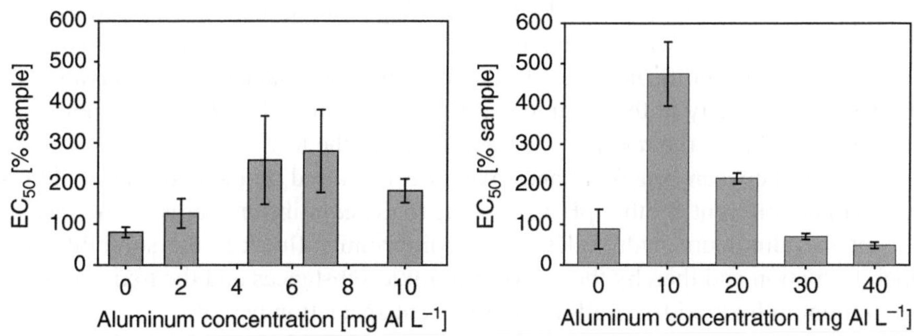

Fig. 2 EC_{50} values and standard deviations for the aliquots from sample III (*left graph*) and sample IV (*right graph*) measured with the algal test. A high EC_{50}-value is equivalent to a low toxic effect

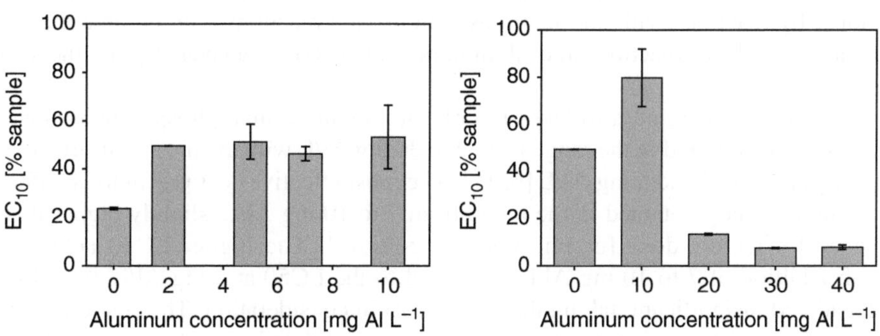

Fig. 3 EC_{10} values and standard deviations for the aliquots from sample III (*left graph*) and sample IV (*right graph*) according to the algal test. The higher the EC_{10}-values, the lower the toxic effect

values and standard deviations of the aliquots from sample III and IV are shown (72 h exposure).

The EC10 values were all below 100% and therefore estimated by interpolation. Figure 3 shows the same tendencies as Fig. 2, that is lower toxic effect for aluminum concentrations between 2 and 10 mg Al/L compared with stormwater runoff without aluminum. As shown, highest EC50 occurs at 10 mg Al Al/L. At 20 mg Al Al/L, the EC10 is lower than the EC10 for the aliquot without aluminum addition. This is not consistent with the EC50 for the same aliquot, where the EC50 is indicating that this aliquot has a lower toxic effect than the aliquot without aluminum addition. In principle, the EC10 and EC50 should show the same tendency. However, due to uncertainties in the tests and also in their interpretation, deviations from this might happen. Due to this uncertainty, it cannot be concluded whether the addition of 20 mg Al Al/L to stormwater runoff has a positive or negative effect on the toxicity. It is therefore, in order to avoid toxic effects, recommended to add not more than 10 mg Al Al/L when treating stormwater runoffs.

The tendency for the daphnia test was the same as for the algae test, that is untreated stormwater runoff had a significant toxic effect as well as runoff samples with addition of 30 and 40 mg Al Al/L; showing an inhibition at 25% and 95% in the undiluted aliquots with 30 and 40 mg Al Al/L, respectively. The other aliquots did not show any significant toxic effects. With the bacterial test, the observation of an increasing toxicity in the aliquots with 30 and 40 mg Al Al/L was not observed – most likely due to the relative poor sensitivity of the test.

The limit between beneficial and toxic effects caused by the addition of aluminum will be different at other pH-values due to the equilibrium between the speciation of the aluminum hydroxides. The equilibrium influences the solubility and flock formation and thereby the removal of toxic substances and the toxic effect of the aluminum hydroxides itself due to different speciation hereof.

At pH 8.2, EC50 for the microalgae Selenastrum capricornutum equal to 460 μg Al/L is reported [10]. The EC50 reported is low compared to the results in this study where aluminum began to show toxic effects at a concentration level of 20,000–40,000 μg Al/L. The large difference is probably due to the nature of the samples tested in this study, resulting in removal of toxic substances and phosphorus by reactions with aluminum as well as side reactions with organic matter. Because of these reactions the aluminum may become unavailable to the test organisms.

Addition of aluminum to lakes in order to obtain optimal phosphorus removal efficiency is reported in the range from 2 to 30 mg Al/L with the most cost-effective dosing between 2 and 5 mg Al/L [6, 11]. The most effective decrease in toxic effect was in this study obtained in the range from 7 to 10 mg Al/L, slightly higher than the cost-effective dose for phosphorous removal. The higher EC50 achieved with addition of 7 to 10 mg Al/L compared to the EC50 at 5 mg Al/L should be considered insignificant, taking the higher cost in consideration. The cost-effective concentration of aluminum for reduction of toxic effects is therefore around 5 mg Al/L. This correspond to a consumption of approximately 300 kg Al/year for the stormwater pond from where the samples originated.

Conclusion

The stormwater runoff analyzed in this study was found to show toxic effects according to two of the three applied toxicity tests. The stormwater was highly toxic to the algae, lesser toxic to the crustaceans and non-toxic to the bacteria – revealing that the bacteria Vibrio fisheri was the least sensitive organism and that the algae Selenastrum capricornutum was the most sensitive organism. Two stormwater runoff samples showed a high toxic effect to the algae with EC50-values at 10% dilution and two samples showed a lower toxic effect with EC50-values at 85% dilution.

To assess the impact of aluminum on the toxic effect of stormwater runoff, aluminum was added to two samples in concentrations ranging from 2 to 40 mg Al/L. The toxicity test using the algae Selenastrum capricornutum showed reduced toxic effect for aliquots with aluminum concentration from 2 to 10 mg Al/L compared with the toxic effect for pure stormwater runoff. That is, the aluminum may reduce the toxic effect in stormwater runoff. At concentration levels above 20 mg Al/L, aluminum itself becomes toxic and the toxic effect is higher than for the pure runoff. This limit between beneficial and toxic effects will depend on a number of water quality parameters, such as the concentration of organic matter, phosphorous, toxic compound, pH etceteras.

The most effective aluminum concentration to reduce toxic effects was in this study found to be 7 mg Al/L. However, the most cost effective aluminum concentration is 5 mg Al/L which corresponds to the interval typically applied for cost-effective phosphorous removal.

Acknowledgements The work was partly financed by PH-Consult ApS, Denmark, under the framework of the Urban Water Technology Research School.

References

1. Göbel P, Dierkes C, Coldewey WG (2007) Stormwater runoff concentration matrix for urban areas. Journal of Contaminant Hydrology 91:26–42
2. Marsalek J, Rochfort Q, Brownlee B, Mayer T, Servos M (1999) An exploratory study of urban runoff toxicity. Water Science and Technology 39:33–39
3. Marsalek J, Rochfort Q (1999) Toxicity of urban wet-weather pollution sources: stormwater and CSOs International Conference on Urban Storm Drainage, 8, Sydney, August 30 – September 3, 1999 Proceedings, vol. I–IV
4. Fisher, D (1995) The acute whole effluent toxicity of stormwater from an international airport. Environmental Toxicology and Chemistry 14:1103–1111
5. Dutka BJ (1988) Priority setting of Hazards in waters and sediments by proposed ranking scheme and battery of tests approach. Zeitschrift für angewandte Zoologie 75:303–317
6. Cooke GD, Welch EB, Peterson SA, Nichols SA (2005) Restoration and Management of Lakes and Reservoirs, 3rd Edition. CRC, Boca Raton, FL. ISBN: 1–56670–625–4
7. Vollertsen J, Lange KH, Nielsen AH, Nielsen NH, Hvitved-Jacobsen T (2007) Treatment of urban and highway stormwater runoff for dissolved and colloidal pollutants. Proceedings of the 6th International Conference on Sustainable Techniques and Strategies in Urban Water Management (NOVATECH), Lyon, France, June 25–28, 2007, 8 pp

8. Hvitved-Jacobsen T, Yousef YA (1988) Analysis of rainfall series in the design of urban drainage control systems. Water Research 22:491–496
9. Pardos M, Benninghoff C, Guéguen C, Thomas, RL, Dobrowolski J, Dominik J (2000) Suspended matter water-elutriate toxicity from water and waste water in Cracow (Poland) evaluated with Microtox® and Selenastrum capricornutum assays. Lakes and Reservoirs: Research and Management 5:67–73
10. Call, 1984 as cited in Gostomski F (1990) The toxicity of aluminum to aquatic species. US Environmental Geochemistry and Health 12:51–54
11. Auvray F, Hullebusch EDv, Deluchat V, Baudu M (2006) Laboratory investigation of the phosphorus removal (SRP and TP) from eutrophic lake water treated with aluminum. Water Research 40:2713–2719

Microbial Contaminants Removal in a Stormwater Pond

Thomas J.R. Pettersson and Johan Åström

Introduction

Stormwater contaminants discharged from urban areas are often considered as threats limited to the receiving waters since the contaminants mainly being viewed as environmental degradation agents rather than, for example raw water degrading agents. To reduce the load of stormwater contaminants the most efficient action in a long-term perspective is the practice of source control, but when acute short-term remediation actions is needed stormwater ponds have showed to be a good measure to reduce contaminants and to achieve cleaner and healthier water bodies [1] favoring the implementation of the European Water Framework Directive where all waters including rivers, lakes, coastal waters and ground waters should reach "good status" by December 2015 [2]. The pollution reduction efficiency of stormwater ponds mainly depend on the physical processes that govern the sedimentation of particulate bound pollutants [1]. This is also valid to the removal of microbial contaminants in stormwater ponds, and of great significance to prevent drinking water sources from pathogens carried in stormwater. To ensure drinking-water safety, microbial as well as chemical and physical hazards need to be assessed. In a public health perspective the microbial contaminants may be of more importance for the society compared to chemical and physical contaminants [3]. Pathogens in natural water catchments may originate from various human and animal sources, and cause waterborne infections through consumption of contaminated drinking water. The management of faecal contaminants from human and animal sources in urban catchment areas depends on the type of sewer system. For separate sewer systems, animal faeces from roads and impervious areas may be transported in the stormwater and enter the receiving water without further treatment. Stormwater is known to carry substantial amounts of faecal bacteria [4, 5] and a few studies have reported the presence and survival of pathogen in wetlands and stormwater ponds [6–8].

T.J.R. Pettersson and J. Åström
Department of Civil and Environmental Engineering, Water Environment Technology, Chalmers University of Technology, Sweden

S. Rauch et al. (eds.), *Highway and Urban Environment*, Alliance for Global Sustainability Bookseries 17,
DOI 10.1007/978-90-481-3043-6_20, © Springer Science+Business Media B.V. 2010

In the assessment of faecal contaminants and pathogens in water environments, indicator organisms have been traditionally used. Total coliforms and *E. coli* are indicator bacteria commonly used for regular monitoring in drinking water sources. Indicator organisms used for parasites are intestinal enterococci and sulphite-reducing clostridia and for viruses bacteriophages is used, which all tend to survive longer in the water environment than *E. coli* making them more useful as indicators for environmentally resistant pathogens [9]. Bacteriophages, of which coliphages are the most common, are considered as model organisms of enteric viruses in water [9], for example to assess the virus removal efficiency for water treatment facilities such as in stormwater ponds.

This paper presents field measurements on the levels of stormwater contamination, and the relative removal of microbial, physical and chemical contaminants in a traditional stormwater pond. Also a discussion on relative removal of virus particles in relation to the human health risk associated with pathogens in raw water sources is included.

Experimental Method

The pond that has been studied in this paper is the Järnbrott stormwater pond situated 5 km from the city of Göteborg. It has been investigated in previous studies [1, 10, 11], and is an off-stream stormwater pond with a permanent pool, and with no constant inflow during dry periods. The pond is divided into three sections with different bottom materials and with water depths between 0.6 and 1.5 m. The catchment area consists of about 160 ha impervious surfaces and is a mix of residential, industrial and traffic areas. For metals and PAH, the major source in the catchment is expected to be the traffic, where an important highway is situated within the catchment. Possible sources of microbial contaminants in the catchment include commercial and residential settlements, where faecal bacteria may originate from domestic and wild animals. Since the catchment only include separate sewer systems, household sewage will not contaminate the stormwater entering the pond.

Three storm events during fall 2004 (22 October, 18 November and 24 November 2004) were sampled using two ISCO 6700 samplers that were installed at the inlet and the outlet of the stormwater pond. The samplers were equipped with flow meters which monitored the flow and triggered flow-weighted sampling during the storm events. The inlet sampler was equipped with a velocity-height (V/H) probe (ISCO 750 module) and the outlet sampler with a pressure probe (ISCO 720) close to the outlet concrete weir. The samplers, equipped with 12 (1 L) plastic and glass bottles, were programmed to take a subsample at every 800 m^3 of stormwater passing the measuring point at the inlet and outlet respectively. Each sample was added to a set of two bottles (totally six sets, each representing 4,800 m^3 of stormwater) and were kept at ambient temperatures (0–10°C) in the sampler. The samples were transported in cold to the laboratory for analysis. Characteristics of the three storm

Table 1 Characteristics of the three storm events at the Järnbrott pond, 2004

Date	Rain depth (mm)	Antecedent dry period (days)	Stormwater volumes (m^3)
10 October	11	3	15,000
18 November	9	2	12,000
24 November	12	1	16,000

events are given in Table 1, including rain depths, dry periods (number of days antecedent the rain event) and stormwater volumes.

Laboratory Analyses

Event mean concentrations (EMC) for the contaminants studied were determined by analyses of samples pooled to one composite sample prior to the analysis. Samples were analyzed for total suspended solids (TSS), volatile suspended solids (VSS) and heavy metals (zinc, copper, lead, cadmium, nickel and cobalt). Nitrogen analysis included measurements on total nitrogen (total-N), nitrite (NO_2–N), nitrate (NO_3–N) and ammonium (NH_4–N), and the phosphorous analyses included total phosphorus (total-P) and phosphate phosphorus (PO_4–P). The organic contaminants of PAH, including the 16 specific US EPA-PAH [12], were also analyzed. Concentrations of these contaminants were given as μg/L.

Microbial analyses included total coliforms, *E. coli* and sulphite-reducing clostridia (SRC) (during three events) as well as intestinal enterococci (IE) and bacteriophages (during two of the events). Total coliforms and *E. coli* were analyzed with the membrane filtration method [13], which also was used in the analysis for intestinal enterococci [14]. Sulphite-reducing clostridia were analyzed according to the standard [15] except for the preheating to exclude vegetative cells. Bacteriophages analysed included somatic coliphages (SC) and F-specific RNA coliphages, and samples were sterile-filtrated (45 μm high-flow filter) to remove debris and bacteria prior to the standard analyses [16]. Microbial concentrations given as colony forming units (CFU/100 mL, bacteria) and plack-forming units (PFU/100 mL, bacteriophages) were presented on a logarithmic scale ($\log_{10}$) and for comparison reasons also the chemical and physical contaminants are presented on a logarithmic scale ($\log_{10}$).

Data Processing

The contaminant loads, based on a mass balance for the inflow and outflow of the pond, were determined for the three events. A mass balance approach was applied to determine the contaminant load (M_i) calculated as the stormwater volume (V_i), registered during the event, times the measured contaminant event mean concentrations (C_i). Site mean concentrations (SMC) for each specific contaminant were calculated as the accumulated load over accumulated stormwater volume, such as $SMC = \Sigma M_i/\Sigma V_i$. Based on the calculated SMC for the inlet (SMC_{in}) and the outlet (SMC_{out}), the removal efficiency R (%) was calculated according to the following expression:

$$R = \frac{SMC_{in} - SMC_{out}}{SMC_{in}} \times 100 \quad (1)$$

Results and Discussion

The concentration of total suspended solids, nutrients, heavy metals and PAH at the inlet and outlet of the Järnbrott stormwater pond is illustrated in Fig. 1 for the three storm events. The measurements exhibited a clear reduction of total suspended solids (TSS) and total phosphorous (Tot-P) was observed, while no reduction was seen for nitrate (NO_3–N) and total nitrogen (Tot-N). As indicated in error bars, the concentrations of nitrate varied highly between the three events. For the metals in Fig. 1b represented by cadmium (Cd), copper (Cu) and lead (Pb), a reduction was observed in the pond, although lower for cadmium compared to the others. The concentration of poly-aromatic hydrocarbons decreased in the stormwater pond, as represented by the accumulated concentrations of the 16 EPA-PAH shown in Fig. 1b.

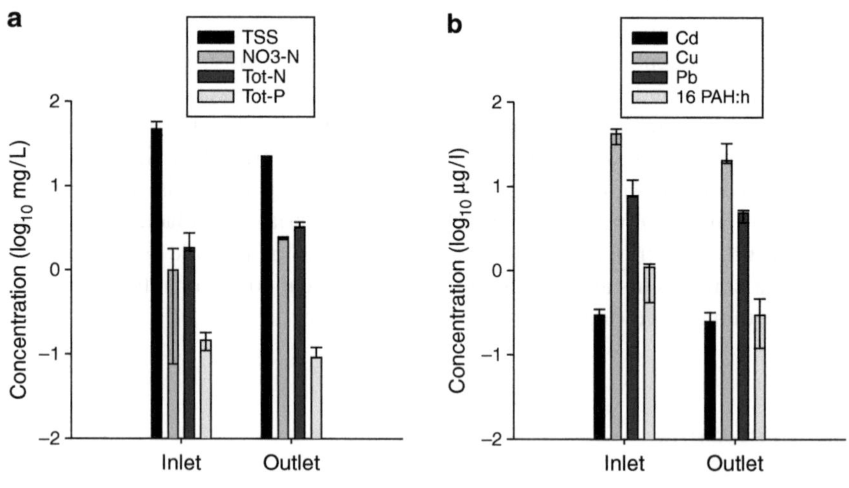

Fig. 1 Event mean concentrations (log-values) of (**a**) suspended solids, nitrogen compounds and total phosphorous, and (**b**) metals and PAH measured at inlet and outlet of the stormwater pond. Median concentrations for the three storm events are given with min and max in *error bars*

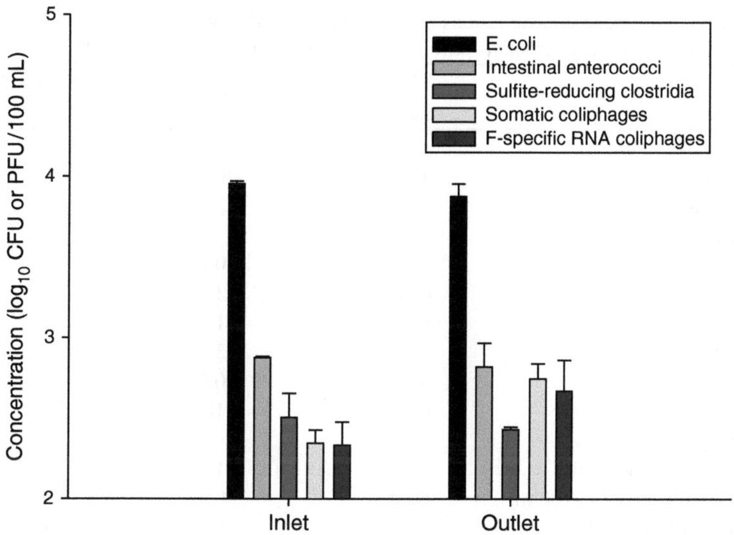

Fig. 2 Event mean concentrations (log-values) of three different indicator bacteria and two types of coliphages. Median concentrations for the three storm events are given with min and max in *error bars*

The stormwater pond reduced heavy metals as well as organic contaminants, and may therefore act as a barrier to prevent the receiving water from chemical contaminants associated with stormwater discharges.

The microbial concentrations, measured at the inlet and outlet of the Järnbrott stormwater pond are shown in Fig. 2. Results for the indicator bacteria show that *E. coli*, intestinal enterococci and sulphite-reducing clostridia were detected at different levels, with *E. coli* levels more than 1 log-unit higher than the others. Comparing concentrations from inlet and outlet samples, the results indicate that the pond did not significantly reduce faecal bacteria. Results for the bacteriophages, measured at two events in the pond, instead exhibited an increase in concentrations from the inlet to the outlet, including a source term in the pond. Comparing inlet and outlet concentrations for the separate events indicated variations between the events, especially for the F-specific coliphages. For one event these were slightly reduced while in the other event a high increase in concentration was observed comparing inlet to outlet of the pond.

Removal Efficiencies

The site mean concentrations of the various contaminants, together with the removal efficiency (%), are calculated for the three storm events (Table 2). The Järnbrott stormwater pond effectively reduced suspended solids (62% and 52% for

Table 2 Site mean concentrations (SMC) and removal efficiency in the stormwater pond for selected contaminants during the analyzed storm events, fall 2004

Parameters	SMC In	SMC Out	Removal (%)
Suspended matter			
TSS (mg/L)	47	18	62
VSS (mg/L)	19.1	9.3	52
Nutrients			
NO2–N (mg/L)	0.02	0.03	–65
NO3–N (mg/L)	0.90	2.39	−164
NH3–N (mg/L)	0.18	0.16	12
Tot-N (mg/L)	2.05	3.39	–66
PO4–P (mg/L)	0.06	0.07	−15
Tot-P (mg/L)	0.14	0,10	29
Indicator organisms			
E. coli (CFU/100 mL)	7,240	7,100	2.0
Tot coliforms (CFU/100 mL)	60,000	54,300	9.4
SRC (CFU/100 mL)	295	251	15
SC (CFU/100 mL)	229	574	−151
F-RNA (CFU/100 mL)	204	499	−145
IE (CFU/100 mL)	749	691	7.8
Heavy metals			
Cd (μg/L)	0.30	0.23	24
Cu (μg/L)	41.4	24.3	41
Pb (μg/L)	9.16	4.67	49
Zn (μg/L)	143	91.3	36
PAH			
16 USEPA PAH [a] (μg/L)	0.94	0.31	67

[a]Method from the US-EPA [12].

TSS and VSS respectively), as well as heavy metals (~25–50%). On the other hand, the reduction was weak for the nutrients and the indicator microorganisms. An increase of the contaminant concentrations in the outflow was observed for nitrite and nitrate, as well as for nitrogen as a total. For the indicator organisms, a removal was observed for the bacteria while an increase was observed for the bacteriophages.

Present results are in sharp contrast to a study performed in another Swedish stormwater pond, Flemingsberg, where a removal of faecal coliforms and enterococci were observed in the range of 97–99.9%. In the same study somatic coliphages were removed to around 70%. The high removal of microbial organisms in that pond was explained with the reduction of suspended solids being the main factor for the bacterial reduction [7].

Factors affecting the removal efficiency in the pond include physical sedimentation, die-off, accumulation and resuspension of the microorganisms. Indicator bacteria as well as bacteriophages may settle in the pond during low flow conditions and re-enter into the water phase later on as a result of high stormwater inflows. A higher survival has been reported for coliphages compared to faecal coliforms and *E. coli*, both in water [17] and soil [18], and coliphages survived the best of the organism groups in the sediment [6]. In contrast, pathogens as Giardia survived best in the water column and least in the sediment, but Giardia was together with Cryptosporidium observed in concentrations one to three orders of magnitude higher in sediment compared to the water phase [6]. The increase observed for the somatic and F-specific RNA coliphages in the present study may result from the high survival characteristics combined with the resuspension of settled microorganisms, or by a point source in the pond vicinity, for example birds swimming in the pond or other animals residing at the waterside.

Results from the current study suggest that stormwater ponds may represent an ineffective barrier for protecting raw water sources from microbial contaminants. After entering the pond, microbial levels are reduced over time, governed by the survival properties of each pathogen or indicator organism. Microbial survival rates in stormwater sediments has been reported in another study [7], and T_{50}-values (time for a 50% reduction) for *E. coli*, faecal enterococci, Clostridium and coliphages of 27, 27, 252 and 370 days, respectively, were reported. In addition, F-specific RNA coliphages has been reported to represent a model for norovirus survival [17]. Bird drops and domestic animals may potentially represent additional sources for faecal contamination in the pond. F-specific RNA coliphages may originate from water fowls such as gulls and goose [19], and at the Järnbrott pond these birds together with wild ducks have been observed, roughly between 20–50 specimens simultaneously. If we assume the increased F-specific RNA coliphage concentration in the pond originates from bird droppings we can calculate an estimate of the number of specimens needed to yield that concentration increase. In a study made on gull faeces [19] the observed F-specific RNA coliphage concentrations were in average $7.5\cdot\times 10^4$ PFU/g of faeces. Since the concentration increase were about 300 CFU/100 mL in the Järnbrott pond, with a dry weather volume of 6,000 m^3, the total counts increased with $18\cdot\times 10^9$. Using the faecal concentration reported on the gull faeces [16], and assuming an average faeces production of 30 g/day per gull and 10 days dry period (in average), this would imply a total production of $2.25\cdot\times 10^5$ coliphages per gull (during 10 days). Using this value with the coliphage increase in the pond, the number of gulls needed to verify this is calculated to 80,000 ($18\cdot\times 10^9/2.25\cdot\times 10^5$) gulls in the pond all day long. This is of course impossible and indicates that the concentration increase in the pond must have another source than droppings of waterfowls.

A high release of coliphages into raw water sources from stormwater ponds therefore represents a human health risk in relation to virus that needs to be further assessed. The composite sampling technique applied in the present investigation gives a representative description for parameters which concentration in water and sediment is well described by a normal distribution. Environmental parameters may however rather be log-normal or poisson-distributed, which may typically be the case for several pathogens [20]. The composite sampling technique may bias the analyzed mean concentration for the sample, representing the entire event, towards the peak concentration in one subsample. As the microbial health risk due to contaminated raw water is governed by short-term peak concentrations, the analysis for composite samples may reflect peak concentrations in sub-samples analyzed for a storm event. Additional flow-weighted samplings of stormwater events, with analysis of discrete sub-samples, are thus needed to describe the variability in stormwater flows for discrete event.

Conclusion

From this study we mainly conclude that stormwater ponds perform well in the reduction of traditional stormwater pollutants including suspended solids, some nutrients and heavy metals. The removal of microorganisms was however negligible for the faecal indicator bacteria. The increase in the pond of somatic and F-specific coliphages, used as a model organisms for virus, show that stormwater ponds does not represent an efficient microbial barrier. Further studies must be carried out to investigate the microbial removal efficiencies in other stormwater ponds and the microbial impact from birds and animals potentially residing in the pond. Analyzing subsamples from the course of the storm events are needed for describing the microbial variability over time.

References

1. Pettersson TJR (1998) Water quality improvement in a small stormwater detention pond. Water Sci Technol 38:115–122
2. EC (2000) Directive 2000/60/EC of the European Parliament and of the Council of 23 October 2000 establishing a framework for community action in the field of water policy. Luxembourg
3. WHO (2004) Guidelines for drinking water quality, 3rd edition, World Health Organization: Geneva. Available at: http://www.who.int/water_sanitation_health/dwq/guidelines/en/index.html. accessed July 3, 2008
4. Geldreich EE, Best LC, Kenner BA, Van Donsel DJ (1968) The bacteriological aspects of stormwater pollution. Water Pollution Control Federation, 40:1861–1872
5. Marsalek J, Rochfort Q (2004) Urban wet-weather flows: sources of fecal contamination impacting on recreational waters and threatening drinking-water sources. J Toxicol Environ Health. Part A, 67:1765–1777
6. Karim MR, Manshadi FD, Karpiscak MM, Gerba CP (2004) The persistence and removal of enteric pathogens in constructed wetlands. Water Res 38:1831–37
7. Stenström TA, Carlander A (2001) Occurrence and die-off of indicator organisms in the sediment in two constructed wetlands. Water Sci Technol 44:223–230
8. Kay D, Wyer MD, Crowther J, Wilkinson J, Stapleton C, P. Glass P (2005) Sustainable reduction in the flux of microbial compliance parameters from urban and arable land use to coastal bathing waters by a wetland ecosystem produced by a marine flood defense structure. Water Res 39:3320–3332
9. WHO (2004d) Microbial fact sheets, in Guidelines for drinking-water quality, 3rd edition, World Health Organization: Geneva. pp. 221–296
10. Pettersson TJR (1999) Stormwater ponds for pollution reduction, in Diss. Department of Sanitary Engineering. Chalmers University of Technology: Göteborg.

11. Pettersson TJR, Lavieille D (2007) Evolution on pollutant removal efficiency in stormwater ponds due to changes in pond morphology, in Highway and Urban Environment: Proceedings of the 8th Highway and Urban environment Symposium, G.M. Morrison and S. Rauch, Editors. Springer: Nicosia, Cyprus 2006. pp. 429–439
12. NEN (1984) Water – determination of the content of six polycyclic aromatic hydrocarbons (PAH) by high-pressure liquid chromatograghy. Document no. NEN 6524:1984 nl
13. ISO (2000) ISO 9308-1, Water Quality - Detection and enumeration of *Escherichia coli* and coliform bacteria-Part 1: Membrane filtration method. International Standardisation organisation: Geneva
14. ISO (2000) ISO 7899-2, Water Quality - Detection and enumeration of intestinal enterococci-Part 2: Membrane filtration method. International Organization for Standardization: Geneva
15. ISO (1986) ISO 6461-2, Water Quality - Detection and enumeration of the spores of sulphite-reducing anaerobes (Clostridia)-Part 2: Membrane filtration method. International Organization for Standardization: Geneva
16. ISO (1995) ISO 10705-1, Water Quality – Detection and Enumeration of Bacteriophages-Part 1: Enumeration of F-specific RNA Bacteriophages. International Organization for Standardization: Geneva
17. Allwood PB, Malik YS, Hedberg CW, Goyal SM (2003) Survival of F-specific RNA coliphage, feline calicivirus, and *Escherichia coli* in water: a comparative study. Appl Environ Microbiol 69:5707–5710
18. Gantzer C, Gillerman L, Kuznetsov M, Oron G (2001) Adsorption and survival of faecal coliforms, somatic coliphages and F-specific RNA phages in soil irrigated with wastewater. Water Sci Technol 43:117–124
19. Cole D, Long SC, Sobsey MD (2003) Evaluation of F + RNA and DNA coliphages as source-specific indicators of fecal contamination in surface waters. Appl Environ Microbiol 69:6507–6514
20. WHO (2004) Microbial aspects, in Guidelines for drinking-water quality, 3rd edition, World Health Organization: Geneva. p. 121–144

Reuse of Solid Waste from the Clearing of Road Basins and Ditches in Civil Engineering: Environmental Characterization, Ways of Treatment and Their Efficiency

Paul-Emile Martin, Laurent Eisenlohr, Guy Bobillon, Pierrick Esnault, and C. Gardet

Introduction

One phase of this work has been the study of an efficient mode of treatment of the carriageway drainage materials, aiming at improving their environment quality. Efficient cleaning up methods exists for chemical treatment (heavy metals, hydrocarbons) as well as for giving a texture adapted to material reuse (screening, washing). These often complex methods are costly, which may form an obstacle for an effective reuse of the cleaning out solid waste.

Experimental Method

Classification of Drainage Solid Waste

Road drainage sediment classification according to their environmental characteristics has been carried out, distinguishing two main sediment categories.

1. **Sediment from ditches**

Studied ditches are solely rassy ditches and curved channels. Concrete curved channel and ditches generate sediment similarly to carriageway dust (compound

P.-E. Martin
Ecole National des Travaux Publics de l'Etat, Lyon

L. Eisenlohr and G. Bobillon
Cete Lyon (Public Work Regional Engineering Office), Lyon

P. Esnault and C. Gardet
SETRA (Technical department for transport, roads and bridge engineering, and road safety)

S. Rauch et al. (eds.), *Highway and Urban Environment*,
Alliance for Global Sustainability Bookseries 17,
DOI 10.1007/978-90-481-3043-6_21, © Springer Science+Business Media B.V. 2010

fraction essentially sandy). Flow speed in these little rough structures is indeed not favorable to fine compound fraction sedimentation.

2. **Sediment from road basins**

This category gathers every basin type. Let us note that some structures commonly called ditches have been included in this category since the sediment type have characteristics closer to basins than to ditches: it is notably the case of every ditch staying under water.

Sediment Environmental and Geotechnics Characterization

As part of this work about 100 samplings have been carried out on basins and ditches in France. Hydrocarbons and metal trace element (MTE) contents have been analyzed. Data from bibliography [1] have been used as well.

Regulation threshold values is not available in France for this kind of sediment before reuse, the proceeding has been as following: for MTE threshold values are VCI usage sensible (impact value for sensible aim in French). These values, used until 2006, correspond to those used for polluted sites and soil in a so-called sensitive context (garden, park, etc.). For hydrocarbons, the used results have been compared to inert waste landfill thresholds. To complete this classification, material geotechnical properties have been studied in detail with the help of geotechnical tests: SETRA's GTR [2] makes it possible to grasp reuse of materials in embankment or sub grades, from various geotechnical parameters. F category organic soil and industrial by-products corresponds the best to studied materials. Their using possibilities in engineering structures depend on their organic matter content (OM%). At this stage we classify this material type in two categories. For organic matter contents above 10%, F12 materials have not characteristics satisfying to be reused in engineering. For organic matter contents from 3% to 10%, F11 materials may be used under conditions similar to mineral matter. To characterize slope geotechnical properties of materials such as road drainage sediment, the following tests have been carried out on three samples:

Grain size analysis
Eighty micrometer fraction analysis
Methylene blue value
Initial moisture
Proctor test

Moisture Analysis

The open air spreading method has been studied more in detail with the help of tests on sediments taken in the Lyon region.

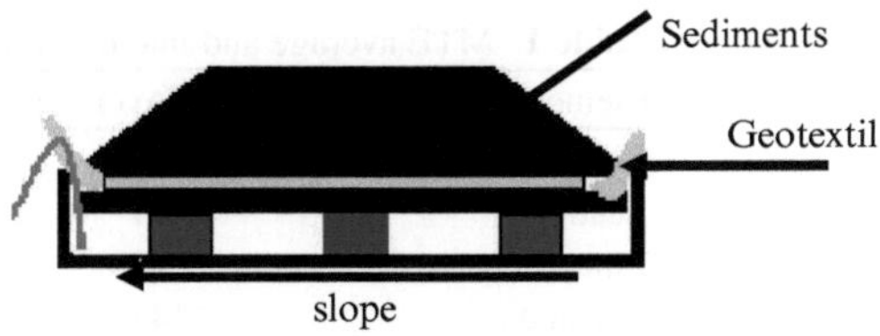

Fig. 1 Experimental device of water loss

Presentation of the Water Loss Experimental Device

To appreciate the dryness evolution by open air spreading, the device presented in Fig. 1 has been realized. It musts be noted that these tests were carried out inside, and consequently sheltered from climatic hazards. Weighing has been carried out with a measure uncertainty in the gram order. Each previously tared (box) weight has been measured until stabilization.

First values may admit some distortions. Indeed this matter put for drying has not the same dryness as the one within the basin. This link with the first measure is not inconvenient facing the experiment objective, which is to determine the time from which dryness is stabilized as well as the final dryness value. A first experiment has been carried out with settling pond sediments. Two tests have been led with 6.5 and 4.5 cm sediments thickness, respectively, at the beginning of the experiment. Weight evolution was regularly followed-up. After 3 weeks a sample was taken and put in a drying oven (150°C during 14 h). Dry soil sample mass was determined, and by knowing sample weight during the course of time, it is possible to deduce the moisture evolution.

To grasp sediment height influence on moisture evolution, other tests have been carried out from B4-B6 basins located on Lyon's East RingRoad. Sediment was put within tubes of variable diameter and height. A test in box similar to the first series was set up as well. The adopted device is presented in Fig. 4.

A drainage and water recovery device was also set up. At last, to prove that a relative uniformity between sediments exists, B4-B6 and B17 sediments draining has been compared for the same sediment height (5.5 cm). Sediments of these two basins have rather different grain sizes, B17 being globally finer than B4-B6.

Result and Discussion

Sediment Environmental and Geotechnic Characterization

Ditches: The only studied variable concerning ditches is traffic. It emerges from the first results that sediments from ditching cleaning out do not contain pollutions significant in metal trace elements (nickel, chrome, cadmium). Concerning hydrocarbons, zinc, copper, and lead, an increasing correlation between hydrocarbon content and traffic supported by the road axis, could be observed (Table 1).

Table 1 MTE average and maximum contents in ditches sediments

Element	Average	Maximum
Cadmium	0.43	1.5
Lead	79	550
Zinc	190	1620
Chrome	43	210
Copper	74	1800
Nickel	28	115

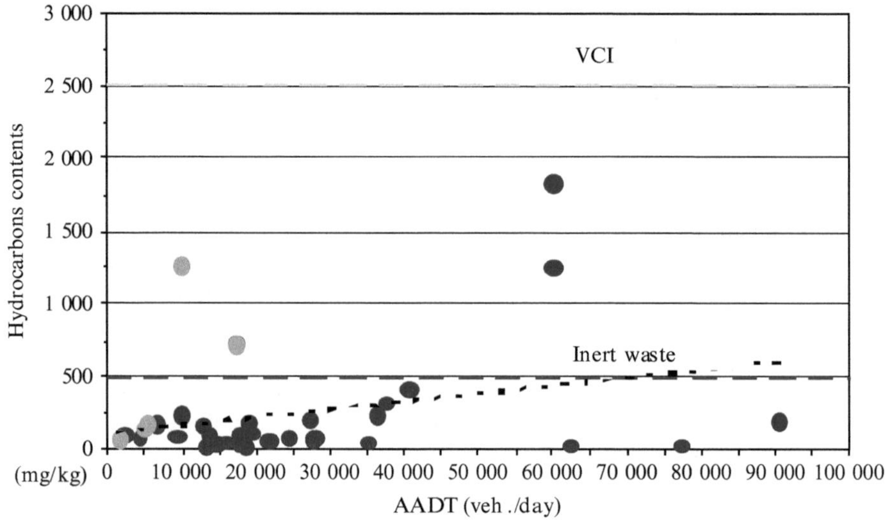

Fig. 2 Variation of sediment hydrocarbons contents with daily traffic

We observed that these pollutants are likely to exceed threshold values for traffic above 20,000 vehicles/day. Fig. 2 presents the observed results for lead and hydrocarbons.

Road basin: Studied parameters concerning road basins are: traffic, size of carriageway, cleaning out frequency. No link could be established on carriageway size parameter: this result seems pertinent as it has been established in a former document [3] that pollutions gathered in runoffs depend more on traffic than on carriageway size. Cleaning out frequency seems to point out that MTE concentration increases with the last cleaning out date: this result seems to be coherent with a renewing car fleer which emits less pollutants (Fig. 3).

Likewise, safety barriers, the main zinc source, tend to emit less zinc over the course of some years [4]. These results are however to interpret prudently: the number of basins where the last cleaning out date is known is limited, and the collected data deals with a short cleaning out duration.

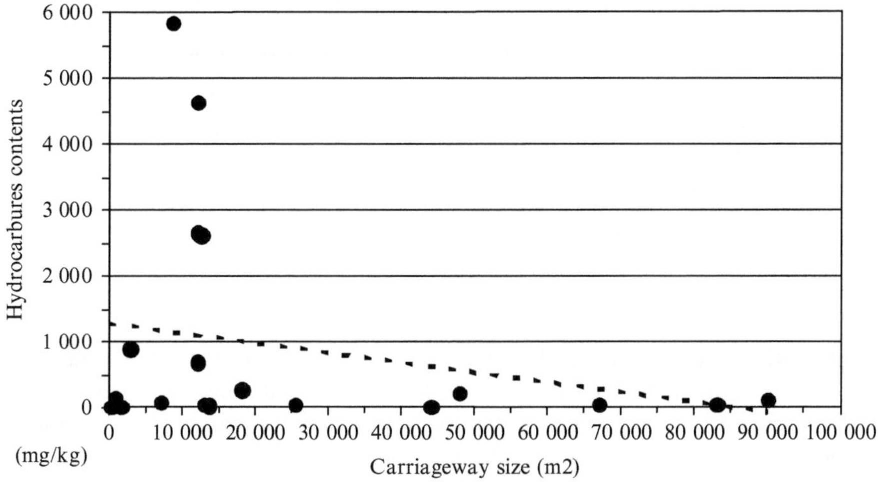

Fig. 3 Influence of size of carriageway on hydrocarbons contents in basins sediments

Table 2 Hydrocarbon average and maximum contents in sediments with different basin typologies and traffics

Hydrocarbon contents	Average	Maximum
Infiltration basin	54.7	350
Stormwater pond, traffic <20,000 vehicles/day	135	900
Stormwater pond, traffic >20,000 vehicles/day	190	1620

Concerning the hydrocarbons content, two basins typologies have been identified: basins under water (stormwater ponds) and dry basins (infiltration basins). Hydrocarbon concentrations in dry areas are very low. This result seems to be logical, as exposure to ambient light and air favor hydrocarbon degradation. Low values in hydrocarbons will, however, have to be confirmed with further samples.

Concerning stormwater ponds; below 20,000 vehicles/day the measured hydrocarbon contents are under threshold-values, but it becomes important and well above threshold-values above 20,000 vehicles/day (Table 2).

For MTE, distinguishing between basins under water and dry basin has not been retained. Cadmium, chrome and nickel are present in non significant quantities in ditches. Zinc, lead and copper are likely present in quantities beyond the threshold values: this result can be explained by the superior efficiency of stormwater ponds and infiltration basins on settling: MTE are indeed mainly linked to fine particles, less easy for settling [5]. Before any reuse in embankment, it is convenient to characterize the zinc, lead, and copper content in sediments. Hydrocarbon characterization is only for traffic intensities superior to 20,000 vehicles/day (Table 3).

Table 3 MTE average and maximum contents in basins sediments

Element	Average	Maximum
Cadmium	043	1.5
Lead	79	550
Zinc	190	1620
Chrome	43	210
Copper		
Nickel		

Geotechnic characteristics: The first parameter making it possible to classify a soil is the organic matter content. Data from literature and sample campaigns show a slight organic matter percentage in sediments from ditches. These sediment characteristics stay about 5% for variability in the range 3.4%–6.6%. Conversely basins present contents in the order of 10% for a variability range from 3% to 37.5%. It is so convenient to foresee a treatment aiming at pull down organic matter before any reuse, by liming or landfarming. The later treatment has moreover the advantage of offering a hydrocarbon degradation possibility.

Grain size, the second main parameter, has been studied on both sediment typologies: materials being likely to present a large fine particle content, only laser grain size analyses are presented here. Grain size composition near to in-site soil in road ditches is observed. The possibilities of reusing these sediments in engineering depend on the local geological context. Sediments from basins all present important percentages of particles inferior to 80 μm (Table 4), indicating a very water sensitive material.

Moreover, variability within samples is very important (Fig. 4). Before any reuse of these materials in engineering, it is convenient to:

- Diminish the organic matter content
- Reduce moisture
- Homogenize the material

As indicated above, one of the two discriminating parameters for reusing sediments in engineering is moisture. The simple device is presented in material and methods. Classically, three phases (Fig. 5) are observed during the experiment:

Table 4 Geotechnical characteristics of different basin sediments

Sample	Particles above 80 μm (%)	Moisture
B6-B4	22.9	40.9
B17	62.7	101.9
B27	23.7	4.7
Cheviré Sud [6]	35.9	–

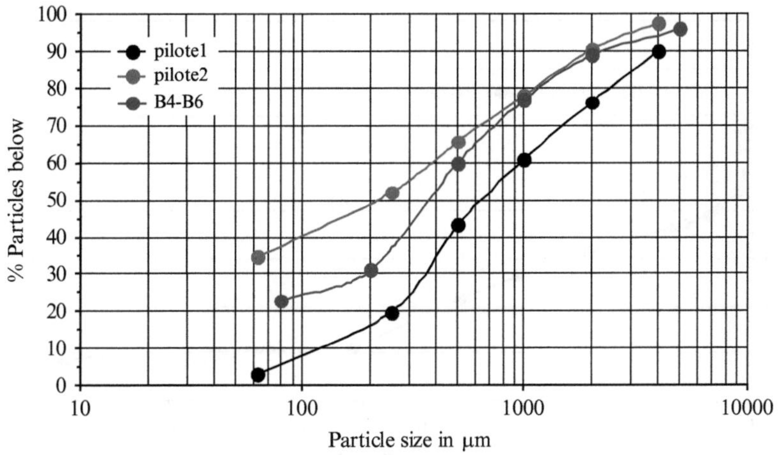

Fig. 4 Grain size analysis for three samples of the same sediment

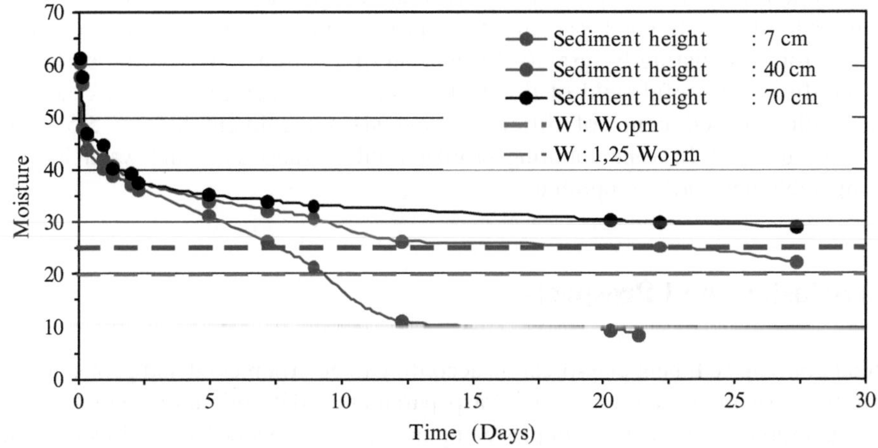

Fig. 5 Moisture reduction with different heights

1. A draining phase, during which water is drained towards geotextil. This phase leads to a quick water loss and lasts about 1 day.
2. A phase where evaporation is relatively intense (until 15 days) and comes to a quasi-linear water loss.
3. A last phase where water loss is very slow (from 15 days).

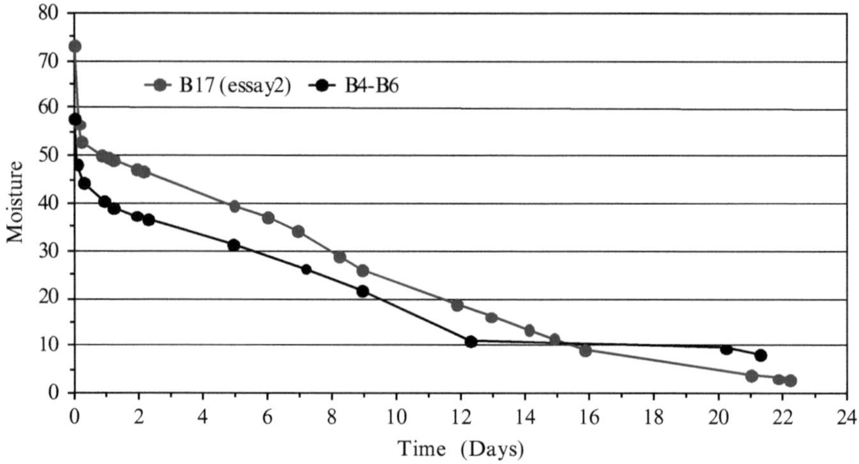

Fig. 6 Moisture reduction with two different grain size sediments

It is recorded that proctor optimum (water content at which soil may be used via a reagent addition) is attained with a value inferior to 20 days, for a sediment layer inferior to 40 cm. Beyond, packing down provoked by sediment seems to hinder water from evacuating by gravity (formation of floating water on the surface). To confirm the useful field of this technology, a comparative test was carried out with two sediments of different grain sizes (sediment height: 5.5 cm). Graphs in Fig. 6 have neighboring profiles until 12 days date, and water loss is comparable until proctor optimum.

Conclusions and Prospects

Road draining sediment classification according to environmental and geotechnical criteria have made it possible to grasp parameters determination, which is an indispensable preliminary to improve engineering. This work has established that the geotechnical properties of the studied material present some homogeneity and that their aptitude for engineering improvements deserves to be analyzed case by case, beginning with the most problematic parameters: grain size and water content.

It appears that simple treatment technologies are able to provide an adapted answer to make studied materials suitable for engineering improving. Materials environmental classification has, for the time being, only been interested in risk concerning direct or indirect material ingestion (using VCI): studies in columns are in progress to evaluate, from sediment composition, potential risks for contamination of water resources by infiltration of water percolating via sediments. Treatment possibilities concerning organic matter and hydrocarbons have been studied in 2007, but the results are not conclusive: a second campaign will be renewed in 2008.

References

1. Setra (1995) entretien des reseaux d'assainissement routiers et pollution des sols. Memorandum, Bagneux
2. Road Earthwork Guide (1992) SETRA
3. Hurtevent (2006) Calculation of the loads of chronic pollution from roadways runoff, information notes, SETRA
4. Pagotto C (1999) Etude sur l'emission et le transfert dans les eaux et les sols des elements trace metalliques etdes hydrocarbures en domaine routier, these, Poitiers University
5. Legret M (2001) *Pollution et impact d'eau de ruissellement de chaussée.* Ed. LCPC, Nantes
6. Auriol JC (2005) *Caractérisation geotechnique en laboratoire des matériaux issus de l'assainissement routier en vue de leur réutilisation éventuelle.* LCPC
7. Martin P-E (2007) Reuse in civil engineering of solid waste from the cleaning of road basin and ditches, characterisation and treatments possibilities, Meeting on Recent Developments in Sediment transport and Natural and Urban Sediments, NOV 28–29, 2007 Lyon, FRANCE

References

[illegible]

Methodology Used for Comparing Two Locations for the New Lisbon Airport Based on the Impacts on Surface Water

Ana Estela Barbosa, João Soromenho Rocha, and João Nuno Fernandes

Introduction

Lisbon airport, named Portela, was constructed in 1940 and needs to be replaced with a more modern and improved infrastructure, to serve future demands. In spite of the fact that the need of a new airport was foreseen four decades ago, only in 1999 a site was formally chosen. Unexpectedly, a private proposal for an alternative location for the New Lisbon Airport (NLA) was announced and published in 2007, starting a public and political discussion about the goodness of the first choice. The Portuguese Ministry of Public Works, Transportation and Communications commissioned the National Laboratory of Civil Engineering (LNEC) to perform a technical comparison of the two alternative locations for the NLA. The study was assigned in June 2007, and deserved a major focus of attention from different sectors of the society – politicians, economists, entrepreneurs, academics and the public in general. LNEC assigned the study to a multidisciplinary team of more than 40 people, including LNEC researchers and consultants from universities and private companies, both national and international. The study took into account the assessment of several environmental, social and economic factors.

This paper presents the study undertaken by the team from LNEC that was responsible for the comparison of the effects of the NLA on surface water resources at the two possible locations. More details of the work can be found in reports published in Portugal [1, 2]. Any new urbanization project may have different proposed locations within a country and/or region, each of one serving some specific social, economical or other objective. The political decision processes concerning alternative locations of major national infrastructures must take into account, amongst other factors, the effects of the project in the environment and the cost of environmental mitigation measures.

A.E. Barbosa (✉), J.S. Rocha, and J.N. Fernandes
National Laboratory for Civil Engineering, Hydraulics and Environment Department, Av. do Brasil, 101, 1700-066, Lisbon, Portugal
e-mail: aestela@lnec.pt

S. Rauch et al. (eds.), *Highway and Urban Environment*, Alliance for Global Sustainability Bookseries 17,
DOI 10.1007/978-90-481-3043-6_22, © Springer Science+Business Media B.V. 2010

Urban development and new infrastructures on a catchment have negative effects on the local water quality and quantity balance. The hydrological regime is changed, with increased volumes of stormwater runoff, higher peak flow rates and flood water levels as well as lower base flows in watercourses, reduction of soil moisture and groundwater recharge. Additionally, there is an increase in pollutant loads diverted into sewers or surface waters. The new demand of water for different uses (domestic and industrial) must be considered – it can place pressure on the local water system (surface or groundwater) or represent an important cost in treating water for consumption, in the case of lack of quality of the local resources.

A new airport will produce two kinds of wastewater: stormwater from the paved areas (runways, parking lots, buildings and roads) and sewage generated by the airport restaurants, toilets and maintenance facilities. These should represent important water pollution sources that need to be managed in accordance with the new demands from the Water Framework Directive.

The Two Airport Alternative Locations

The projected NLA area was estimated to be a rectangle of around 2,000 ha. The distance between the two alternative locations for the new airport is about 40 km, equal to the distance of each place to Lisbon. In Fig. 1 the two alternative locations

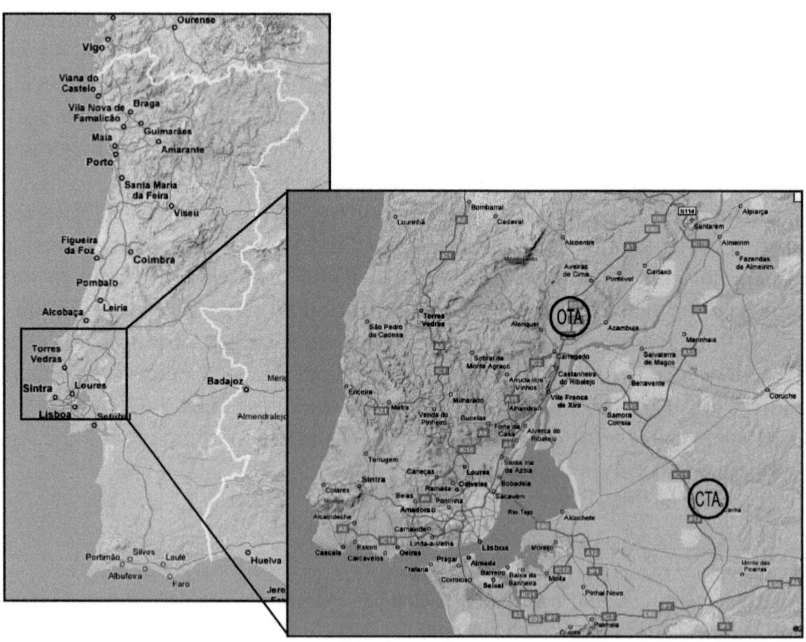

Fig. 1 Alternative locations

are presented, namely Ota (first choice) and Campo de Tiro de Alcochete (CTA), the proposed location. Both locations are within the huge river Tagus Basin and in the vicinity of the large Tagus estuary.

Methodology

Objectives and Constraints

A major constraint of the study was the time limit of 6 months and the fact that not much field studies were developed for the locations, in particular for the new one. The first alternative location (Ota) presented a set of preliminary studies and a layout of the runways that although incipient, could support the analysis to a much higher degree compared to the most recent proposal, *Campo de Tiro de Alcochete* (CTA). Therefore the working methodology needed to be accurate, effective and make the most out of the existing information.

The team responsible for the analysis of the surface water resources was twice at the location of CTA, collecting information and doing field work, and once at the location of Ota. These visits and the experience of the working team were important assets for the development of the study. Part of the LNEC team was involved in the preliminary studies of the Ota location, providing terms of reference for water resources studies.

An infrastructure with the complexity of a new major national airport concerns also the construction of new transportation structures, and changes the urban development in its vicinity. Neither of these aspects was analysed. The study was focused on surface water resources of the areas directly affected by the impermeabilization due to the construction on the airport platform.

Selection of Characterization Indicators

In order to support the analysis, a set of characterization indicators have been established. Their purpose was to describe in an objective manner the characteristics and conditions of the surface water, or other factors that may affect it. The following characterization indicators have been assigned to the study:

1. Percentage of catchment impermeabilization
2. Length of rivers that are affected
3. Floods
4. Soil erosion factors
5. Pollution sources
6. Water quality and uses
7. Local availability of water for the airport uses

Establishment of Decision Indicators

The characterization indicators are tools to describe in a similar way the two locations in appreciation. In order to support the (political) decision the best methodology is to aggregate the relevant information provided by the characterization indicators into decision indicators. Therefore this study commenced for deciding to use two decision indicators, namely: (A) "Flow under flood conditions", and (B) "State and integrated management of surface water resources". The decision indicator (A) aggregates in a direct or indirect manner the information provided through the characterization indicators 1–3. The decision indicator (B) intends to agglutinate factors related to the water quality and uses (concerns the characterization indicators 5–7). Both were developed and applied to the two locations [2]. It was then understood that the decision indicator (B) was not able to distinguish, in a significant way, the two locations under analysis. Therefore it was decided that the most relevant decision indicator should be "Flow under flood conditions". The assessment of the costs for environmental mitigation measures was carried out, concerning the scope of this indicator. A similar analysis concerned the opportunities that the new infrastructure could represent for the local water resources management and protection.

Comparison of the Two Locations

Tables 1 and 2 present a summary of the characterization of the seven selected indicators for the two locations. They were based on all existing information, provided by the representants of the two project locations [3–5], and field work, including assessment of water levels and pollution sources, and measurements with a multimetric water quality probe. Historical water quality measurements and the characteristics of the sanitation systems, at each location, have been analyzed as well [6, 7]. Other relevant information was provided by the Environment Transport and Planning [8] and the river Tagus Catchment Plan [9]. Flood discharges for the Ota location was calculated by FCG [4].

The available information concerning the CTA was scarce, therefore simple calculations of flood discharges have been done with the rational method (for instance, in [10]). In that method the runoff coefficient C, that is, the proportion of the total rainfall that will reach the storm drains, was estimated following the recommendations in [11, 12].

Concerning the analysis of "Flow under flood conditions", there is a large difference between the two sites, being their relative position in the watershed the most relevant discrepancy. The geographical location of the two sites, presented in Fig. 1, and the variability of the precipitation are not relevant factors for distinguishing the cases.

Table 1 Description of the seven characterization indicators for the Ota location

Characterization indicator	Description
Catchment impermeabilization (%)	3.5% of the area of the watershed of Ota River and 25% of the watershed of the tributary Alvarinho
Length of rivers that are affected (km)	Approximately 25 km
Floods	Historically flood site; needs huge constructions to deal with the flood discharges including one dam
Soil erosion factors	Soil uses are not sensitive but there are steep slopes
Pollution sources	Relevant domestic, industrial and agricultural (arable, cattle and swine) pollution sources
Water quality and uses	Bad water quality reported for the last 10 years. Most frequent parameters exceeding water quality limits are: total suspended solids, biochemical oxygen demand; phosphorus and nitrogen; *faecal coliform*. Surface water used for irrigation
Local availability of water for the airport uses	Surface water is not a good source of water due to its very bad quality

Table 2 Description of the seven characterization indicators for the CTA location

Characterization indicator	Description
Catchment impermeabilization (%)	Five percent of the area of the watershed of Vale Cobrão Stream
Length of rivers that are affected (km)	Approximately 26 km
Floods	No historical floods reported. Flood discharge calculations indicate no need for major constructions to change the rivers path
Soil erosion factors	The factors indicate small soil erosion. Mild slopes
Pollution sources	Only diffuse pollution sources: agriculture (arable, cattle and sheep)
Water quality and uses	Evidences of historical water eutrophication at a local reservoir, kept as a reserve for fire combat
Local availability of water for the airport uses	Surface water is not a good source of water due to its scarcity. Groundwater is the main water source

A Geographical Information System, ArcGis®, was used to define each watershed, therefore allowing different measurements and calculations (e.g., area, perimeter and length of the affected rivers).

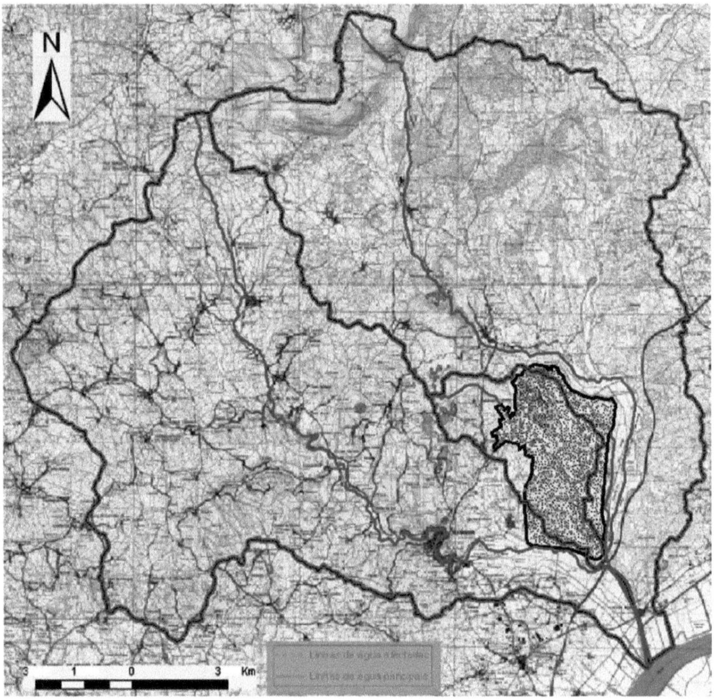

Fig. 2 Ota location

In the case of Ota, shown in Fig. 2, the airport platform is placed almost at the end of the Ota and Alenquer watersheds. It covers almost two third of the Alvarinho watershed (65%), a sub-basin of the Ota River. This position leads to an increase of the drainage area and to a large inundation area. Besides, the high slopes contribute to a reduction of the time of concentration and to an increase of the flood discharges. In addition, the proposed platform is located at the confluence of three rivers and in a zone where the tide influence occurs. The combination of flood events with high tide levels prevents the water flow and, consequently, extended areas are flooded. This happens frequently, therefore, in the case of choice of this location, the accessibility to the airport would have severe constraints. According to this scenario, there is a demand for significant flood control constructions, including dikes and regularization works on river Ota, and a dam and deviation pipes for the Alvarinho River.

The CTA option is located at the headwaters of the Vale do Cobrão Watershed (Fig. 3) where the slopes are mild. For that reason the flood discharges are not as important as at the Ota location. The construction works needed to ensure the flow of the floods are not very complex, and can be implemented with relatively small costs. Summing up, the ratio of the flood discharges at the two locations is about one to three. In CTA the flood discharge for a return period of 100 years is approximately 70 m^3/s, and for Ota about 226 m^3/s. Moreover, the topography is much

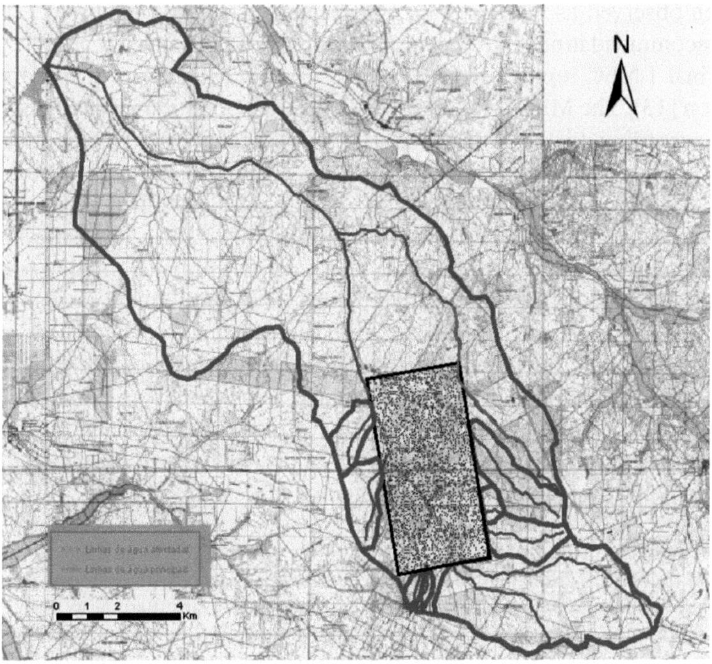

Fig. 3 CTA location

more compatible with flood drainage in CTA than in Ota. This situations leads to much easier construction works for the CTA case.

Comparison of the Costs of Measures for Impact Mitigation

The costs of the intervention on the transformation of river network at both locations are mainly related with the flood drainage.

At Ota, the main work is to change the course of the Alvarinho River, by building a large dam upstream of the airport platform and transferring the water from the Ota River to the Alenquer River. The intervention in CTA is quite simple; it requires the construction of channels following the contour of the airport platform at the East and West sides. Consequently, it was estimated that the Ota flood water works would cost about 13.4 M€ while the cost for CTA would reach 0.9 M€.

Conclusions and Recommendations

The study undertaken supplied objective quantitative and qualitative data that indicated that the location CTA is significantly better than the Ota alternative, concerning surface water resources analysis. During the course of the study, some issues have been observed as matters that could be solved only by the project itself, and several recommendations have been made in order to guide it.

The final LNEC report, with the global assessment, in fact reached the same conclusion [13]. The Ministry of Public Works Transportation and Communications decided according to the LNEC report results, and the study has already been through public consultation and discussion. Although several sectors of the Portuguese society that defended the Ota location are not satisfied, there is a general agreement upon the reliability of the study and its results. It is believed that the methodology used is objective, reliable, and easy to implement and to provide the desirable level of conclusions, capable of supporting a political decision. For the case related here there were important time and information constraints, nevertheless the methodology fitted and is sufficiently flexible to be implemented for cases with two or more locations in appreciation, and also when there are different levels of information, concerning the sites and the projects development. The projected NLA area was estimated to be of around 2,000 ha, but it is considered that the methodology may be applied to various spatial scales.

Acknowledgements This study was developed in the context of a large team work, therefore the material presented here benefited from the discussions that took place during the meetings. In particular, the authors are grateful to Professor Maria do Rosário Partidário from the Lisbon Technical University, who provided the insight into the methods for the strategical environmental assessment, and to Eduarda Beja Neves from LNEC, responsible for the coordination. The authors also acknowledge the kind help and all the information provided by the *Campo de Tiro de Alcochete* (CTA), and the Center for Technical and Military Formation of the Air Force, CFMTMA (Ota).

References

1. Barbosa AE, Rocha JS, Fernandes JN (2007) Location of the New Lisbon Airport at the Campo de Tiro de Alcochete. Surface water resources study. (In Portuguese) Report 311/2007-NRE, Laboratório Nacional de Engenharia Civil, 17 pp
2. Rocha JS, Fernandes JN, Barbosa AE (2007) Technical compared analysis of the alternative locations for the New Lisbon Airport. Surface water resources study. (In Portuguese). Final Report, Report 409/2007-NRE, Laboratório Nacional de Engenharia Civil, 36 pp
3. http://www.naer.pt Preliminary Environmental Impact Study of the Ota location for the NLA- NAER
4. FCG, Prosistemas (2004) Estudo Hidrológico e Hidráulico do Novo Aeroporto na Ota (Etapa B) – Estudo Prévio, Modelação hidráulica e Infra-Estruturas, Memória descritiva e justificativa
5. IDAD, Instituto do Ambiente e Desenvolvimento (2007) Localizações alternativas para o Novo Aeroporto de Lisboa. Projecto elaborado para Confederação da Indústria Portuguesa. EEP 04.07-07/17, 96 pp
6. http://snirh.pt/ Surface water quality data
7. http://insaar.inag.pt. Data concerning national water and wastewater systems
8. Environment Transport and Planning (2003) Estudo de contaminação de solos e águas subterrâneas no perímetro militar do Campo de Tiro de Alcochete. Relatório elaborado para o Campo de Tiro de Alcochete. P-1110-POR, 26 de Dezembro de 2003, 46 pp
9. http://www.inag.pt Basin Plan for the Tagus river Catchment Vol. III October 2000

10. Chow VT, Maidment DR, Mays LM (1988) Applied hydrology. McGraw Hill Book Company, New York
11. Portela M, Silva A, Melim C (2000) O efeito da ocupação urbana nos caudais de ponta de cheias naturais em pequenas bacias hidrográficas, 5? Congresso da Água, Lisboa
12. Portela M, Hora G (2002) Aplicação da fórmula racional à análise de cheias em Portugal continental: valores do coeficiente C, 6.? Congresso da Água, Porto
13. LNEC (2008) Technical compared analysis of the alternative locations for the New Lisbon Airport, at the Ota site and the Campo de Tiro de Alcochete site. 2nd Stage – Compared evaluation of the two locations. (In Portuguese) Report 2/2008 – DT, Laboratório Nacional de Engenharia Civil, 324 pp

Determination of Polycyclic Aromatic Hydrocarbons (PAHs) in Surface Waters and Runoff

Olalekan S. Fatoki, Titus Van Ree, and Joyce Nakhavhembe

Introduction

PAH are ubiquitous environmental pollutants of natural and anthropogenic origins. Microbial synthesis, synthesis by higher plants and volcanic activities are major natural source contributions but they contribute a very small amount compared to anthropogenic sources [1]. They are important priority pollutants emanating anthropogenically from incomplete combustion of organic matter mainly coal- and oil-burning, plants and vehicle emissions as combustion products. Polycyclic organic matter that contains benzo [α] pyrene and other PAH containing materials is released from forest and agricultural fires and they are most likely absorbed onto smoke particles settling on all kinds of surfaces where they are transferred by rainfall into aquatic environment. [2–4]. PAH are slow to degrade in the environment, and sediments in particular are "sinks" where chemicals tend to concentrate [5, 9].

Their mutagenic and carcinogenic nature has been widely recognized by several researchers [2, 6, 7]. They are classified as carcinogenic because they are metabolized to dihydrodiols by hydrocarbon hydroxylase, which is present in the liver and these dihydrodiols and their epoxide derivatives bind to DNA and proteins and start mutagenic processes in the cells [11]. PAH can enter human beings via inhalation through lungs when they breathe air which is contaminated by PAHs from cigarette smoke, wood smoke, coal smoke and smoke from industrial sites [12, 13]. PAHs can also enter the human body through drinking water contaminated with PAHs, or swallowing food, soil or dust particles containing PAHs. These compounds can also enter the human body through skin contact. In the body, PAHs are stored in all

O.S. Fatoki (✉)
Cape Peninsula University of Technology, Cape Town, South Africa
e-mail: fatokiO@cput.ac.za

T.V. Ree and J. Nakhavhembe
University of Venda, South Africa

S. Rauch et al. (eds.), *Highway and Urban Environment*,
Alliance for Global Sustainability Bookseries 17,
DOI 10.1007/978-90-481-3043-6_23, © Springer Science+Business Media B.V. 2010

organs that contain fats, like kidneys and liver, where they changed into different substances that may even be more harmful than the original PAHs [13].

Due to the toxicity of PAHs, particularly their carcinogenic and mutagenic character, they are listed by Environmental Protection Agency (EPA) as priority pollutants because they have negative impact and also affect quality of life negatively [14]. Due to their environmental importance, PAHs have been studied extensively in mussels, surface runoff, sediment and soil [10, 15–17]. The increasing presence of these compounds necessitates their regular monitoring in the environment. In South Africa however, particularly in the study area, the status of PAHs is not clearly known because of paucity of data (these compounds are rarely studied) which might be due to analytical problem of recovery from complex matrices. Due to the complexity of the matrix and the very low concentrations they are found to be present in aqueous media, PAH determination in environmental samples is often a difficult task even after fractionation of sample extracts, therefore good chromatographic selectivity both in separation and detection are required during environmental analysis [2, 18]

This study investigates the levels of PAHs (pyrene, azulene, anthracene, dibenzothiophene and fluoranthene) in some rivers and surface runoffs in Venda, Limpopo Province of South Africa. This relatively poor region of South Africa is subjected periodically to environmental degradation because of poor water sully and infrastructure. Many people in this region are still using water directly from streams and rivers in the area for household purposes.

Materials and Method

The study area in this work was the Vhembe district in the Limpopo Province of South Africa (Fig. 1). Thohoyandou is the major town in the area and it is the capital city of the former Venda homeland.

Quality Assurance Studies

Validation of liquid-Liquid extraction procedure for analysis of PAHs in water samples was done by extracting spiked doubly distilled water with dichloromethane. As indicated in the tables below, 1 mL of 1,000 mg/L standard mixture, consisting of the six PAHs mixtures, were added to 500 mL of doubly distilled water. This was first pre-extracted thrice with 30 mL dichloromethane to serve as blank sample. Spiking of sediment samples was done by adding 1 mL of 1,000 mg/L PAH standard mixtures to 10 g of dried sediment (also pre-extracted with 30 mL of dichloromethane). The spiked water and sediment samples were then extracted and analyzed using the methods described below for water and sediment samples.

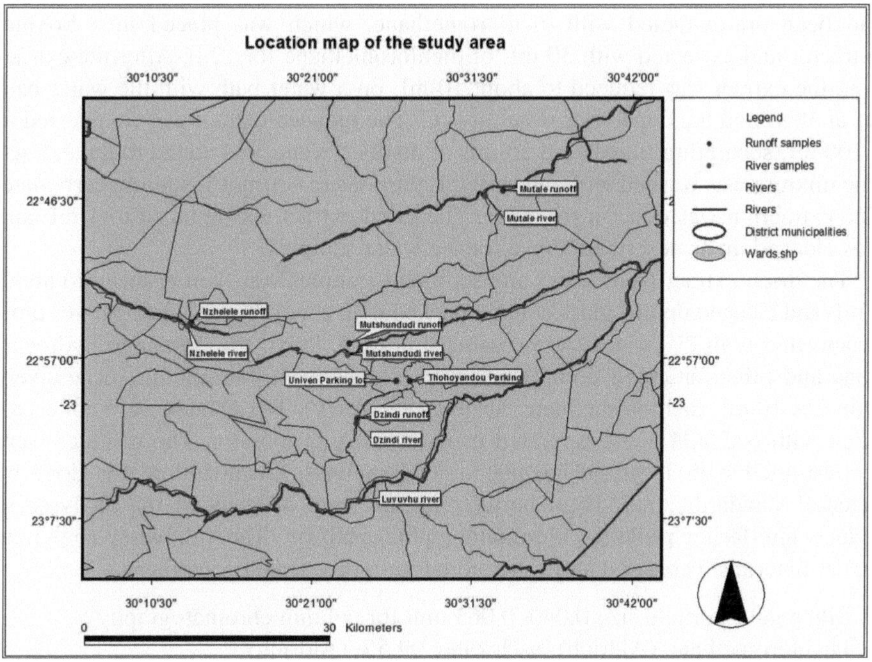

Fig. 1 Map of the sampling area

Sampling and Analysis of Water and Sediment Samples

Water samples were collected in 2.5 L bottle from rivers around Thohoyandou – Nzhelele, Luvuvhu, Mutale, Mutshundudi and Dzindi rivers (Fig. 1) between September 2005 and January 2006. Sediment samples were also collected from the rivers. Run-off samples and runoff sediments were collected at riverbanks at points where run-off enters the rivers, and at roadsides and parking lots around Thohoyandou Town. Water samples were stored at < 4°C and analyzed as soon as possible after collection. A 500 mL aliquot of water sample was saturated with 75 g of NaCl. The mixture was transferred into a separating funnel and extracted three times by shaking vigorously with 30 mL dichloromethane. The three extracts were pooled together and washed three times with 30 mL of 0.1 M sodium carbonate in a 100 mL separating funnel. The organic layer was separated and dried with anhydrous Na_2SO_4. The dried extract was reduced by evaporation and the extraction was done in triplicate.

Soxhlet extraction was used for the extraction of sediment samples since it is recommended by the US Environmental Protection Agency for the extraction of semi-volatile and non-volatile organics from solid matrices. Sediment samples were air dried, crushed with a mortar and pestle, and sieved through a 0.5 mm mesh sieve. 10 g of sediment samples were weighed into an extraction thimble, which had been pre-extracted with dichloromethane, which was placed in a Soxhlet extractor and extracted with 30 mL of dichloromethane for 12 h. After the extraction, the extract was reduced to about 10 mL on a water bath with the water bath set at 38°C and the condenser water at 4°C. The reduced extract was transferred to a 100 mL separating funnel and 20 mL of distilled water was added to the extract. The mixture was washed with three 10 mL portions of 0.1 mol/L sodium carbonate. The extraction was done in triplicate. The dried extract was reduced to 1 mL and was cleaned up as described above for the water samples.

The dried extract from water and sediment samples was then reduced to about 1 mL and cleaned up in a glass column packed with about 25 g of silica gel (40 μm), deactivated with 2% water (w/w), using n-hexane. Polycyclic aromatic hydrocarbons and other saturated compounds were eluted from the column successively with 2 × 10 mL dichloromethane: hexane (85:15 v/v). The eluates were collected, dried with Na2SO4 and evaporated using a rotary evaporator. The residues were reconstituted with 1 mL of hexane for GC analysis. Quantization was done by external standardization. High purity reagents were used in all the analyses to reduce interfering peaks in chromatographic analysis. The following analytical grade materials were used and were stored in tightly sealed containers:

- Silica gel - particle size 0.040–0.063 mm for column chromatography
- Dichloromethane (Aldrich); *n*-Hexane, 99.5% (Aldrich).

GC Conditions

Perkin Elmer Clarus 500 Gas Chromatograph, with FID detector and capillary column (Perkin Elmer SA (Pty). The gas chromatograph had an auto-injection, dual column system for both FID and ECD. The chromatograms were handled by a total Chrom Work Station programmed temperature with He as carrier gas. The operating conditions were: temperature programme; hold for 30 min. at 70°C, heat at 12°C/min and hold at 330°C; column diameter, 250 μm, and a column packing of ELHE; He carrier gas flow rate, 540 mL/min.

Results and Discussion

Results of the quality assurance studies are shown in Table 1. Due to the low solubility of PAHs in water and the subsequent low concentrations they are found in aqueous media, most the methods developed so far for their analysis include an extraction/concentration step, followed by elution with an appropriate solvent and chromatographic analysis [18]. Some of these methods include liquid-liquid extraction (LLE), solid phase extraction and micro-sold phase extraction for water samples and Soxhlet extraction (SE), microwave-assisted extraction (MAE), micro-solid phase extraction and ultrasonic extraction methods for solid samples [8, 19–22].

Table 1 Percentage recoveries of PAHs from spiked water and sediment samples

PAH standard	Recoveries (%) Water	Recoveries (%) Sediment	Retention time (min)
Indene	76	74	5.86
Azulene	67	61	9.53
Dibenzothiophene	74	72	14.23
Anthracene	102	77	14.56
Fluoranthene	68	60	17.79
Pyrene	95	90	18.62

LLE has been successfully applied to a wide range of environmental matrices, such as surface water, precipitation, sea water and waste water samples [19]. Determination of PAHs in marine samples of Siokolo Fishing Settlement was done using LLE [23] and recoveries ranged from 64.78%–91.94%.

SE is recommended by the US Environmental Protection Agency (USEPA) for the extraction of semi-volatile and non-volatile organics from solid matrices [8, 21, 22]. SE is a standard method and is preferred because it is easily standardized, with high recoveries when compared to other matrix dependent techniques like microwave extraction, pressurized liquid extraction and supercritical fluid extraction [19, 21]. PAHs in marine water samples of Siokolo Fishing Settlements were determined [23], and extraction of water sediment samples was based on SE and recoveries ranged from 90.58%–118%.

PAHs are commonly analyzed by gas chromatography coupled with either mass spectrometry (GC-MS) or flame ionization detector (GC-FID), or by liquid chromatography coupled with mass spectrometry (LC-MS) or UV-diode array detection (LC-UV-DAD) (19, 24). In principle, GC-FID is simple and more direct for the determination of PAHs compared to other methods, if a suitable column is used. The use of capillary column in GC as employed here has brought a high level of excellent reproducibility, high sensitivity and high resolution of PAHs on routine basis [8]. Thus the use of GC-FID was the method of choice in this study.

The results of recovery experiments by liquid-liquid (LLE) and Soxhlet extraction (SE) methods with determination by GC-FID, for water and sediment samples are shown in Table 1. The recoveries varied between 68%–102% for spiked water samples and from 60%–90% for spiked sediments. The results were adjudged satisfactory and the methods were therefore used for analysis of environmental water and sediment samples.

Concentration of PAHs in water samples varied between 0.1–53.5 (μg/L) in river water samples (Table 2) and between 22.8–9,870 (μg/kg) in river sediments (Table 3). PAHs levels in runoffs varied between 1.0–2,500 (μg/L) (Table 4) while levels in runoff sediment ranged from 72.5–34,000 (μg/kg) at the Univen parking lot (Table 5). This might be due to high levels of vehicles parked here.

Generally the levels of PAHs in sediments samples are consistently higher than those in water samples. The higher concentrations in sediments were expected because of possible accumulation of PAHs on sediments. The sediments could act

Table 2 PAH concentrations in river water samples (μg/L)

Rivers	Sampling date	Indene	Azulene	Dibenzo-thiophene	Anthr-acene	Fluor-anthene	Pyrene
Nzhelele	Aug 09, 2005	2.5	ND	1.9	6.7	4.2	4.6
Nzhelele	Aug 27, 2005	1.9	ND	3.3	28.6	24.2	0.1
Nzhelele	Sept 23, 2005	1.2	ND	2.2	8.2	8.3	11.1
Mutshundudi	Aug 09, 2005	0.8	ND	0.9	53.5	23.5	4.6
Mutshundudi	Aug 27, 2005	1.8	ND	7.4	16.7	19.9	ND
Mutshundudi	Sept 23, 2005	3.3	ND	2.0	13.7	6.8	52.4
Mutale	Aug 27, 2005	1.5	ND	0.3	32.2	19.9	0.5
Dzindi	Sept 24, 2005	1.1	ND	9.4	34.2	3.6	9.4
Levhuvhu	Sept 24, 2005	1.6	ND	6.7	11.3	7.6	11.0

Table 3 Concentrations of PAHs in river sediment samples (μg/kg)

Rivers	Sampling date	Indene	Azulene	Dibenzo-thiophene	Anthr-acene	Fluor-anthene	Pyrene
Nzhelele	Aug 09, 2005	67.8	ND	50.0	1,460	588	24.6
Nzhelele	Aug 27, 2005	27.4	ND	401	9,190	5,110	26.6
Nzhelele	Sept 23, 2005	161.3	ND	89.8	2,320	1,520	69.3
Mutshundudi	Aug 09, 2005	156.0	ND	463.0	7,050	505	122.0
Mutshundudi	Aug 27, 2005	81.1	ND	1,650	9,870	2,420	22.8
Mutshundudi	Sept 23, 2005	114	ND	1,980	2,200	7,560	941
Mutale	Aug 27, 2005	162.6	ND	38.1	1,170	4,360	100.4
Dzindi	Sept 24, 2005	53.4	ND	73.5	1,360	1,030	42.3
Levhuvhu	Sept 24, 2005	110	ND	17.9	1,210	3,150	1,490

Table 4 Concentrations of PAHs in runoff water samples (μg/L)

Rivers	Sampling date	Indene	Azulene	Dibenzo-thiophene	Anthr-acene	Fluor-anthene	Pyrene
Mutale	Dec 01, 2005	1.3	ND	1.0	37.7	78.5	10.3
Mutale	Dec 21, 2005	1.6	ND	3.3	195.0	81.3	25.2
Nzhelele	Dec 21, 2005	5.0	ND	31.5	17.1	3.4	940
Nzhelele	Jan 10, 2006	6.1	83.1	14.0	8.4	7.1	620
Mutshundudi	Dec 21, 2005	4.3	134	67.5	118.0	241.0	2,500
Mutshundudi	Jan 10, 2006	10.1	17.2	36.1	13.1	14.9	1,960
Dzindi	Jan 10, 2005	3.6	82.1	0.6	6.7	4.2	192.0
Thohoyandou	Dec 01, 2005	2.0	ND	7.1	81.5	10.8	10.2
Thohoyandou	Dec 19, 2006	1.9	10.5	10.5	230.0	251	7.2
University Parking lot	Dec 19, 2005	3.2	9.5	25.8	196.0	9.2	14.9

as "sink" for the PAHs as they bind on sediments [16, 24–26]. Runoff levels of PAHs were also consistently higher than the values found in waters from nearest rivers. The higher levels of PAHs in the runoffs than in river water samples showed the runoffs to be possible sources of PAHs in the receiving water bodies. Anthracene was the major PAH in both river and runoff water samples and in sediment samples.

Table 5 Concentrations of PAHs in runoff sediments (μg/kg)

Rivers	Sampling date	Indene	Azulene	Dibenzo-thiophene	Anthr-acene	Fluor-anthene	pyrene
Mutale	Dec 01, 2005	72.5	469	877	313	2,350	4,620
Mutale	Dec 21, 2005	106	558	298	322	8,490	530
Nzhelele	Dec 21, 2005	166	ND	118	–	419	636
Nzhelele	Jan 10, 2006	176	ND	140	1,730	414	2,610
Mutshundudi	Dec 21, 2005	96.4	ND	2,530	18,200	5,170	4,180
Mutshundudi	Jan 10, 2006	123	ND	31,500	14,500	1,500	9,890
Dzindi	Jan 10, 2005	159	ND	341	6,740	15,100	17,800
Thohoyandou	Dec 01, 2005	67.5	ND	1,370	112	7,010	617
Thohoyandou	Dec 19, 2006	179	ND	2,700	16,400	12,700	3,280
University Parking lot	Dec 19, 2005	124	ND	270	23,400	3,570	34,000

Probable sources of PAHs in the water and sediment samples might be contributions from traffic density coupled with the surface wash off of petroleum spillages from gas stations and auto garages. Other sources might be from indiscriminate refuse dumps and leakages from these dumps and from partially treated waste water effluents from the town's water treatment plants which drains into the rivers. PAHs which have anthropogenic origin are formed from either combustion of organic matter or as a discharge of petroleum related materials. Petroleum derived residues contain high levels of 2–3 ring PAH compounds, that is petrogenic, while high molecular weight PAHs.are formed in high temperature combustion processes [27]. High levels of pyrene and fluoranthene as shown relatively in this study were indicative of PAH pollution from incomplete combustion, pyrolysis of fuel and oil burning (i.e., they have pyrolytic origin) [28]. Most of the PAHs identified in this study are those usually found in environmental samples some of which are carcinogenic [29, 30].

Generally the values in this study were higher than those reported for some developed countries [31]. These elevated levels are probably due to indiscriminate disposal of untreated industrial waste water into the environment, washout of spills from auto garages, burning of fossil fuel and vegetation and exhausts from vehicles which is common practice in the study area [31]. However, the levels of PAHs found in the water samples are lower than the maximum allowable concentrations of 3 mg/L in water samples [13]. Thus the levels of PAHs in the water samples would not pose any serious health threat to downstream users of water from the rivers.

Conclusion

Liquid/liquid extraction (LLE) and Soxhlet extraction (SE) methods have been used to determine PAHs in water and sediment samples, respectfully. Levels in surface runoff were higher than those in river water samples and levels in sediment samples were higher than those in water samples. Generally levels of PAHs found in river water samples were lower than the maximum allowable concentration in water samples and therefore would not pose any health threat to downstream users of water from the rivers investigated.

References

1. Edwards NT (1983) Polycyclic aromatic hydrocarbons (PAHs) in terrestrial environment – a review, J. Environ. Qual. 12:427–440
2. Bruzzoniti MC, Sarzanini C. Mentasti E. (2000). Preconcentration of contaminants in water analysis – a review, J. Chromatogr. A 902:289–309.
3. Kolahgar B, Hoffman, C, Heiden, AC. (2002). Application of stir sorptive extraction to the determination of polycyclic aromatic hydrocarbons in aqueous samples, J Chromatogr. A: 225–230
4. Eirora AA, Blanco EV, Mahia PL, Lorenzo SM, Rodriguez DP, Fernandez EF (2000) Determination of polycyclic hydrocarbons in a complex mixture by second-derivative constant energy synchronous spectrofluorimetry, Talanta, 51, 677
5. Environment Canada (2001) Toxic chemicals in Atlantic Canada – polynuclear aromatic hydrocarbons (PAHs), http://www.ns.ec.ca/epb/envfacts/pah.html
6. Oanh NTK, Reutergardh LB, Dung NT (1999) Emission of polycyclic aromatic hydrocarbons and particulate matter from domestic combustion of selected fuels, Environ. Sci. Technol. 33:2703–2709
7. Ghauch A, Rima J, Fachinger C, Suptil J, Martin-Bouyer M (2000) Room temperature phosphorescence analyses of PAHs using an imaging sensing system combined with bifurcated optical fibre and a cooled charge coupled device detector, Talanta 51:807–816
8. Bjorseth A (1977) Analysis of polycyclic aromatic hydrocarbons in particulate matter by glass capillary gas chromatography, Anal. Chim. Acta 94:21–27
9. Ging PB, VnaMetre PC, Callender E (2001) Bottom sediments of Lorence Creek Lake, San Antonio, Texas reflect contaminant trends in an urbanizing watershed. http://water.usgs.gov/pubs/FS/Fs14999
10. Butler AC, Sibbald RR (1986) Isolation and gas chromatographic determination of saturated and polycyclic aromatic hydrocarbons mussels, Bull. Environ. Toxicol. 37:570–578
11. Liguori L, Heggstad K, Hove TH, Julshamn K (2006) An automated extraction for isolation twenty four polyaromatic hydrocarbons (PAHs) from various marine matrixes, Anal. Chim. Acta 573–574:181–188
12. Deutsch-Wenzel RP, Brune H, Grimmer G, Dettbarn G, Misfeld J (1983) Experimental studies in rat lungs on the carcinogenicity and dose-response relationships of eight frequently occurring environmental polycyclic aromatic hydrocarbons, J. Natl. Cancer Inst. 71:539–544
13. Agency for Toxic Substances and Disease Registry, ATSDR: 2006 Toxicological profiles CD-ROM, ATSDR, Washington, USA
14. Salinas-Castilo A, Fernandez-Sanchez JF, Segura-Carreterro A, Fernandez-Guiterrez A (2005) Solid-surface phosphorescence characterization of polycyclic aromatic hydrocarbons and selective determination of benzo[a]pyrene in water samples, Anal. Chim. Acta 550:53–60
15. Wust W, Kern U, Herrmann R (1992). Street washoff behaviour of heavy metals, poly-aromatic hydrocarbons and nitrophenols, in the Fourth International Symposium Highway Pollution Proceedings, Middlesex Polytechnic, UK, 265 pp
16. Wilcocock RJ, Corban GA, Nothcott GL, Wilkins AL, Langdon AG (1996) Persistence of polycyclic aromatic hydrocarbons of different molecular size and water solubility in surficial sediment of an intertidal sandflat, Environ. Toxicol. Chem. 15:670–676
17. Conte P, Zena A, Pilidis G, Piccolo A (2001) Increased retention of polycyclic hydrocarbons in soils induced by soil treatment with humic substances, Environ. Pollut. 112:27–31

18. Algara M, Radin C, Fornier de Violet Ph, Lamotte M, Garrigues Ph, Hardy M, Gillard R (2000) Direct fluorometric analysis of PAHs in water and in urine following liquid solid extraction, J. Fluoresc. 10:366–359
19. Busetti F, Heitz A, Cuomo M, Badoer S, Traverso P (2006) Determination of sixteen polycyclic aromatic hydrocarbons in aqueous and solid samples from an Italian waste water treatment plant, J. Chromatogr. 1102:104–115
20. Rezaee M, Assadi Y, Hossein MM, Aghee E, Ahmadi M, Berijini S (2006) Determination of organic compounds in water using dispersive liquid-liquid microextraction, J Chromatogr. 1116:1–9
21. Banjoo DR, Nelson PK (2005) Improved ultrasonic procedure for the determination of polycyclic aromatic hydrocarbons in sediments, J Chromatogr. A 1066:661–668
22. Helaleh MIH, Ali-Omair A, Nisar A, Gevao B (2005) Validation of various extraction techniques for quantitative analysis of polycyclic aromatic hydrocarbons in sewage sludge using gas chromatography-ion trap mass spectrometry, J. Chromatogr. 1083:153–160
23. Anyakora C, Ogbeje A, Palmer P, Coker H (2005) Determination of polynuclear aromatic hydrocarbons in marine samples of Siokolo Fishing Settlement, J. Chromatogr. 1073:323–330
24. Maldonado C, Bayona J, Bodineau L (1999) Sources, distribution and water column process of aliphatic and polycyclic aromatic hydrocarbons in northwestern Black Sea water, Environ. Sci. Technol. 33:2693–2702
25. Kim GB, Maruya KA, Lee RF, Lee JH, Koh CH, Tanabe S (199)Distribution and sources of polycyclic aromatic hydrocarbons in sediments from Kyeonggi Bay, Korea, Mar. Pollut. Bull. 38:7–15
26. Southwood GR (1979) The environmental load and cycle of polycyclic aromatic hydrocarbons, Sci. Total Environ. 6:239–250
27. Xu J, Yu Y, Wang P, Guo W, Dai S, Sun H (2007) Polycyclic aromatic hydrocarbons in surface sediment from Yellow River, China. Chemosphere 67:1408–1414
28. Feng J, Guo Z, Chan KC, Fang M (2007) Properties of organic matter – $PM_{2.5}$ at Changdao Island, China – a rural site in the transport path of the Asian continental outflow, Atmos. Environ. 41:1924–1936
29. Tate CH, Env D, Arnos KF (1990) Water Quality and Treatment a Handbook of Community Water Supplies, McGraw Hill, New York, 4th edition, 106 pp
30. Bjorseth A, Knutzen J, Skei J (1979) Determination of polycyclichydrocarbons in sediments and mussels from Saudafjord, W. Norway by glass capillary gas chromatography, Sci. Total Environ 13:71–86
31. Ogunfowokan AO, Asubiojo OI, Fatoki OS (2003) Isolation and determination of polycyclic aromatic hydrocarbons in surface runoff and sediments, Water, Air Soil Pollut. 147:245–261

Use of Absorptive Media to Monitor Stormwater Contamination in Small Urban Drains

Stephen Marshall, Vincent Pettigrove, Matthew Potter, and Tony Barrett

Introduction

Urban streams are frequently contaminated with heavy metals and petroleum hydrocarbons [1]. Industrial catchments in Melbourne contribute disproportionately high concentrations of these pollutants to stormwater [2]. Major sources of stormwater pollution from industrial areas include building materials, paved surfaces and/or poor work practices. Councils in the Melbourne region are responsible for ensuring that appropriate premises and work practices are used by industries within their region. One of these councils, the City of Kingston, implemented a stormwater Education and Enforcement (E&E) program to improve the quality of runoff from industrial estates in their region. This program involved auditing stormwater management practices in industrial catchments, together with community education and enforcement where necessary, to improve any poor practices.

The City of Kingston wanted to know whether their program improved the quality of runoff from their industrial areas. Practical evaluation of an E&E program requires estimation of prevailing runoff water quality pre-and post-program. However, it is difficult to reliably measure the quality of runoff water from small (<100 ha) urban sub-catchments [3]. Continuous water quality samplers may provide a good indication of pollutant concentrations given sufficient time and adequate flow [4]. However, for surveying more than a few sub-catchments, the installation and maintenance of automatic sampling equipment is prohibitively expensive. Sediment quality assessment is an alternative approach for identifying

S. Marshall
Centre for Environmental Stress and Adaptation Research (CESAR), The University of Melbourne, ParkvilleVictoria, 3010

V. Pettigrove and M. Potter
Research and Technology, Melbourne Water Corporation, PO Box 4342 Melbourne, Victoria 3001

T. Barrett
City of Kingston, PO Box 1000 Mentone, Victoria 3194

S. Rauch et al. (eds.), *Highway and Urban Environment*, Alliance for Global Sustainability Bookseries 17,
DOI 10.1007/978-90-481-3043-6_24, © Springer Science+Business Media B.V. 2010

major pollution sources [8], but this is not practical in concrete-lined channels, designed to minimize sediment accumulation.

We developed and trialed an alternative passive sampling method based on solid-media. The method uses polythene mesh bags of granulated activated carbon (GAC) media to absorb pollutants present in industrial runoff. We applied this method to determine the effectiveness of the City of Kingston's E&E program in reducing the amount of pollution discharged from industries into receiving waters.

Method

This study was conducted at in industrial areas within the City of Kingston, 25 km south-east of Melbourne, Australia. Sub-catchments with clearly defined drainage to a single outfall were initially selected from a drainage plan. Sub-catchments containing land zoned residential, traversed by major roads, or with significant areas currently under construction were excluded. Visual inspection excluded a further two drains which were either submerged or could not be accessed safely, leaving 20 sub-catchments for assessment, each draining to a single outfall.

Polyethylene mesh bags measuring 400 mm × 300 mm edged with 40 mm hook-and-loop fastening strips were filled with 500 mL of granular activated carbon (GAC). Polyvinyl chloride (PVC) mounting plates fitted with 40 mm hook-and-loop strips were fastened to drain surfaces with construction adhesive. Bags of media were then secured to PVC mounting plates with hook-and-loop strips either within or immediately below each drain. After exposure for 1 week (including five business days), bags were sampled and replaced with fresh media. Concentrations of accumulated hydrocarbons (C6-C36) and heavy metals (As, Cd, Cr, Cu, Fe, Ni, Pb, Zn) were determined by Ecowise Environmental (Scoresby, Victoria) according to methods WSL-030 and WSL-032 respectively.

Daily rainfall records were provided by Melbourne Water for the Clayton Retarding Basin, Braeside Park, and Mentone gauges. Weekly rainfall was estimated as mean 24 h totals for the three gauges, summed for the week preceding collection. Weekly rain frequency was calculated as the sum of days on which mean rainfall exceeded 0.2 mm.

Where concentrations were below detection limits (DL), values were imputed by Helsel's robust method [5], using UnCensor v5.1 [6]. Data were $\log_{10}$ transformed prior to parametric statistical tests. After three sampling collections, principal component analysis was conducted on heavy metal and TPH media concentrations to identify redundant parameters. Principal components were extracted from a correlation matrix, retained where eigenvalues were over 1. Parameters highly correlated with Cu (Fe, Ni, Cr) or consistently below detection limits (As, Cd, Pb) were not determined after the third week.

Sub-catchments were paired by two-step cluster analysis of catchment characteristics and pollutant concentrations collected during the first sampling stage only.

Clustering used a log-likelihood distance measure, with number of clusters set at ten to divide sub-catchments into matched pairs. Sub-catchment pairs were then randomly assigned to control and treatment groups. To identify homogeneous groups of sub-catchments, a second two-step cluster analysis was performed with number of clusters determined by Schwarz's Bayesian Information Criterion. Differences in mean pollutant concentrations between treatment and control sub-catchments were evaluated by paired *t*-tests, with pairs matched by cluster and collection date. Pollutant correlations with catchment characteristics, rainfall amount and frequency were evaluated by Spearman's rank test. All statistical tests were performed in SPSS v15 [7].

Results

Of 69 samples analyzed after 3 weeks monitoring, 100% of As, 66% of Cd and 88% of Pb concentrations were below analytical detection limits, therefore these metals were excluded from subsequent data analysis. Principal component analysis extracted three factors representing 88% of variation in the remaining six parameters: The first factor correlated with Fe, Ni, Cr, and Cu, the second factor correlated with Zn, and the third factor correlated strongly with TPH concentrations. Therefore, parameters other than Cu, Zn and TPH were redundant and could be dropped from analysis without losing substantial information.

Prior to the commencement of the E&E program, 150 GAC samples were collected from 20 sub-catchments over approximately 8 weeks. Bag recovery was 93.8%. High water caused by heavy rain prevented collection from four sub-catchments on Nov 8, and six sub catchments on Nov 15. Heatherton drain outfalls were submerged prior to collection on Nov 8, although accumulated pollutant concentrations at these four sub-catchments were similar to previous weeks.

After the E&E program finished, 114 GAC samples were collected from 20 sub-catchments over approximately 6 weeks. Bag recovery was 95%. Sample bags were missing from sub-catchment H1 on Oct 24, and H4 on Nov 11 and Dec 6. Heavy rain in the week preceding Dec 6 also caused the loss of bags from sub-catchments D11, D4, and S1.

Cluster analysis, using the data collected prior to the commencement of the E&E program, separated the 20 sub-catchments into ten control-treatment pairs, summarized in Table 1. Four homogeneous sub-groups were separated, corresponding roughly to the spatial location of sub-catchments. Settlement and Heatherton groups corresponded with sub-catchments draining into the Settlement and Heatherton drains. Sub-catchments serviced by the Dunlops drain were separated into two groups, Dunlops A and Dunlops B. Sub-catchments in Dunlops A were generally larger than Dunlops B, with higher levels of hydrocarbon and lower levels of Zn pollution (Table 1).

As both Dunlops A and Settlement groups had an odd number of members, sub-catchment D6 was included in the Settlement group for between group comparisons.

Table 1 Sub-catchment pairs (C = control; T = treatment) and homogeneous groups determined by cluster analysis. Catchment characteristics and pollutant concentrations are means for groups of sub-catchments. Pollutant concentrations refer to the initial survey period prior to the commencement of the E&E program only

Sub-catchment	Exp Group	Pair	Catchment group	Sub-catchment age (years)	Number of properties	Cu	Zn	TPH
H3B	C	7	Heatherton	50	21.7	11.8	103	414
H2	T	7						
H1	C	8						
H4	T	8						
D6	C	6	Settlement	29	13.4	13.9	179	508
S5	T	6						
S3	C	9						
S1	T	9						
S2	C	10						
S6	T	10						
D3	C	5	Dunlops A	17	29.4	18.2	113	2,080
D11	T	5						
D9	C	1						
D4	T	1						
D1	C	2	Dunlops B	18	17.5	17.2	165	563
D2B	T	2						
D8	C	3						
D7	T	3						
DE1	C	4						
D5	T	4						

Trends in rainfall and pollutant accumulation over time are illustrated in Fig. 1. Rainfall was more frequent in 2007 than 2006. Although mean rainfall was higher in 2006, this was dominated by the week ending Nov 6, when 19.7 mm of rain fell in the preceding week. The lowest rainfall recorded in 2007 was 0.33 mm in the week ending Dec 12, whereas 3 of 8 weeks in 2006 recorded rainfall of 0.2 mm or less.

Pollutant concentrations were generally higher in 2007 than in 2006 (Fig. 1). Cu, Zn and TPH were positively correlated with rainfall frequency (Spearman's rho = 0.14, $p = 0.019$; rho = 0.39, $p < 0.001$; and rho = 0.16, $p = 0.01$ respectively). Correlations were similar whether rainfall quantity or frequency was considered, hence only results for frequency are presented. Zn was more strongly associated with rainfall than either Cu or TPH.

Mean concentrations of Cu, Zn and TPH accumulated by media in the surveys before and after the commencement of the E&E program are summarized in Table 2. Concentrations of pollutants accumulated were in the order: Cu < Zn < TPH. Zn concentrations ranged from 13 to 800 mg/kg, while Cu concentrations ranged from < 5 to 130 mg/kg.

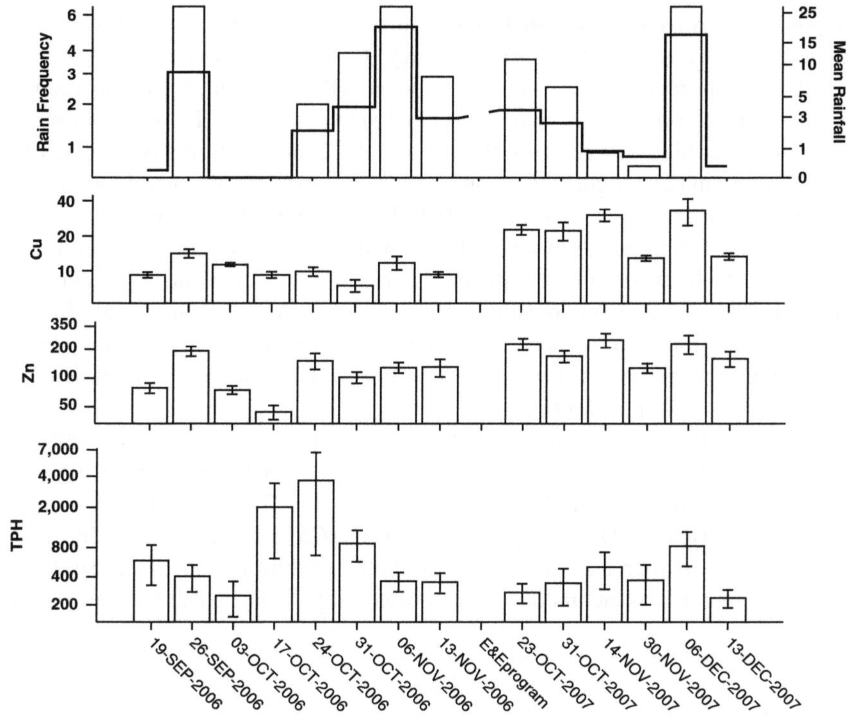

Fig. 1 Trends in rainfall and accumulation of pollutants (in mg/kg) by GAC before and after delivery of the E&E program. Cu, Zn and TPH bars represent mean concentrations ± one standard error. Rain frequency (*bars*) represents the number of days mean rainfall exceeded 2 mm across the sampling area, while mean rainfall (*line*) is the sum of daily rainfall totals in mm for the period sample bags were exposed

Table 2 Summary of Cu, Zn and TPH concentrations accumulated by solid media. All concentrations are in mg/kg

Variable	N	Median	Mean	Std error	Minimum	Maximum	<DL (%)
Cu	264	11.0	15.3	0.8	<5	130	2.7
Zn	264	110.0	147.1	7.7	13.0	800	0
C6-C9	264	13.3	64.0	13.9	<20	2,800	60.2
C10-C14	264	45.0	200.3	80.1	<20	19,000	23.9
C15-C28	264	64.5	479.5	173.2	<50	41,000	43.2
29-C36	264	24.9	66.6	8.0	<50	1,300	77.7
TPH	264	212.6	810.4	254.0	<140	60,135	15.6*

*Proportion of samples where all hydrocarbon fractions were below their respective detection limits. Hence, at least one hydrocarbon fraction was above detection limits in 84.4% of samples.

Hydrocarbons were predominately medium-weight (C10–C14; C15–C28), with occasional spikes in light (C6–C9) and heavy (C29–C36) fractions.

Hydrocarbon fractions were co-correlated, so TPH (the sum of all four fractions) was generally a good summary of individual size fractions. TPH distribution was skewed towards high values; with the mean TPH concentration (810 mg/kg) four times the median (212 mg/kg). Cu concentrations correlated negatively with catchment age (Spearmans $\rho = -0.12$, $p = 0.009$, before E&E program only).

The effect of the education and enforcement (E&E) program was obscured by large variations between sub-catchment groups. Mean Cu, Zn and TPH concentrations grouped by catchment are summarized in Fig. 2. A downward trend in Zn and TPH concentrations was observed in treatment sub-catchments in three groups (Fig. 2). Cu concentrations were significantly lower in Dunlops A treatment sub-catchments ($t_9 = 5.4$, $p < 0.001$).

Zn concentrations were significantly lower in Dunlops A ($t_9 = 3.4$, $p = 0.008$), and marginally significantly lower in Heatherton ($t_8 = 2.3$, $p = 0.055$) treatment sub-catchments. TPH was significantly lower in Heatherton ($t_8 = 2.9$, $p = 0.021$), Dunlops A ($t_9 = 2.3$, $p = 0.046$) and Dunlops B ($t_{17} = 2.5$, $p = 0.025$) treatment sub-catchments.

In contrast, an increase in media pollutant concentrations in the Settlement group of treatment sub-catchments was identified (Fig. 2). Zn and TPH concentrations

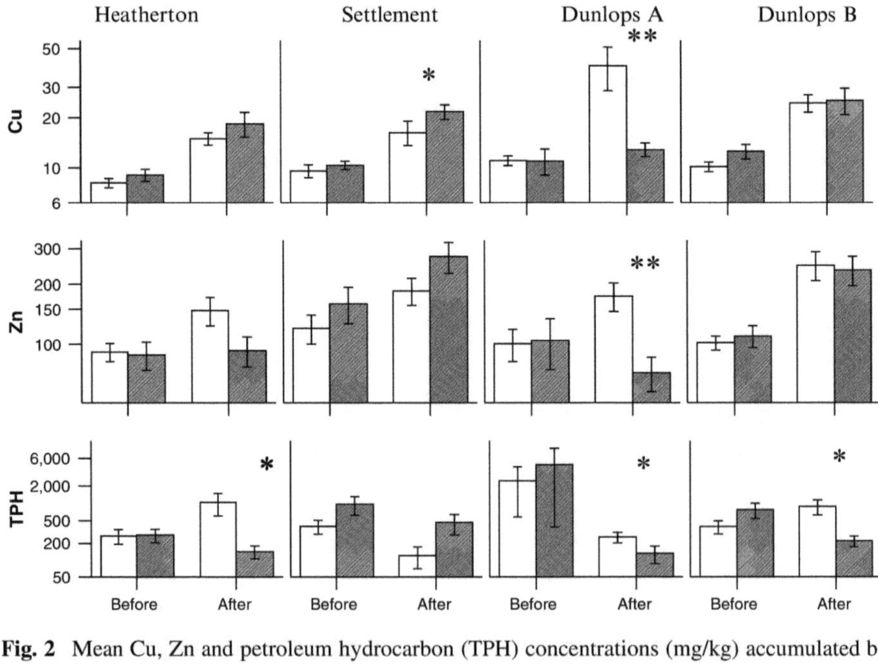

Fig. 2 Mean Cu, Zn and petroleum hydrocarbon (TPH) concentrations (mg/kg) accumulated by GAC media grouped by sub-catchment. *Bars* represent mean ± 1 standard error. Control sub-catchments in *white*, treatment sub-catchments shaded. Difference between groups significant at p = 0.05*, p = 0.01**

were higher and Cu concentrations significantly higher ($t_{16} = -3.0$, $p = 0.009$) in treatment sub-catchments after the E&E program.

Discussion

No reduction in accumulated pollutants was apparent after immersion by storm flows at the Heatherton sub-catchments on Nov 1 and 15. Although further testing is required for performance validation, this observation suggests monitoring during moderate rainfall is feasible.

The association between Cu and catchment age suggests an indirect relationship with vehicle use. Cu pollution in urban catchments has been linked to vehicle use [8], hence higher Cu concentrations in new sub-catchments may be related to the prevalence of larger property parcels with bigger car parks.

Increased Cu concentrations after the E&E program in the Settlement treatment group of sub-catchments was influenced by high Cu concentrations at sub-catchment S6. Stormwater pipes draining areas outside this sub-catchment were discovered after sampling had commenced, casting doubt on the boundary integrity of this sub-catchment. If properties outside the E&E program delivery boundary contributed stormwater to this outfall, Cu concentrations at S6 may not reflect the E&E program's performance in this sub-catchment.

The positive association with rainfall and even distribution of Zn concentrations between sub-catchments suggests the major inputs of this metal were structural rather than point-source. Galvanized roofing, atmospheric deposition and road surface materials have all been implicated as sources of Zn in urban runoff [9, 10]. Therefore, more frequent rainfall was a likely factor contributing to increased pollutant concentrations in 2007, particularly Zn.

Although hydrocarbon pollution has been linked with catchment features such as roads and parking lots [11], diffuse sources cannot reasonably account for the degree of hydrocarbon pollution observed in this study. TPH pollution was episodic, with low base loads punctuated by occasional pollution events. This observation is consistent with poor stormwater management such as machinery wash-down to stormwater, inadequate work shop waste storage, or even deliberate dumping of waste oil. The E&E program was specifically designed to address stormwater management practices. The reduction in TPH concentrations in most treatment sub-catchments, particularly those with the highest TPH loads, suggests it was at least partly effective.

The reduction in TPH compared to Zn and Cu suggests that changes in property-level stormwater management had a greater impact on pollution by hydrocarbons than heavy metals. The reduction in Zn concentrations in Heatherton and Dunlops A treatment catchments suggest that changing stormwater management practices can reduce heavy metal pollution, although not to the same extent as hydrocarbons. The association of Zn with rainfall is consistent with evidence that structural features such as galvanized iron roofing can contribute significant quantities of Zn to stormwater [9]. At the time of follow-up sampling not all E&E program recommendations to businesses had been implemented due to significant construction and planning costs (e.g., construction of bunding around a wash-off area), hence a reduction in heavy metal pollution due to these measures may only become apparent in future surveys.

Limiting the range of pollutants analyzed to Zn, Cu and TPH reduced chemical analysis costs considerably, while maintaining the ability to discriminate between relatively clean and heavily polluted sub catchments. Further testing with alternative media is recommended to refine the sampling method, as targeted media could provide sensitive detection of a wider range of pollutants [12].

Conclusion

Passive sampling with solid media was an affordable alternative to traditional stormwater monitoring approaches. Our results suggest the E&E program was more effective in reducing stormwater pollution by hydrocarbons than by heavy metals. Significant differences in accumulated pollutant concentrations between treatment and control sub-catchments demonstrate changing community attitudes to stormwater management can have an immediate effect on the quality of urban runoff.

References

1. Makepeace DK, Smith DW, Stanley SJ (1995) Urban stormwater quality – summary of contaminant data. Critical Reviews in Environmental Science and Technology 25:93–139
2. Davis AP, Shokouhian M, Ni SB (2001) Loading estimates of lead, copper, cadmium, and zinc in urban runoff from specific sources. Chemosphere 44:997–1009
3. Lee H, Stenstrom MK (2005) Utility of stormwater monitoring. Water Environment Research 77:219–228
4. Thomson NR, McBean EA, Snodgrass W, Mostrenko I (1997) Sample size needs for characterizing pollutant concentrations in highway runoff. Journal of Environmental Engineering-ASCE 123:1061–1065
5. Helsel DR (1990) Less than obvious – statistical treatment of data below the detection limit. Environmental Science and Technology 24:1766–1774
6. Newman DMC (2007) UnCensor v5.1. University of Georgia, Savannah River Ecology Laboratory, P.O. Drawer E Aiken, S.C. 29801
7. SPSS (2006) SPSS for Windows, release 15.0.0. 233 S. Wacker Drive, 11th floor Chicago, Illinois 60606
8. Davis AP, Shokouhian M, Ni SB (2001) Loading estimates of lead, copper, cadmium, and zinc in urban runoff from specific sources. Chemosphere 44:997–1009
9. Brown JN, Peake BM (2006) Sources of heavy metals and polycyclic aromatic hydrocarbons in urban stormwater runoff. Science of the Total Environment 359:145–155
10. Pitt R, Lalor M (2000) The Role of Pollution Prevention in Stormwater Management, Monograph 9 in: Models and Applications to Urban Water Systems. James W (ed), Geulph, Ontario. pp. 1–20
11. Scoggins M, McClintock NL, Gosselink L, Bryer P (2007) Occurrence of polycyclic aromatic hydrocarbons below coal-tar-sealed parking lots and effects on stream benthic macroinvertebrate communities. Journal of the North American Benthological Society 26:694–707
12. Madrid Y, Camara C (1997) Biological substrates for metal preconcentration and speciation. TrAC Trends in Analytical Chemistry 16:36–44

Effects of Atmospheric Salt Deposition on Highway Runoff Characteristics – A Pilot Case Study

Pedro Baila Antunes and Ana Estela Barbosa

Introduction

Monitoring studies of road runoff, from two different sites (highways A25 and IP6), located on coastal areas of Portugal [6–8] detected high levels of salinity and Carbon Oxygen Demand (COD). Table 1 compares results for these two sites, displaying interesting similarities between the parameters connected to salinity and COD. On the other hand, pollutants typical for highway runoff characterization, such as total suspended solids (TSS) and the heavy metals zinc (Zn), copper (Cu) and lead (Pb) clearly indicate different levels of traffic pollution. [5, 7] proposed that such evidences should be related to the vicinity of the Atlantic Ocean, since the non-coastal Portuguese roads did not show this same pattern.

Multivariate exploratory statistical techniques, applied to a set of coastal and non-coastal Portuguese roads indicated different patterns for the two groups, in terms of salinity, conductivity and chlorides, as expected, but also of COD concentrations [7].

According to [16] the marine aerosol can be transported to large distances. In areas with high salinity and where evaporation is systematically superior to the precipitation, the salts crystallization at the top surface of pavement materials tend to provoke a significant degradation of the road pavement, with structural consequences relevant to road engineering[17]. The bituminous material is lifted and its adhesion to the road base is destroyed.

In coastal areas, under climate conditions such as the Portuguese ones, maritime salts transported by the atmosphere are deposited and may remain in the road pavement and crystallize during dry periods between rainfall events, eventually, leading

P.B. Antunes (✉)
Environment Department, Higher School of Technology of the Higher Institute for Technology of Viseu, Campus Politécnico, 3504-510 Viseu, Portugal
e-mail: baila@estv.ipv.pt

A.E. Barbosa
Hydraulics and Environment Department, LNEC – Portuguese National Laboratory for Civil Engineering, Av. do Brasil 101, 1700-066, Lisboa, Portugal

S. Rauch et al. (eds.), *Highway and Urban Environment*,
Alliance for Global Sustainability Bookseries 17,
DOI 10.1007/978-90-481-3043-6_25, © Springer Science+Business Media B.V. 2010

Table 1 Data comparison between IP6 and A25 pollutants (adapted from [5, 7])

Parameter	Average concentration (mg/L)		SMC (mg/L)		Pollutants loads (kg/ha year)	
	IP6	A25	IP6	A25	IP6	A25
Conductivity (μS/cm)	384.3	305.1	–	–	–	–
Salinity	183.0	161.0	175.2	266.0	1580.6	2287.2
Chlorides	35.0	51.3	36.5	48.2	329.0	460.8
TSS	235.6	49.3	224.7	44.7	2027.1	427.0
COD	173.2	84.1	195.9	83.0	1766.8	793.2
Zn	0.077	0.194	0.08	0.205	0.69	2.0
Cu	0.032	0.0242	0.03	0.014	0.29	0.13
Pb	0.010	0.0032	0.01	0.005	0.08	0.05

to a certain level of degradation and extraction of the superficial layer of the pavement. Such phenomenon may increase the road pavement potential for environmental pollution, for example contributing to increase the refractory organic matter levels.

To provide a consistent answer to the question: "Which are the particular characteristics of road runoff in coastal areas?" and to establish a conceptual model for coastal roads pollutants are two of the tasks of the project *Guidelines for Integrated Road Runoff Pollution Management in Portugal* (Project G-Terra), a 3 years study that started in January 2008, and is funded by the Portuguese Science and Technology Foundation. Given the relevance of this issue and the novelty of it, it will be the subject of a Ph.D. thesis to be developed within the G-Terra project.

The diagram presented in Fig. 1 describes a first approach to the conceptual processes and the leading variables that could be more relevant in controlling the phenomena under study. The analysis of G-Terra case studies monitoring results will permit to select the most relevant variables and processes, and establish mathematical/empirical relationships between them.

Among the methodologies that will be used for data analysis, there are simple and advanced statistical analyses.

Possible correlations among road runoff quality parameters namely, salinity, chlorides, TSS, CQO, Total Organic Carbon (TOC), total hydrocarbons and oil and grease and independent variables will be studied. The independent variables belong to three different groups: (i) rainfall event characteristics; (ii) air masses transport and salt deposition, and (iii) road site characteristics.

Within the G-Terra 5 case studies, the A25 site monitored by [5] in 2003 and 2004, was chosen as the pilot case study, concerning the effects of atmospheric salt deposition on highway runoff characteristics. Four monitoring periods are planned at the A25 site during 2008 and 2009. It is expected to gather data from around 40 different rainfall events, under various climacteric circumstances, and approximately 300 runoff samples.

Table 2 reports the principal characteristics of the studied site A25 and Fig. 2 shows the location of the area in Portugal, and satellite view of the site. Table 3

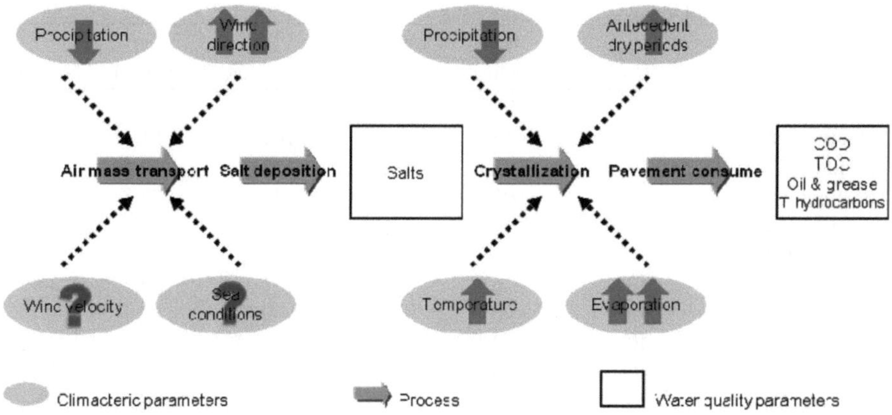

Fig. 1 The conceptual processes and the leading variables

Table 2 Main characteristics of the highway site studied [5]

Location	Next to Aveiro city (75,000 inhabitants)
Distance to the Atlantic Ocean	5.6 km
Monitored catchment area	250 m^2 (totally impervious)
Annual average daily Traffic	27,000 vehicles (2003) 2,935 vehicles (May, 2008)
Other relevant characteristics	The highway crosses a few meters above a coastal lagoon (ecological sensitive), with high salinity

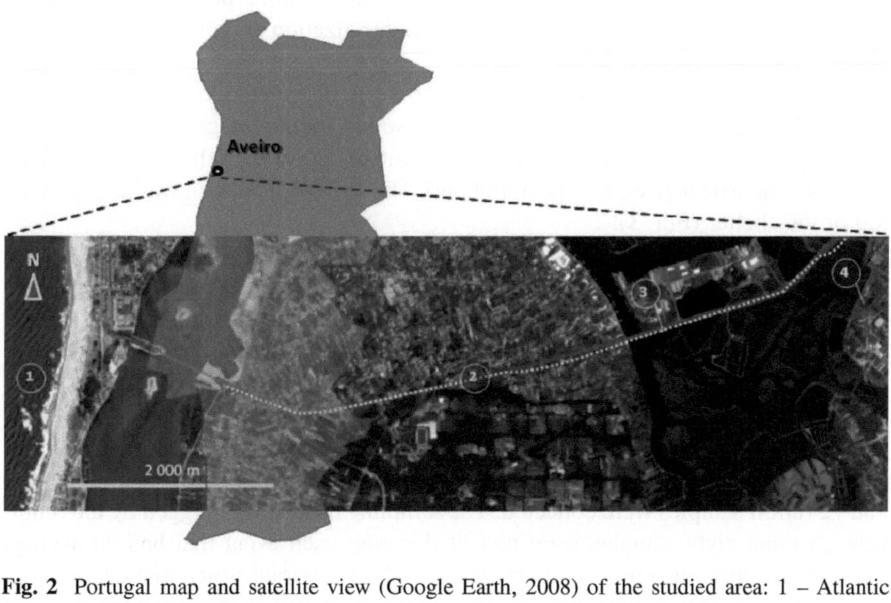

Fig. 2 Portugal map and satellite view (Google Earth, 2008) of the studied area: 1 – Atlantic Ocean; 2 – Highway A25; 3 – Monitoring site location; 4 – Meteorological Station

Table 3 Results from the 2003/2004 monitoring study at the A25 site [5]

Parameter	No. samples	Range of values Min.	Max.	Average	Median	St. Dev.
Conductivity (μS/cm)	40	67.3	1,790.0	305.1	156.5	354.4
Salinity (mg/L)	40	36	950	161	81	188
Turbidity (FNU)	40	1.2	152.7	55.9	47.2	39.4
TSS (mg/L)	40	<5.0	295.0	49.3	33.8	52.6
Total hardness (mg $CaCO_3$/L)	40	28.3	181.8	91.0	91.4	48.9
Chlorides (mg/L)	40	6.6	139.5	51.3	52.3	38.8
NH_4^+ (mg/L)	40	<0.1	1.2	<0.1	<0.1	–
Total P (mg/L)	40	<0.16	0.72	0.25	0.23	0.15
COD (mg/L)	40	<0.3	170.0	84.1	84.2	53.1
BOD_5 (mg/L)	25	<2	66	25	12	13

contains results of the previous characterization of A25 highway runoff that will be considered in the present study, as well.

Methodologies

In February 2008 the field monitoring system was installed at the site, in a shelter. It consisted of a rain-gauge, a flowmeter (ISCO 730 Bubbler flow module), a bubbler flow and a Thel-Mar volumetric weir (fit to a 400 mm pipe), and an automatic water sampler (ISCO 6700) working in synchronization. Figure 3 presents such system, the same as used by [5].

An innovation for the G-Terra studies is the use of a methodology based on [7], named the wet candle device, selected as a sound methodology to determine the atmospheric chloride deposition rate (amount of chlorides salts deposited from the atmosphere on a given area per unit time). The wet candle device was placed at the top of the shelter (Fig. 3).

The Meteorological Station of Aveiro University is located approximately 2 km to the East from the monitoring site, as can be seen in Fig. 2. A second wet candle device will be placed there and meteorological and atmospheric data will be collected and analyzed. This station continuously records meteorological and atmospheric data, useful to explain the air mass transport and salt deposition. The saline aerosol production is induced by sea conditions. Data concerning wave height, direction and period, will be gathered in the research.

Between February and April 2008, ten different rainfall events were monitored and 72 runoff samples were collected. The sampling routine was trigged by the runoff flow. Around eight samples were collected during each event that had an average duration of approximately two hours. The samples were transported to the laboratory as soon as possible, and the preparation, conservation and analysis procedures were

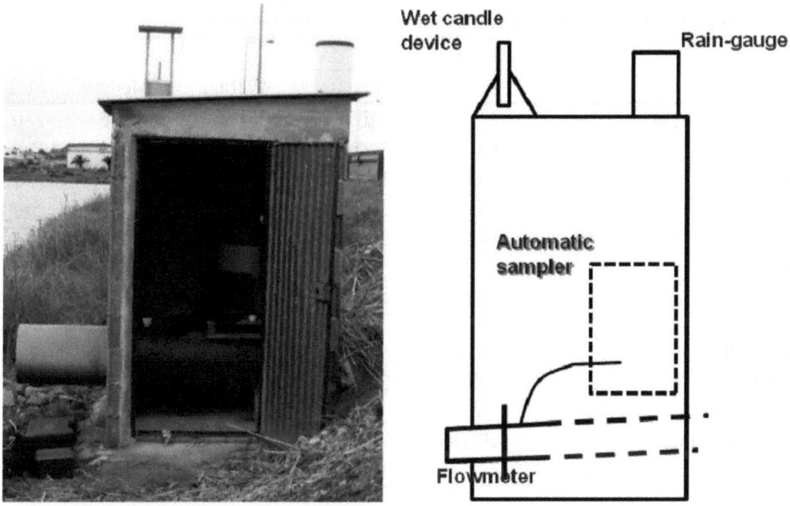

Fig. 3 Photo and scheme of monitoring system at the A25 site

followed in accordance with international standards procedures [4], based on an adequate Quality Assurance and Quality Control plan establish for G-Terra Project. Among other quality parameters were: COD, TOC, oil and grease, salinity and chlorides. Other common parameters determined in the runoff water samples are: temperature, pH, conductivity, turbidity, Total Suspended Solids (TSS), total hardness, Total Kjeldahl Nitrogen (TKN), total phosphorus, Biochemical Oxygen Demand (BOD_5), total Fe, total Zn, total Cu, total Pb, total Cd and total Cr.

Results and Discussion

The results concerning the first 2008 monitoring period at the A25 pilot case study are summarized in Table 4, and the salt deposition rate is presented in Table 5.

These preliminary results are similar to the other highway runoff studies in costal zones made in Portugal. It is common, namely the high levels of salinity, organic matter and related parameters, as the comparison between Tables 4 and 1 can confirm. It is also observed higher levels for nitrogen and phosphorus.

Considering the A25 case (see Tables 3 and 4) the levels of conductivity, salinity, chlorides and COD, although of similar magnitude, show higher concentrations in 2008. The 2008 Average Daily Traffic (ADT) at A25 (29350 vehicles in May 2008) compared to the value for 2003 (27,000 vehicles), may explain the higher values for the parameters related to traffic sources. Nevertheless, it is not expected that salinity and parameters such as nitrogen and phosphorus are associated with the ADT.

Table 4 Summary of pollutant concentrations in A25 highway runoff

Parameter	No. samples	Range of values Min.	Max.	Average	Median	St. Dev.
Conductivity (μS/cm)	72	113.0	970.0	388.3	405.5	230.2
Salinity (mg/L)	72	60	520	205	216	120
Turbidity (FNU)	72	5.8	118.0	43.6	30.4	28.4
TSS (mg/L)	72	<5.0	250.0	67.7	45.0	60.1
Total hardness (mg $CaCO_3$/L)	72	36.0	208.0	85.3	78.0	44.8
Chlorides (mg/L)	72	15.8	370.9	108.2	81.4	92.6
TKN (mg/L)	24	<0.2	6.2	1.9	1.0	2.2
Total P (mg/L)	72	<0.05	7.38	1.14	0.38	1.78
COD (mg/L)	72	25.0	375.0	109.1	105.5	56.9
COT (mg/L)	72	6	91	31	27	17
Oil and grease (mg/L)	32	<0.184	7.260	1.359	0.810	1.789
BOD_5 (mg/L)	72	<2	90	14	6	16

Table 5 Salt deposition rate at A25 site using a wet candle device

Period of determination	Feb 19 until Apr 1, 2008	Apr 2 until May 26, 2008
Salt deposition rate (mg/m² day)	59	84

In terms of general climacteric conditions that may influence the salinity/organic matter levels, the two monitoring periods took place during the autumn and the winter seasons. May the higher values of organic matter be correlated with the high values of salinity? At this stage of the research, it is not possible to answer the question, although the facts indicate that such relation is possible.

The BOD_5 concentration is independent from the salt deposition, and although showing a higher maximum value in 2008 (90 mg/L), average and median (14 and 6 mg/L) values are lower compared to the 2003/04 study (25 and 12 mg/L).

The 2008 monitoring study includes the COT determination, a parameter directly associated with the organic matter. Comparing the COT results obtained at A25 with the values found in the literature [9–11], they indicate that highway runoff from A25 has high levels of COT, meaning higher levels of organic matter.

The A25 oil and grease values showed a very high standard deviation (7,260 mg/L) and a minimum concentration below 0.184 mg/L. The average and median are of 1.359 and 0.810 mg/L, respectively. The oil and grease are not commonly measured in Portugal. Concentrations obtained in other studies were summarized in Table 6. As can be observed, the levels of concentrations are much below the measurements at A25. Future sampling at this site, and other G-Terra case studies, may help to understand the sources of oil and grease and their processes within the road environment.

All the oil and grease samples concentrations are below the Portuguese law for wastewater point discharges levels. However, concerning the SST, COD and BOD_5

Table 6 Oil and grease averages determined in highway runoff in Portugal

Road	No. samples	ADT	Average (mg/L)	Reference
A1	5	30,299	19.2	
A2	7	16,344	3.1	[12]
A6	23	2,918	11.2	
IP6	8	6,539	0.5	[7]
A23	20	9,855	0.58	[1]

Table 7 Salt deposition rate (chloride deposition) measure in a variety of studies in different environments, using wet candle device

Marine environment/country	Distance from the sea (m)	Salt deposition rate (mg/m^2 day)	Reference
Atlantic Ocean/Portugal	5,600	59–84	Antunes and Barbosa, (working paper)
Atlantic Ocean/Nigeria	45–365	800–140	[2]
Indian Ocean/Malaysia	30–400	400–100	[15]
Mediterranean Sea/Spain	20–170	111–7	[14]
Atlantic Ocean/Brazil	10–1,100	460–6	[13]

concentrations, 35%, 14% and 3% of samples, respectively, were above the discharge limits of this law. The atmospheric chloride deposition rate values (Table 5), when compared to data from in several studies reported in Table 7, show clear evidences of the presence of salt.

These results were expected, considering: the sea proximity and the A25 road platform altitude, nearly corresponding to the sea level, and the absence of significant obstacles to the transport of the marine aerosols.

Final Remarks

In 2005 and 2006 researchers from Portugal realized that coastal roads showed a specific pattern for some pollutants in highway runoff, namely the TSS, COD, and chlorides and salinity [5, 7]. A 3 year project, G-Terra, funded by the Portuguese Science and Technology Foundation, intends to study among others, this issue. One important task of the G-Terra is to gather monitoring data at different road sites. The study started in January 2008.

The first results concerning monitoring were obtained at the A25 coastal zone pilot case study. They are represented by the sampling of a total of ten events. The site has been previously monitored in 2003/2004 [5]. Procedures for monitoring highway runoff included automatic water sampling, and rain and flow measurements.

The wet candle device, based on [3], was selected for the G-Terra studies as a sound methodology to determine the atmospheric chloride deposition rate. The results for the salinity accumulation rates, with a weighted average of 74 mg/m^2 day, clearly show the presence of atmospheric salt at the A25 site.

The initial outcomes seem to confirm a high level of salinity, organic matter and related parameters in the A25 highway runoff. It was also observed higher levels for nitrogen and phosphorus, when compared to the 2003/2004 monitoring study. The oil and grease concentrations showed a very high standard deviation (1.789 mg/L) and a minimum concentration below 0.184 mg/L. These results are not understood yet, and future monitoring studies within the G-Terra project will certainly help to comprehend the sources of oil and grease and their processes within the road environment.

All the oil and grease samples concentrations are below the Portuguese law for wastewater point discharges levels. However, concerning the SST, COD and BOD_5 concentrations, 35%, 14% and 3% of samples, respectively were above the discharge limits of this law. This sort of knowledge is relevant for the establishment of guidelines for road runoff pollution management, the final objective of G-Terra project.

These results are still being analyzed. It is believed that the adopted methodologies are adequate to the study. These and the data from the three next monitoring periods at the A25 site will provide enough data for simple and advanced statistical analysis that will support the identification of the particular characteristics and processes, concerning road runoff in coastal areas.

If the phenomena associated with the high salinity in highway runoff in coastal areas are better understood and quantified, a new and valuable contribution to the knowledge will provide information for highway runoff management in countries like Portugal, with a significant coast line.

References

1. Albuquerque AM (2006) Study of the performance of the treatment pond of the highway A23 – Ligação à Covilhã. Master thesis dissertation (in Portuguese), Beira Interior University
2. Ambler HR, Bain AJ (1955) Corrosion of metals in the tropics. Journal of Applied Chemistry 5:437–467
3. American Society for Testing and Materials (ASTM) G 140-02 (2002) Standard test method for determining atmospheric chloride deposition rate by wet candle method. Annual Book of ASTM Standards, Philadelphia
4. American Water Works Association (2005) Standard Methods for Examination of Water and Wastewater, 21st edn
5. Antunes PA, Barbosa AE (2005) Highway Runoff Characteristics in Coastal Areas – A case Study in Aveiro, Portugal. Tenth International Conference on Urban Drainage, Copenhagen, 6 pp
6. Barbosa AE, Antunes PB (2004) Águas de Escorrência de Estradas Sistemas para Minimização de Impactes, 2° Relatório, Relatório 128/04-NRE/DHA, Laboratório Nacional de Engenharia Civil, 66 pp
7. Barbosa AE, Henriques MJ, Fernandes JN (2006) Quality of Highway Runoff in Coastal Areas: Special Cases? Eigth Highway and Urban Environment Symposium, Nicosia, Cyprus, 10 pp

8. Barbosa AE, Fernandes JN, Henriques MJ (2006) Características poluentes duma estrada costeira e avaliação do sistema de tratamento das suas escorrências. 12? Encontro Nacional de Saneamento Básico, Cascais, 15 pp
9. Barrett ME, Zuber RD, Collins ER, Malina JF, Charbeneau RJ, Ward GH (1995) A Review and Evaluation of Literature Pertaining to the Quantity and Control of Pollution from Highway Runoff and Construction. Centre for Research in Water Resources, Austin, 186 pp
10. Driscoll ED, Shelley PE, Strecker EW (1990) Pollutant loadings and impacts from highway stormwater runoff volume III: analytical investigation and research report, FHWA-RD-88–008. Federal Highway Administration, USA, 160 pp
11. Kayhanian M, Suverkropp C, Ruby A, Tsay K (2006) Characterization and prediction of highway runoff constituent event mean concentration. Journal of Environmental Management 85:279–295
12. Leitão TE, Barbosa AE, Henriques MJ, Ikävalko V, Menezes, JT (2005) Avaliação e Gestão Ambiental das Águas de Escorrência de Estradas, Relatório Final Relatório 109/05 – NAS, Laboratório Nacional de Engenharia Civil, 243 pp
13. Meira GR (2004) Chloride aggressiveness in marine atmosphere zone connected with corrosion problems in reinforced concrete structures. Ph.D. Thesis (in Portuguese), Federal University of Santa Catarina
14. Morcillo M, Chico B, Mariaca L, Otero E (2000) Salinity in marine atmospheric corrosion: its dependence on the wind regime existing in the site corrosion science 42:91–104
15. Mustafa MA, Yusof KM (1994) Atmospheric chloride penetration into concrete in semitropical marine environment. Journal of Cement and Concrete Research 24:661–670
16. Warneck P (1999) Chemistry of the Natural Atmosphere. 2nd edn. Academic Press, New York
17. Woodbridge MS, Obika B, Freer-Hewish R, Newill D (2002) Salt damage to bituminous surfacings: results from road trials in Botswana. Sixth Conference on Asphalt Pavements for Southern Africa, Cape Town

Size Fractionation of Heavy Metals in Highway Runoff Waters

Beatrice Béchet, Bertrand Durin, Michel Legret, and Pierre Le Cloirec

Introduction

Highway runoff waters may contain pollutants that have accumulated on the carriageway. Numerous field surveys have demonstrated that heavy metals (Cd, Cr, Cu, Ni, Pb, Zn) are common pollutants in highway runoff waters that degrade the quality of natural water bodies [1–6]. These hazardous pollutants have been found to be attached to particles, especially finer particles than 100 μm, usually quantified by dissolved, suspended and settleable solids measurements [7]. Clay minerals, very fine silts, metallic oxy-hydroxides and organic matter were identified as main components of the composite particulate matter [8, 9]. The knowledge of partitioning of trace metals between the different solids fractions is crucial for stormwater management. The pollutant size distribution determines the efficiency of runoff water treatment by settling in detention basins. Moreover, the toxicity and bioavailability metal characteristics also depend on the speciation of pollutants that could be discharged into natural water bodies. Partitioning of pollutants is also challenging nowadays due to issues of colloid-mediated transfer of heavy metals, as it was already shown for roadside soils impacted by runoff waters [10].

In the case of stormwater treatment in detention-infiltration basins, previous studies have demonstrated that the settling of particles contributes to the formation of contaminated sediments [11, 12]. Even if earlier field data assess that groundwater is seldom contaminated [13, 14], recent studies have pointed out potential

---

B. Béchet, B. Durin, and M. Legret
Laboratoire Central des Ponts et Chaussées, Division Eau et Environnement, Route de Bouaye, BP 4129, 44341, Bouguenais cedex, France
DDE 69/SERDD/MPE, 33 rue Moncey, 69421, Lyon CEDEX 03, France

P. Le Cloirec
Ecole de Chimie de Rennes (ENSCR), Campus de Beaulieu, Avenue du Général Leclerc, 35700, Rennes, France

S. Rauch et al. (eds.), *Highway and Urban Environment*, Alliance for Global Sustainability Bookseries 17,
DOI 10.1007/978-90-481-3043-6_26,

downward colloid-transport of heavy metals [15, 16]. Seasonal and temporal variations in runoff quality, such as spreading of de-icing salts may have an important impact on the colloidal matter in the runoff waters and in the sediment (aggregation, adsorption/desorption of pollutants). Several studies highlight the partitioning of trace metals in runoff waters between particulate and dissolved-bound fractions, the "dissolved" fraction being operationally defined by filtration through a 0.45-μm pore size membrane [7, 17]. But, only few studies have begun recently to investigate the colloidal fractions of metals in runoff waters (nanometer to micrometer size range) [9, 18].

A peri-urban experimental site was chosen to sample highway runoff waters. The objective was to study the distribution of selected trace metals (Cd, Cr, Cu, Ni, Pb, Zn) among dissolved, colloidal and particulate fractions, in relation to potential constituents of colloidal and particulate matter (Al, Fe, Mn, Si, organic C). The relationship between physico-chemical water characteristics and chemical elements was investigated and the fractionation of trace metals was quantified by filtration and ultrafiltration.

Experimental

Highway Study Site

The experimental site is located on the south bypass motorway of Nantes (Loire-Atlantique, France), where it crosses the Loire River at the Cheviré Bridge [14]. Opened in 1991, the Cheviré Bridge supports nowadays an average total daily traffic of 90,000 vehicles. The bridge is 1,500 m long and 24.6 m wide. Carriageways consist of three lanes in each direction, separated by a central reservation and the pavement consists of a conventional asphalt surface. The runoff waters of the 19,000 m^2 contribution area are collected in gulleys and flowed in aluminum collectors. Then, they are gathered in a detention-infiltration basin. Nantes has an oceanic climate with a mean annual rainfall of 820 mm over the last 50 years. Between August 2004 and May 2006, runoff waters were sampled in a concrete pipe, at the inlet of the basin, with an automatic sampler to take mean samples according to time during rain events. Twenty one samples were taken over the year (three events in spring, seven in summer, nine in autumn and two in winter).

Analysis of Runoff Waters

The physico-chemical characteristics (pH, conductivity, suspended solids (SS)) were measured according to French and EU standard methods, either on site or after sampling at the laboratory on mean composite samples. Total concentrations of major chemical elements and of trace metals were obtained from raw samples by digestion with nitric acid, ultrasonic stimulation and at least 3 h of rest. Samples were filtrated through a 0.45-μm pore size membrane (Millipore) and total concentrations were measured in the filtrate. Dissolved concentration was obtained by analysis of the filtrates of the raw samples through a 0.45-μm pore size membrane and acidification of the filtrates at pH < 2 with nitric acid. All solutions were kept at 4°C in a refrigerator prior to analysis.

Major elements (Na, Mg, K, Ca, Si, Al, Fe, Mn) and Zn concentrations were determined by inductively-coupled plasma-atomic emission spectrometry (ICP-AES) (Varian Liberty 220). All results were established from duplicate analysis, the detection limit for Zn was 10 μg/L and precision ranged from 3% to 5%. Pb, Cu, Cd, Cr and Ni concentrations were determined by electrothermal atomic absorption spectrometry (AAS) (Varian SpectrAA 300 or SpectrAA 220Z). All results were established from duplicate analysis, the detection limits for Pb, Cu, Cd, Cr and Ni were 2, 2, 0.1, 0.5 and 2 μg/L, respectively, and precision ranged from 3% to 5%. An internal quality control solution was incorporated into each series. When the deviation exceeded 5% for ICP-AES or 10% for AAS, the series were started over again. The total organic carbon (TOC) was measured by oxidation and quantification of CO_2 by infra-red spectrometry (NF EN 1484). The detection limit was 2 μg/L. The ultra-pure water employed was deionized using ion exchange columns (Elga Maxima). All reagents were of analytical grade (Merck, Suprapur or Pro Analysis) and all glassware in contact with the samples were washed with 10% nitric acid and then rinsed with deionized water.

Size Fractionation of Runoff Waters

Size fractionations were implemented for four samples (one sample for each season), using in series filtration and ultrafiltration, to characterize the distribution of trace metals between particulate, colloidal and dissolved fractions. The upper size limit of colloids in runoff waters is defined to be around 10 μm, as determined by [19]. The lower limit is considered to be 5 kDa referred to the limit between colloids and compounds such as dissolved species and very small natural organic colloids (fulvic acids). A 2-L feed sample was used for in series fractionation. Large particles were removed by settling during 24 h. Experiments were carried out in a pressured stirred batch cell (Amicon cell). The membranes selected were Millipore nitrocellulose membranes with 8, 1.2, 0.45, 0.22 and 0.05-μm pore sizes for filtration and Vivascience polyethersulfone membranes with molecular weight cut-offs of 100, 30, and 5 kDa for ultrafiltration. The membrane cut-offs were chosen according to the procedure of [9, 18], and depending on the available membrane sizes. The protocol was adapted from one sample to the next one, depending on the results obtained for each filtration step. If the elemental concentrations were not significantly different between two filtration steps, the corresponding cut-off was suppressed in the next sequence. Mass balance was calculated for each step of the fractionation to verify the efficiency of the protocol. The suspensions were analysed for major chemical elements, organic carbon and heavy metals after acidification at pH < 2 with nitric acid. The 8-µm and 5 kDa cut-offs were considered for synthetic presentation of the results.

Results and Discussion

Physico-Chemical Characteristics

During the data collection period, the pH was rather constant with a mean value of 6.6 (Table 1). Variations of conductivity observed all over the year were hardly related to rain event characteristics or season. The conductivity was well correlated ($r^2 = 0.96$) with the mineralization of runoff waters, estimated by calculation of the ionic strength from dissolved concentrations of Cl, SO_4, NO_3, Na, Ca, Mg and K. The suspended solid concentrations exhibited a great standard deviation, due to some extreme values above 1,000 mg/L. Except for these data; SS concentrations were in relation with the flow intensity. Even if only few TOC concentrations were available, TOC correlated with SS ($r^2 = 0.77$). The two extreme values of SS could be explained by particle aggregation: the first one due to very high dissolved concentrations of Na (386 mg/L) and Cl (870 mg/L), reflecting the contribution of de-icing salt spreading, the second one linked to a high concentration of organic matter (TOC 41 mg/L). Compared to data from A11 highway sampling site near Nantes [17], we observed that pH and conductivity median values were very close. Very high values of conductivity related to de-icing salt spreading were not measured for runoff waters at the Cheviré site but some, up to 24 mS/cm, were measured in the water retained in the basin [16] as well as at the A11 site. The SS concentrations were very different but the influence of the characteristics of the sampling sites should be considered; the traffic intensity and contribution area being more important at the Cheviré site.

Table 1 Results of the water analysis. Total concentrations of major elements

2	pH	Cond.	TOC	SS	Al	Fe	Mn	Ca	K	Mg	Na	Si
		µS/cm		mg/L								
Event number	21	21	7	17	17	17	17	17	17	17	17	14
Mean	6.6	352	18.4	402	1.7	4.7	0.2	33.4	14.9	3.2	67.2	2.3
Median	6.6	312	20.2	284	1.4	3.8	0.1	29	9.1	2.4	34.5	2.1
Min.	6.2	53	3.5	74	dl	0.2	dl	8.4	1.7	0.8	5.1	0.8
Max.	7.1	748	41	1.340	5.5	16.7	0.6	85	46.1	9.8	563	6.0
St. Dev.	0.24	202	13.2	368	1.45	4.3	0.18	21.7	13.1	2.3	129	1.3

dl: Below detection limit

Total Concentrations of Major and Trace Elements

Runoff elemental concentrations, named "total concentrations", include the bulk concentrations that desorbed from dispersed particulate matter exposed to nitric acid (surface-bound constituents) as well as a part of particles themselves which are dissolved by acidification process, plus the constituents present initially as dissolved species.

Al, Fe and Mn concentrations varied widely between rain events, but no relationship could be established between these elements and SS, probably due to the incomplete dissolution of minerals and to the contribution of organic matter. High variations of total trace elements concentrations were measured over the data collection period (Table 2). A comparison with the literature showed that the concentrations of heavy metals were in the same range as measured 10 years ago at the Cheviré site [14]. Values were also in the same range as current measurements in England [5].

Despite the removal of lead from fuel, the constant concentration of Pb could be attributed to the increase of traffic and to the presence of other sources of lead than leaded gasoline. Ca and Na were the main major elements in runoff waters, as generally observed for surface waters. The quite high concentration of K could

Table 2 Results of the water analysis. Total concentrations of trace elements

Event date	Zn	Cd	Cr	Cu	Ni	Pb
			μg/L			
Aug 24, 2004	275	0.08	0.9	47	3.0	2.0
Sept 10, 2004	1,618	0.21	6.5	49	7.8	25.9
Nov 10, 2004	1,900	0.39	5.2	24	9.0	28.3
Oct 10, 2004	1,837	0.90	14.9	169	10.1	68.7
Apr 10, 2004	2,744	0.38	17.5	186	15.1	50.8
Apr 10, 2004	2,052	0.20	7.1	153	8.3	15.4
Nov 26, 2004	815	0.44	14.8	163	5.5	41.0
Dec 16, 2004	3,136	1.94	22.1	544	23.8	44.6
Jan 17, 2005	nd	nd	nd	nd	nd	nd
Jan 19, 2005	1,455	1.02	16.4	194	10.2	96.5
Apr 23, 2005	3,147	1.60	37.6	945	26.2	38.6
June 06, 2005	nd	nd	nd	nd	nd	nd
July 27, 2005	744	0.23	8.8	146	3.3	46.6
Aug 08, 2005	nd	nd	nd	nd	nd	nd
Aug 08, 2005	693	0.30	6.0	121	5.0	40.0
Sept 25, 2005	1,410	0.90	10.0	243	8.0	70.0
Sept 29, 2005	2,214	1.20	12.0	345	11.0	94.0
Oct 28, 2005	nd	nd	nd	nd	nd	nd
Oct 28, 2005	545	0.20	16.0	128	7.2	35.3
July 07, 2005	386	<0.1	12.9	91	3.6	42.4
May 16, 2006	4,688	<0.4	22.1	179	13.8	126

originate mainly from the geological context of Nantes area. As conductivity and concentrations of Ca, Na, Mg and K were well correlated, the elements could be present mostly as dissolved species in runoff waters.

Distribution of Metals Among Particulate, Colloidal and Dissolved Fractions

The partitioning of major chemical elements between particulate and dissolved fractions, according to conventional 0.45-μm pore size filtration, indicated that Fe and Al were predominantly associated (>80%) with particles greater than 0.45 μm (Fig. 1). Mn was equally distributed in both the dissolved and particulate fractions. Si and organic carbon were mostly in the <0.45 μm size fraction (80%). The trends were observed whatever the seasons as indicated by the low standard deviation of the partitioning for these major elements (35–20%). Ca, Na, Mg and K were predominantly dissolved in runoff waters. All the trace metals were mostly in the >0.45 μm fraction: from 50% (Ni) to 90% (Pb). The variability of the distribution for trace metals was high (SD 20–30%), except for lead (SD 9%). Correlation coefficients between particulate concentrations of heavy metals and Fe, Al, Mn-particle associated showed the affinity of all the trace metals, except Pb, for Fe, Al, Mn-bearing phases. Coefficients varied between 0.78 and 0.97 for Cd, Cr, Cu and Ni in increasing order of correlation whereas lower correlation for Zn was noticed ($r^2 = 0.74 \pm 0.04$). Moreover, high correlations between particulate Cu and Ni and particulate Si ($r^2 = 0.8$) were found. With the exception of a moderate correlation

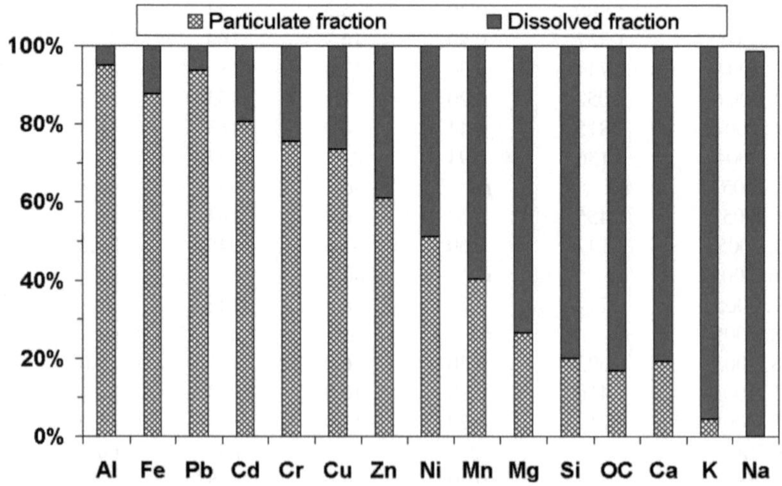

Fig. 1 Partitioning of chemical elements between particulate and dissolved fractions (cut-off 0.45 μm) – Mean values of 17 rain events

of Pb with the particulate fraction of organic carbon ($r^2 = 0.42$), Pb did not correlate with any element. These results indicated that most of the trace metals could be associated to particles such as metallic oxy-hydroxides and for Ni and Cu, clays minerals or siliceous particles. Pb seemed to be linked partially to particulate organic matter.

As filtration of runoff through a 0.45 µm filter can result in colloidal materials being measured as dissolved, in series partitioning was performed on four samples. The whole protocol described above was applied to the two first samples. No significant drop in concentration for trace metals was observed applying the 0.22-µm, 100 and 30 kDa cut-offs. The 0.05-µm cut-off seemed to slightly influence the concentration of Zn and organic carbon but could hardly be related to a specific particle size. Finally, the applied protocol was composed of four cut-offs (8, 1.2 and 0.45 µm and 5 kDa). Three fractions were considered to compare the results: the dissolved fraction <5 kDa (dissolved species, very small colloids such as fulvic acids); the colloidal fraction with sized particles between 5 kDa and 8 µm and the particulate fraction composed of particles above 8 µm. Size fractionation for the May 16, 2006 rain event was selected for illustration because the mass balance was especially accurate for trace metals (Fig. 2). Al, Fe, Cr and Pb were predominantly associated with particles greater than 8 µm in size. Colloidal-bound metals represented only a few percent of the distribution. Zn, Ni, Cu were also mostly particulate-bound but 3–15% of these trace metals was colloid-associated. Na, Ca and K were present in runoff waters mainly as dissolved species (75–90%), Mg being equally fractionated between the fractions <5 and >5 kDa.

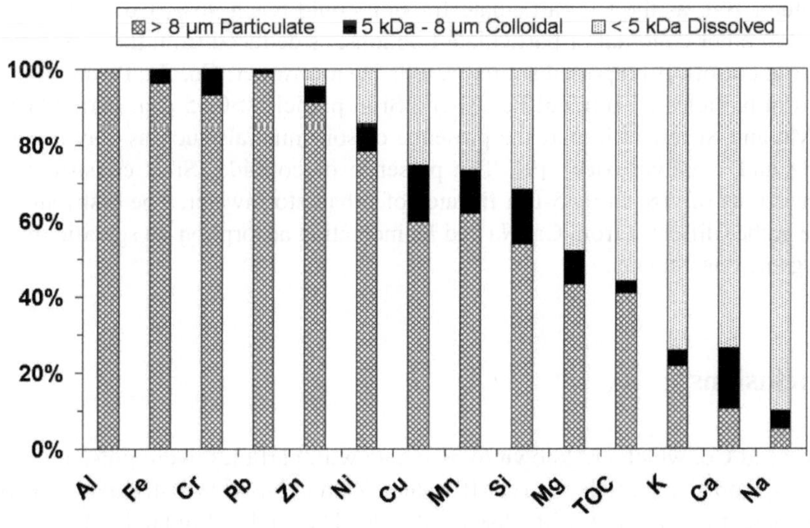

Fig. 2 Distribution of major and trace elements in the particulate, colloidal and dissolved fractions for a spring rain event (May 16, 2006)

Table 3 Distribution range (%) over the four samples for the three fractions

Element	Particulate fraction	Colloidal fraction	Dissolved fraction
Al	97–100	0–3	0
Fe	96–100	0–4	0
Cd, Cr, Pb	85–100	0–15	0
Mn	52–100	0–11	0–41
Zn	65–94	2–15	4–21
Ni	72–98	2–7	0–26
Si	52–74	1–14	26–42
TOC	0–41	3–31	44–70
Mg	33–78	3–9	2–59
Cu	22–96	2–73	3–26
Ca, Na, K	5–69	2–16	20–92

The distribution of major and trace elements was quite similar for the four samples, except for Cu, Mn and TOC. No sample was collected after a de-icing salt spreading which could probably change the distribution of trace metals due to aggregation phenomenon. Four groups of chemical elements resulted from the calculation of percentage of distribution between the fractions (Table 3).

The results of in series partitioning highlight that Al, Fe and trace metals such as Cd, Cr and Pb were mostly present in runoff waters as particle-associated with a maximum of 15% of colloidal species. Both Zn and Ni were primarily in the >8 μm size fraction but the dissolved fraction could reach up to 30%. Their distribution was very close to those of Mn. The dissolved fraction of Si and TOC was rather constant, between 30% and 70%. The similar colloidal fraction of Cu and organic carbon showed the link between this metal and organic particles, as reported in the literature. But, as the Cu particulate fraction could reach 96%, Cu could be also linked to other colloidal or particulate fractions, as deduced from the conventional 0.45-μm cut-off. [9] reported similar results in stormwater: Cu, Zn, Pb and Cr were found in particles >5 μm and Fe, Al and Si in particles >0.45 μm. Particulate Fe, Al, Mn and Si reflect in part the presence of soil minerals such as clay minerals, quartz and oxy-hydroxides [9]. The presence of colloidal Si is consistent with observations of [18] in 0.45-μm filtrates of urban stormwater. The distribution of Mg is rather different from Ca, Na and K, indicating adsorption on specific mineral or organic constituents.

Conclusions

Pb, Cr and Cd, which are known to associate with particles, were present entirely in the >8 μm size particles of runoff waters. Zn and Ni were distributed mostly in particulate fractions but with dissolved (<5 kDa) and colloidal fractions up to 30%. The distribution of Cu is quite equal to Zn and Ni, with up to 70% of colloidal Cu. Even if colloidal metals were relatively small compared to other size fractions, total mobility of metals could be enhanced by facilitated transfer of the colloids during infiltration of runoff waters. The association between Fe and Al-bearing phases and Cd, Cr, Zn and Cu was confirmed, just as the link between Cu and organic matter in the <0.45 μm fraction. Mn-bearing phases were pointed out, as the association of Ni with metallic oxy-hydroxides and siliceous particles. This study revealed also that Pb could be more linked to particulate organic matter than adsorbed on metallic oxy-hydroxides. These observations underscore the need for observation of metal-bearing solids phases and characterization of organic phases in runoff waters.

Acknowledgements This work has partially benefited from financial support of the ANR ECCO (France).

References

1. Harrison RM, Wilson SJ (1985) The chemical composition of highway drainage waters. I: Major ions and selected trace metals. Sci Tot Environ 43:63–77
2. Stotz G (1987) Investigations of the properties of the surface water run-off from Federal Highways in the FRG. Sci Tot Environ 59:329–337
3. Hewitt LN, Rashed MB (1990) An integrated budget for selected pollutants for a major rural highway. Sci Total Environ 93:375–384
4. Wu JS, Allan CJ, Saunders WL, Evett JB (1998) Characterization and pollutant loading estimation for highway runoff. J Environ Eng 124:584–592.
5. Crabtree B, Moy F, Whitehead M (2005) Pollutants in highway runoff. Proc. 10th Int Conf Urban Drainage, Copenhagen, Denmark
6. Flint KR, Davis AP (2007) Pollutant mass flushing characterization of highway stormwater runoff from an ultra-urban area. J Environ Eng 133:616–626
7. Sansalone JJ, Buchberger SG (1997) Partitioning and first flush of metals and solids in urban highway runoff. J Environ Eng Div ASCE, 123:134–143
8. Roger S, Montrejaud-Vignoles M, Andral MC, Herremans L, Fortune JP (1998) Mineral, physical and chemical analysis of the solid matter carried by motorway runoff water. Wat Res 32:1119–1125
9. Tuccillo ME (2006) Size fractionation of metals in runoff from residential and highway storm sewers. Sci Tot Environ 355:288–300
10. Norrström AC, Jacks G (1998) Concentration and fractionation of heavy metals in roadside soils receiving de-icing salts. Sci Tot Environ 218:161–174.
11. Barbosa AE, Hvitved-Jacobsen T (1999) Highway runoff and potential for removal of heavy metals in an infiltration pond in Portugal. Sci Tot Environ 235:151–159
12. Durand C, Ruban V, Amblès A (2004) Mobility of trace metals in retention pond sediment. Environ Technol 8:881–888
13. Yousef YA, Lin LY (1992) Potential contamination of groundwater from Cu, Pb and Zn in wet detention ponds receiving highway runoff. J Environ Sci Health A27:1033–1044
14. Legret M, Le Marc C, Demare D, Colandini V (1995) Pollution par les métaux lourds dans un bassin de décantation recevant des eaux de ruissellement d'origine routière. Environ Technol 16:1049–1060
15. Winiarski, T, Bedell J-P, Delolme C, Perrodin Y (2006) The impact of stormwater on a soil profile in an infiltration basin. Hydrogeol J 14:1244–1251
16. Durin B, Béchet B, Legret M, Le Cloirec P (2007) Role of colloids in heavy metal transfer trough a retention-infiltration basin. Wat Sci Technol 56:91–99

17. Legret M, Pagotto C (1999) Evaluation of pollutant loadings in the runoff waters from a major rural highway. Sci Tot Environ 235:143–150
18. Grout H, Wiesner MR, Bottero J-Y (1999) Analysis of colloidal phases in urban stormwater runoff. Environ Sci Technol 33:831–839
19. McKenzie ER, Wong CM, Green PG, Kayhanian M, Young TM (2008) Size dependent elemental composition of road-associated particles. Sci Tot Environ, doi: 10.1016/j.scitotenv.2008.02.052

Ecotoxicological Impacts on Brown Trout (*Salmo trutta* L.) Exposed to Runoff from a Four Lane Motorway in Norway

Sondre Meland, Bjørn Olav Rosseland, and Brit Salbu

Introduction

In the last decades there has been a growing concern regarding highway runoff as a potential hazard to receiving water bodies. This has led to increasing use of various treatment systems. A lot of the pollutants are believed to be associated with particles, and most of the treatment systems are therefore based on removing the pollutants by sedimentation. However, treatment of particle bound pollutants might be inappropriate in protecting the aquatic biota because toxicity, at least for metals, is mostly related to the dissolved fraction which includes colloids as well as the low molecular mass positively charged metal species [1]. In northern countries, like Norway, road salt is frequently used in the winter maintenance to achieve good friction on the road. The sodium and chloride concentrations can thus be rather high during runoff episodes and might lead to increased mobility, remobilization and finally increased bioavailability of the metals [2–5].

Pollutants often found in elevated concentrations in highway runoff such as nickel (Ni), cadmium (Cd), zinc (Zn), copper (Cu), lead (Pb) and polycyclic aromatic hydrocarbons (PAH) are in several studies shown to potentially disturb normal homeostasis in fish, for example increased enzymatic activity (as well as inhibition), osmoregulatory- and respiratoric dysfunctions [1, 6–8].

This paper gives an overview of results obtained from one out of four simulated runoff episodes entering a highway retention pond system, and describes the chemical exposure concentrations and physiological responses in the native fish species brown trout (*Salmo trutta* L.).

S. Meland, B.O. Rosseland, and B. Salbu
Department of Plant and Environmental Sciences, Norwegian University of Life Sciences, P.O. Box 5003,N-1432 Ås, Norway

S. Rauch et al. (eds.), *Highway and Urban Environment*, Alliance for Global Sustainability Bookseries 17,
DOI 10.1007/978-90-481-3043-6_27, © Springer Science+Business Media B.V. 2010

Experimental Methods

Field Experiment Setup

The experimental setup was located close to a detention pond just outside the City of Oslo in Norway (early May 2007). This pond receives runoff from a four lane motorway (E6) with an average daily traffic around 45,000 vehicles. In brief the pond has a closed pre sedimentation basin in front of the main pond, which has a wet volume of approximately 800 m^3 and is heavily vegetated from the midpoint and towards the outlet [9]. Veterinary approved brown trout (age 1+), weighing 67.2 ± 22.8 g and with a condition factor (K) 1.0 ± 0.1 was supplied from a nearby hatchery (OFA). The fishes were placed in five experimental tanks (70 L, biomass 30 g/L), maintained in continuously flowing stream water pumped from upstream the pond outlet from the River Ljanselva (6 L/min) to acclimate before the experiment (5 days). The reference stream water had a temperature of 9.7 ± 0.2°C and a pH of 7.6 ± 0.1.

To simulate a realistic episode the fish was exposed for a period of 24 h to runoff water pumped from the inlet (after the pre sedimentation basin) and to the outlet of the pond, as well as to mixtures of these waters. Thus, the five tanks had the following pollutant loadings: (1) SW; stream water (control), PI; pond inlet (untreated runoff), PO; pond outlet (treated runoff), PI + SW; pond inlet + stream water (mix ratio 50:50), PO + SW; pond outlet + stream water (mix ratio 50:50).

Water Chemistry

Water from each tank was sampled and fractionated in situ with a peristaltic pump and 0.45 μm membrane filter (Millipore). Two fractions were used throughout the analyses; particulate fraction (total – 0.45 μm filtrate) and dissolved fraction (0.45 μm filtrate). Determinations of the metals were done by ICP-MS and ICP-OES. Total organic carbon (TOC), polycyclic aromatic hydrocarbons (16PAH) and hydrocarbons (C10-C40) were analyzed by ALS Scandinavia. Conductivity, pH, temperature and oxygen were measured with a multi water quality sensor (W21SDI).

Fish Physiology

After 24 h exposure, six fishes from each tank were sampled randomly. The fish was killed by a single blow to the head and sampled according to the EMERGE protocol [10]. Blood was immediately sampled from the caudal vein and analyzed for a total of nine blood parameters; (I-Stat, Abbot), sodium (Na), potassium (K),

Chloride (Cl), total CO_2 (TCO_2), glucose, hematocrit, pH, partial pressure CO_2 (pCO_2) and bicarbonate (HCO_3^-).

The second right gill arch and a part of the liver were dissected for analysis of accumulated metals (ICP-MS): Cd, Cu, Zn, Ni, Pb, Fe and Al.

A second part of the liver was snap frozen in liquid nitrogen and stored at −80°C prior to enzymatic response analysis of superoxide dismutase (SOD, Fluka kit) and catalase (CAT, Cayman kit). SOD and CAT are known to be important enzymes detoxifying free radical induction [11]. In addition, analyses of metallothionein (MT) and total protein content were performed. MT, a cystein rich protein complex able to bind various heavy metals, was analyzed by the Cd-/Zn chelex method [12, 13] and the protein concentration was analyzed by the Bradford assay [14].

Results and Discussion

Water Quality

The accumulated precipitation of 8.6 mm was registered over a period of 6 days prior the exposure experiment. However, the pollutant loadings were rather low compared to earlier studies [15]. As expected the highest concentrations of most of the pollutants were observed in the PI and only minor differences were observed between the pond outlet water (PO) and the stream water (SW) (Table 1). The dissolved fraction appeared to be dominating for Ni, Cu, and Zn and to some extent for Cd, while Pb, Al and Fe were more associated with particles.

The high chloride levels indicated that the runoff water was considerably affected by road salting, being almost 37 times higher in the PI compared to the SW (Table 1). The elevated salt concentrations might explain the higher fraction (%) of dissolved metals species in the untreated runoff compared to the stream water, as increased ion strength can aggregated particles and sedimentation can occur. Furthermore, high levels of salt can mobilize colloidal- or organically bound metals [3].

Metal Accumulation in Gill Tissue

No mortality occurred during the 24 h exposure experiment. Despite the small differences between the various treatment tanks regarding Pb and Al concentrations in the water, the concentration of these two elements appeared significantly higher in trout gills from PI compared to the other groups (gill Pb in SW not significant different from PI, however four values were <LOD, n = 2) (Fig. 1).

The high salt concentrations in the untreated inlet water (PI) led to a substantial and immediate increase in ionic strength in the mixed water of PI + SW and PO + SW. Since both Al and Pb in gills were strongly and significant correlated with the

Table 1 Key water quality parameters and total and estimated fractions (in %) of metals in water sampled on May 7 from the various exposure groups, (n = 1)

Pollutant	Fraction (unit)	SW	PI	PI + SW	PO	PO + SW
Temp	(°C)	9.9	9.8	9.8	11.0	10.5
pH[a]		7.5	7.2	7.2	7.3	7.4
Oxygen	(mg/L)	10.1	5.6	7.0	9.2	9.3
Conductivity	(mS/m)	37.9	652	390	149	87.6
DOC[b]	(mg/L)	6.00	8.97	6.63	6.2	5.84
Calcium	Total (mg/L)	32.3	55.6	47.0	31.7	33.5
Sodium	Total (mg/L)	35.0	1,178	721	235	144
Chloride	Total (mg/L)	59.6	2,210	1,290	442	263
Hydrocarbons[c]	Total (μg/L)	74	130	80	<50	<50
16PAH	Total (μg/L)	n.d.	n.d.	n.d.	n.d.	n.d.
Cd	Total (μg/L)	0.08	0.02	0.09	0.02	0.05
	Part. (%)	35	8	64	54	48
	Dissolved (%)	65	92	36	46	52
Ni	Total (μg/L)	1.9	3.6	3.3	2.6	2.2
	Part. (%)	19	8	16	19	14
	Dissolved (%)	81	92	84	81	86
Cu	Total (μg/L)	6.5	12.8	8.5	6.7	5.5
	Part. (%)	47	29	31	16	17
	Dissolved (%)	53	71	69	84	83
Zn	Total (μg/L)	27.5	43.3	35.2	11.7	19.9
	Part. (%)	32	23	19	29	31
	Dissolved (%)	68	77	81	71	69
Pb	Total (μg/L)	0.8	0.8	0.8	0.3	0.5
	Part. (%)	88	96	89	74	78
	Dissolved (%)	12	4	11	26	22
Al	Total (mg/L)	0.18	0.17	0.14	0.05	0.10
	Part. (%)	87	81	93	77	83
	Dissolved (%)	13	19	7	23	17
Fe	Total (mg/L)	0.36	0.45	0.37	0.14	0.21
	Part. (%)	79	75	82	82	79
	Dissolved (%)	21	25	18	18	21

SW, stream water; PI, pond inlet; PO, pond outlet.
[a]pH presented as the average of three similar episodes due to failure in pH-meter during the present episode.
[b]DOC analyzed as TOC in 0.45 μm filtrate.
[c]Hydrocarbons measured as C10–C40.

chloride concentrations in the water (Fig. 2), an unstable mixing zone was created, and due to ion exchange processes and polymerization [16–18] formed gill reactive metals species will accumulate on available gill surfaces. Several of the elements found in the gills, although in low concentrations, were significantly correlated with each other, indicating co-precipitation (e.g. Pb gill vs. Al gill r = 0.90, Cu gill vs. Ni gill r = 0.83, Cu gill vs. Fe gill r = 0.93).

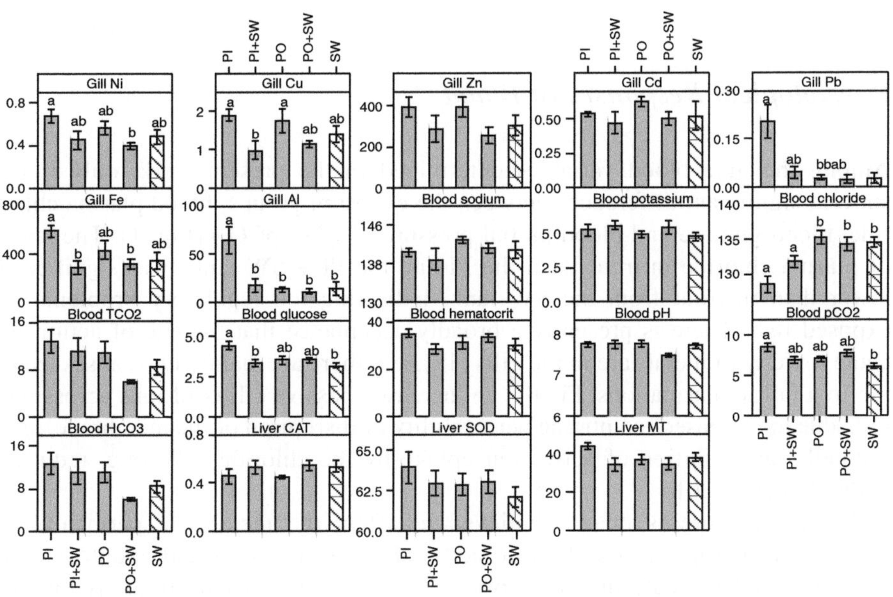

Fig. 1 Physiological responses in brown trout exposed to road runoff (*filled bars*) and stream water (*hatched bars*) for 24 h (mean ± S.E. n = 2–6). Metals in gill tissue expressed in μg/g dw, glucose, sodium, chloride and HCO_3^- expressed in mmol/L, TCO_2 and pCO_2 expressed in mmHg, hematocrit and SOD inhibition rate expressed in %, CAT expressed in μM/min mg/protein and MT in nmol/g ww. *Bars* with a different letter represent a significant difference ($p < 0.05$). SW stream water, PI pond inlet, PO pond outlet

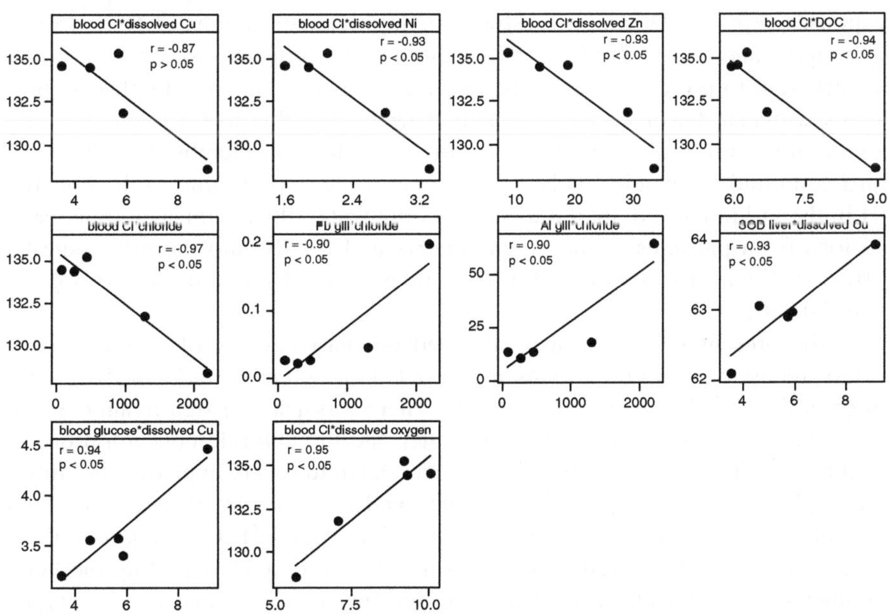

Fig. 2 Correlation between water quality and physiology parameters. (n = 5). The physiological parameters are based on the average of six fishes

Physiological Responses in Blood

Among the nine blood variables determined in the present work, only three showed significant differences between the exposure groups; blood plasma chloride, blood glucose and blood partial pressure of CO_2 (pCO_2) (Fig. 1). The level of plasma chloride in trout from the PI and the PI + SW tanks was significant lower than in trout from the SW tank, indicating ion regulatory problems in exposed fish. There is presently a broadly acceptance that the site of action of acute toxicity of metals are in the gills [19], and there are mainly two causes leading to a fall in plasma ions. Firstly, an increased passive efflux of ions across the gills due to increased membrane permeability or disruption of the membrane, and secondly an inhibition of active ion uptake by the chloride cells (e.g. reduced Na/K-ATPase) [20].

Metal exposure is mainly believed to affect the membrane permeability rather than disrupting the chloride cells, although necrosis and apoptosis are also observed in acute metal exposed fish [21]. Cu concentrations as low as 12.5 μg/L were found to lead to substantial loss of plasma ions in rainbow trout (*Oncorhynchus mykiss.*) [22], even in water at neutral pH and high Ca concentrations. In addition to a loss in plasma ions, an increase in the glucose levels was observed. Such responses are also found in salmonids exposed to several other elements, for example Pb [23] and Al [16]. A similar pattern was observed in the present study, where both chloride and glucose showed a strong relationship with several metals in the dissolved fractions, indicating multiple stress responses (Fig. 2). This is interesting; the high concentrations of DOC (5.84–8.97 mg/L) and Ca (31.7–55.6 mg/L) measured in all the exposure tanks should be sufficient to protect the trout against metal toxicity due to complexion processes between dissolved metal ions and DOC, and due to Ca being a strong competitor with high binding strength to biological membranes [24, 25]. Particle bound Cu would likely be a source of Cu toxicity due to the lowered pH in the gill mucus compared to surrounding water, thus favoring a transfer of Cu ions from particles to the mucus matrix and finally into the gill tissue by diffusion. Such processes could also be relevant for Cu complexed to organic material [26].

In the present study we also observed decreased levels of oxygen in the waters, being lowest in the PI tank measured after 24 h (5.6 mg/L, ca. 50% saturation at the actual temperature, Table 1). There was a strong significant correlation between dissolved oxygen in the water and the chloride concentrations in trout blood (Fig. 2). Blood acidosis is often related to severe hypoxic conditions, with high pCO_2 (hypercapnia) in the water, and with specific changes with lowered blood pH and chloride, and increased TCO_2 and HCO_3. However, there were no significant differences between the exposure tanks regarding the latter parameters (Fig. 1), and they were not significantly correlated to the oxygen concentration.

Superoxide Dismutase-, Catalase- and Metallothionein Response

There were no significant differences in liver enzymatic activity measured as SOD and CAT or concentration of MT between the various exposure regimes (Fig. 1). However, as for glucose in blood, dissolved Cu showed a significant positive relationship with SOD inhibition activity, indicating the presence of reactive oxygen species (ROS) (Fig. 2). Cu, together with Fe, are known to be inducers for free oxygen radicals [11, 27, 28] and both were found in elevated concentrations in gill tissue from fish exposed to the pond inlet (PI) water. Free radicals and formation of ROS can bring about a series of responses in exposed organisms, such as changing gene regulation or expression, stimulating DNA-repair mechanisms, impacting on cell apoptosis, as well as interacting with important biomolecules such as enzymes and proteins [29, 30]. This indicates that gill accumulated Cu and other metals could induce free radicals and thereby disrupt cellular functions like Na/K-ATPase [31] in gills [32].

Conclusions

The present study has showed that highway runoffs, even during low concentration episodes, have the potential to disturb normal homeostasis in brown trout. We registered that several metals in the water heavily affected by road salt was mobilized and present as dissolved metal species. This is probably the underlying mechanisms for the elevated concentrations of metals in gill tissue from the trout exposed to untreated (PI) and treated water (PO). Following the accumulation of metals in the gill tissue, biological responses could also be expected. Therefore, the observed loss of chloride ions in the blood plasma together with increased values of glucose and pCO_2 is attributed to the accumulation of gill-reactive metal species produced in the highway runoff.

References

1. Lydersen E, Lofgren S, Arnesen RT (2002) Metals in Scandinavian surface waters: effects of acidification, liming, and potential reacidification. Critical Reviews in Environmental Science and Technology 32:73–295
2. Lofgren S (2001) The chemical effects of deicing salt on soil and stream water of five catchments in southeast Sweden. Water Air and Soil Pollution 130:863–868
3. Bäckström M, Karlsson S, Bäckman L, Folkeson L, Lind B (2004) Mobilisation of heavy metals by deicing salts in a roadside environment. Water Research 38:720–732
4. Ruth O (2003) The effects of de-icing in Helsinki urban streams, Southern Finland. Water Science and Technology 48:33–43
5. Hindar A, Henriksen A, Kaste O, and Torseth K (1995) Extreme acidification in small catchments in southwestern Norway associated with a sea salt episode. Water Air and Soil Pollution 85:547–552

6. Spry DJ, Wiener JG (1991) Metal bioavailability and toxicity to fish in low-alkalinity lakes – a critical review. Environmental Pollution 71:243–304
7. Reynaud S, Deschaux P (2006) The effects of polycyclic aromatic hydrocarbons on the immune system of fish: a review. Aquatic Toxicology 77:229–238
8. van der Oost R, Beyer J, Vermeulen NPE (2003) Fish bioaccumulation and biomarkers in environmental risk assessment: a review. Environmental Toxicology and Pharmacology 13:57–149
9. Vollertsen J, Åstebøl SO, Coward JE, Fageraas T, Madsen HI (2006) Highway and urban environment: Proceedings of the 8th Highway and Urban Environment Symposium. In Highway and Urban Environment Symposium, Morrison GM and Rauch S, (eds). Cyprus, Springer, pp. 499–509
10. Rosseland BO, Massabuau J-C, Grimalt J, Hofer R, Lackner R, Raddum G, Rognerud S, Vives I, (2001) Workpackage 5: Fish ecotoxicology. The EMERGE fish sampling manual for live fish. pp. 7
11. Halliwell B, Gutteridge JMC (2007) Free radicals in biology and medicine (Oxford: Oxford University Press)
12. Bartsch R, Klein D, Summer KH (1990) The Cd-Chelex assay – a new sensitive method to determine metallothionein containing zinc and cadmium. Archives of Toxicology 64:177–180
13. Olsvik PA, Gundersen P, Andersen RA, Zachariassen KE (2001) Metal accumulation and metallothionein in brown trout, Salmo trutta, from two Norwegian rivers differently contaminated with Cd, Cu and Zn. Comparative Biochemistry and Physiology C-Toxicology and Pharmacology 128:189–201
14. Bradford MM (1976) Rapid and sensitive method for quantization of microgram quantities of protein utilizing principle of protein–dye binding. Analytical Biochemistry 72:248–254
15. Åstebøl SO (2004) Overvåkning av rensebasseng for overvann fra E6 Skullerudkrysset i Oslo, 2003–2004 (In Norwegian). In Vegdirektoratet Utbyggingsavdelingens rapportserie. pp. 29, Statens vegvesen: Oslo
16. Rosseland BO, Blakar IA, Bulger A, Kroglund F, Kvellstad A, Lydersen E, Oughton DH, Salbu B, Staurnes M, Vogt R (1992) The mixing zone between limed and acidic waters – complex aluminum chemistry and extreme toxicity for salmonids. Environmental Pollution 78:3–8
17. Teien HC, Salbu B, Heier LS, Kroglund F, Rosseland BO (2005) Fish mortality during sea salt episodes - catchment liming as a countermeasure. Journal of Environmental Monitoring 7:989–998
18. Lydersen E, Poleo ABS, Nandrup Pettersen M, Riise G, Salbu B, Kroglund F, Rosseland BO (1994) The importance of "in situ" measurements to relate toxicity and chemistry in dynamic aluminium freshwater systems. Journal of Ecological Chemistry 3:357–265
19. Paquin PR, Zoltay V, Winfield RP, Wu KB, Mathew R, Santore RC, Di Toro DM (2002) Extension of the biotic ligand model of acute toxicity to a physiologically-based model of the survival time of rainbow trout (*Oncorhynchus mykiss*) exposed to silver. Comparative Biochemistry and Physiology C-Toxicology and Pharmacology 133:305–343
20. Bonga SEW, Lock RAC (1992) Toxicants and osmoregulation in fish. Netherlands Journal of Zoology 42:478–493
21. Rosseland BO, Staurnes M (1994) Physiological mechanisms for toxic effects and resistance to acid water: an ecophysiological and ecotoxicological approach. In Acidification of Freshwater Ecosystems: Implications for the Future, Steinberg CEW and Wright RW, (eds). Wiley, New York, pp. 227–246
22. Lauren DJ, McDonald DG (1985) Effects of copper on branchial ionoregulation in rainbow-trout, Salmo-Gairdneri Richardson – modulation by water hardness and pH. Journal of Comparative Physiology B-Biochemical Systemic and Environmental Physiology 155:635–644
23. Rogers JT, Patel M, Gilmour KM, Wood CM (2005) Mechanisms behind Pb-induced disruption of Na+ and Cl– balance in rainbow trout (*Oncorhynchus mykiss*). American Journal of Physiology-Regulatory Integrative and Comparative Physiology 289:R463-R472

24. Hollis L, Muench L, Playle RC (1997) Influence of dissolved organic matter on copper binding, and calcium on cadmium binding, by gills of rainbow trout. Journal of Fish Biology 50:703–720
25. Grosell M, Gerdes R, Brix KV (2006) Influence of Ca, humic acid and pH on lead accumulation and toxicity in the fathead minnow during prolonged water-borne lead exposure. Comparative Biochemistry and Physiology C-Toxicology and Pharmacology 143:473–483
26. Tao S, Liang T, Liu CF, Xu SP (1999) Uptake of copper by neon tetras (*Paracheirodon innesi*) in the presence and absence of particulate and humic matter. Ecotoxicology 8:269–275
27. Martinez-Alvarez RM, Morales AE, Sanz A (2005) Antioxidant defences in fish: Biotic and abiotic factors. Reviews in Fish Biology and Fisheries 15:75–88
28. Mason AZ, Jenkins KD (1995) Metal detoxification in aquatic organisms. In Metal speciation and bioavailability in aquatic systems, Volume 3, Tessier A and Turner DR, (eds). Wiley, New York, pp. 479–608
29. Lushchak VI, Bagnyukova TV (2006) Effects of different environmental oxygen levels on free radical processes in fish. Comparative Biochemistry and Physiology B-Biochemistry and Molecular Biology 144:283–289
30. Landis WG (2003) Twenty years before and hence; ecological risk assessment at multiple scales with multiple stressors and multiple endpoints. Human and Ecological Risk Assessment 9:1317–1326
31. Sampaio FG, Boijink CDL, Oba ET, dos Santos LRB, Kalinin AL, Rantin FT (2008) Antioxidant defences and biochemical changes in pacu (*Piaractus mesopotamicus*) in response to single and combined copper and hypoxia exposure. Comparative Biochemistry and Physiology C-Toxicology and Pharmacology 147:43–51
32. Dobrota D, Vachova MM, Kurella EG, Boldyrev AA (1999) Na/K-ATPase under oxidative stress: molecular mechanisms of injury. Cellular and Molecular Neurobiology 19:141–149

Potential of Road Surface for a Non-point Source of Pollutants – Result of Nationwide Survey in Japan

Norihide Tamoto, Takeyoshi Matsui, and Takashi Sakakibara

Introduction

There are many water areas in Japan where the quality standards for closed waters are not met. Eutrophication and its damages in closed waters, such as red tide and water bloom, are still serious [1].

In order to achieve the environmental water quality standards for closed waters, various measures have been implemented in Japan. For example, the Comprehensive Basin-wide Plans for Sewerage Works (CBPSW) are required to be drawn up when determining the amount of works and the treatment levels for satisfying the standards, which involves estimating the amount of pollutants that flow from the land into the water [2]. However, there is little available data on the effects of pollutants from non-point sources, and thus plans should be drawn up based on rough assumptions.

Road surface is a major non-point source of pollutants particularly in cities. Pollutants, such as nitrogen and phosphorus, which cause eutrophication, and heavy metals and chemicals, which may adversely affect people's health and ecosystems [3, 4] have been widely detected in storm water runoff from road surface. However, due to difficulties of predicting rainfall and high costs, only few surveys have been conducted and thus there is little available data. Therefore, data is insufficient for designing and taking countermeasures. In Japan, data that was collected several decades ago is still used, and thus acquisition of latest data is an important topic.

In this study, pollutant loads from road surfaces were surveyed at 36 points in four cities in Japan in 2006 and 2007. The objectives were to: (1) prepare new criteria on pollutant load from road surface in Japan, and (2) analyze the runoff characteristics

---

N. Tamoto and T. Sakakibara
National Institute for Land and Infrastructure Management, Ministry of Land, Infrastructure, Transport and Tourism, Government of Japan

T. Matsui
Japan Institute of Wastewater Engineering Technology

S. Rauch et al. (eds.), *Highway and Urban Environment*,
Alliance for Global Sustainability Bookseries 17,
DOI 10.1007/978-90-481-3043-6_28, © Springer Science+Business Media B.V. 2010

and changes in pollutant loads by comparing data collected in the past. Quantitative data of pollutant load from road surface will provide planners of watershed management with basic information for establishing precise the CBPSW and deciding the priority order of measures for controlling urban nonpoint sources (Table 1).

Method

Surveys were conducted in four cities in Japan twice in each site in 2006–2007 by sprinkling water. In each city, residential, commercial and industrial districts were extracted, and survey was conducted at three mutually close points in each sites. All the survey points were one-lane road each way and were set at where the nearby land use was uniform and there were no other pollutant sources, such as greenery and parks.

Equipment Used for the Survey

In the nationwide survey, mobile sprinkling equipment was prepared and used to reproduce rainfall events. The equipment has nozzles at the top from which water is sprinkled to reproduce rainfall artificially. Since pollutants are highly probable to be unevenly distributed on road surfaces, pollutants on the entire lane were decided to be collected. Thus, the sprinkling equipment was designed so as to change the length according to the width of the road. In a preliminary test, the equipment was verified to be capable of uniformly sprinkling water over road surface. The equipment is shown in detail in Fig. 1a and b. Moreover, the concentrations of the pollutants in runoff water were found to be high soon after the start of sprinkling but converged into certain values in 20 min after the start of spraying.

Methods of the Sprinkling Survey

The survey was conducted by first checking that the antecedent precipitation at the target site had been zero over at least 4 days and then spraying water at 30 mm/h for 60 min.

A preliminary test was also performed to compare data with that when the artificial rainfall intensity was set at 50 mm/h, in which almost no difference was shown. The runoff water from the road surface was collected in the street inlet and was all pumped into a container using a small pump. Specimens were sampled and subjected to quality analysis. Then, the road surface was scrubbed using a nylon brush for 20 min while spraying water at 30 mm/h to wash out pollutants. The scrubbing force was adjusted to be uniform in all survey points by installing metal

Table 1 Target points and date of the survey

City	Site (land use)	Point	Date No. 1	Antedecent precipitaiton	Spraying area (m^2)	Date No.2	Antedecent precipitaiton	Spraying area (m^2)
Chiba	Residential	Point 1	June 2, 2006	4	11.8	Oct 13, 2006	5	11.7
		Point 2	June 2, 2006	4	11.5	Oct 13, 2006	5	10.5
		Point 3	June 2, 2006	4	9.3	Oct 13, 2006	5	10.9
	Commercial	Point 1	Mar 9, 2006	5	13.6	Oct 12, 2006	4	13.7
		Point 2	Mar 9, 2006	5	13.4	Oct 12, 2006	4	13.7
		Point 3	Mar 9, 2006	5	13.6	Oct 12, 2006	4	14.0
	Industrial	Point 1	June 4, 2006	6	17.2	Oct 15, 2006	7	16.7
		Point 2	June 4, 2006	6	18.6	Oct 15, 2006	7	16.3
		Point 3	June 4, 2006	6	19.1	Oct 15, 2006	7	16.6
Yokohama	Residential	Point 1	Aug 3, 2006	9	8.8	Oct 19, 2006	11	8.8
		Point 2	Aug 3, 2006	9	7.5	Oct 19, 2006	11	7.5
		Point 3	Aug 3, 2006	9	8.2	Oct 19, 2006	11	7.9
	Commercial	Point 1	Aug 7, 2006	13	11.6	Oct 21, 2006	13	11.1
		Point 2	Aug 7, 2006	13	11.3	Oct 21, 2006	13	11.0
		Point 3	Aug 8, 2006	14	10.9	Oct 22, 2006	14	10.9
	Industrial	Point 1	Aug 2, 2006	8	13.5	Feb 4, 2007	8	13.9
		Point 2	Aug 2, 2006	8	14.2	Mar 10, 2007	4	14.2
		Point 3	Feb 4, 2007	8	14.4	Mar 10, 2007	4	14.4

(continued)

Table 1 (continued)

City	Site (land use)	Point	Date No. 1	Antedecent precipitaiton	Spraying area (m^2)	Date No.2	Antedecent precipitaiton	Spraying area (m^2)
Osaka	Residential	Point 1	Sep 22, 2006	7	10.6	Jan 23, 2007	5	10.5
		Point 2	Sep 22, 2006	7	10.3	Jan 23, 2007	5	10.3
		Point 3	Sep 22, 2006	7	8.5	Jan 23, 2007	5	8.5
	Commercial	Point 1	Aug 29, 2006	5	16.9	Jan 16, 2007	5	16.7
		Point 2	Aug 29, 2006	5	15.9	Jan 16, 2007	5	15.9
		Point 3	Jan 24, 2007	6	17.3	Jan 16, 2007	5	17.3
	Industrial	Point 1	Aug 28, 2006	4	18.1	Jan 15, 2007	4	18.1
		Point 2	Aug 28, 2006	4	18.6	Jan 15, 2007	4	18.6
		Point 3	Aug 28, 2006	4	18.8	Jan 15, 2007	4	18.8
Kobe	Residential	Point 1	Sep 27, 2006	8	9.9	Jan 12, 2007	4	10.0
		Point 2	Sep 27, 2006	8	8.7	Jan 12, 2007	4	8.8
		Point 3	Sep 27, 2006	8	8.3	Jan 12, 2007	4	7.9
	Commercial	Point 1	Sep 26, 2006	5	15.3	Nov 7, 2006	14	15.5
		Point 2	Sep 26, 2006	5	16.0	Nov 7, 2006	14	16.0
		Point 3	Sep 26, 2006	5	15.7	Nov 7, 2006	14	15.9
	Industrial	Point 1	Sep 28, 2006	7	15.8	Dec 2, 2006	4	16.1
		Point 2	Sep 28, 2006	7	15.1	Dec 7, 2006	4	15.1
		Point 3	Sep 28, 2006	7	14.9	Dec 22, 2006	4	16.0

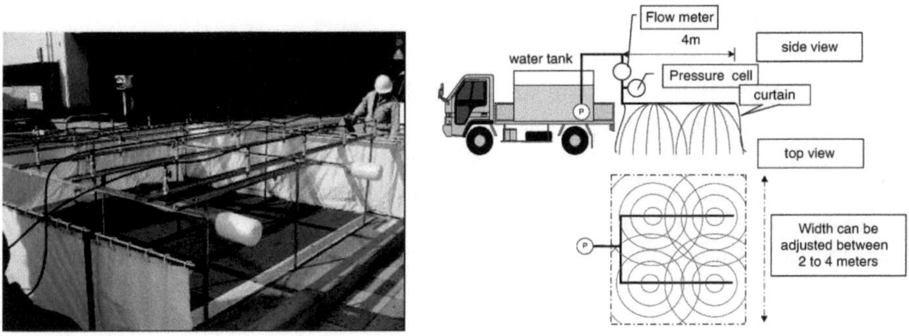

Fig. 1 (**a**) Mobile sprinkling equipment. (**b**) Schematic of the mobile sprinkling equipment

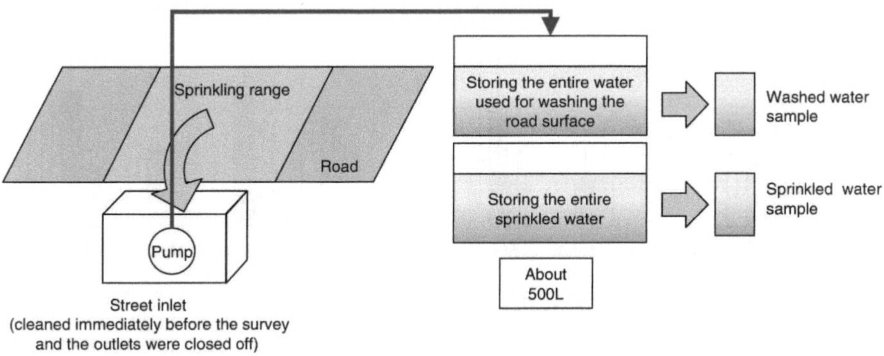

Fig. 2 Water sampling method

Table 2 Results of analysis of variance (Runoff load)

Source of variation	BOD	COD	SS	TN	TP
Land use		**		*	
City		**		**	
Land use × city		**			

*5% Significance.
**1% Significance.

weights. The runoff water from the road surface was also collected first in the street inlet, from which all the water was collected in another container using the pump. Specimens were sampled and subjected to quality analysis. The water sampling method is in Fig. 2. Points and dates of the survey are listed in Table 2.

Water quality constituents analyzed were BOD, COD_{Mn}, SS, Total Nitrogen (TN) and Total Phosphorus (TP).

In this paper, the pollutant load that were discharged just by spraying water is called "runoff load", that discharged by scrubbing is called "scrubbed load", and the sum of the two is called "potential load". Finally, the unit load is defined as the amount of pollutant discharged per unit area.

Results and Discussion

The runoff loads in the cities are summarized in Fig. 3 for each water quality constituent. The error range lines show the maximum and minimum values of the collected data. The mean unit loads for each land use are shown in Fig. 4. They were all calculated for the runoff and scrubbed values separately. Regardless of land use, the mean SS unit load was about 4 kg/ha. The scrubbed load was 8–34 times larger than the runoff load. The other water quality constituents also showed scrubbed

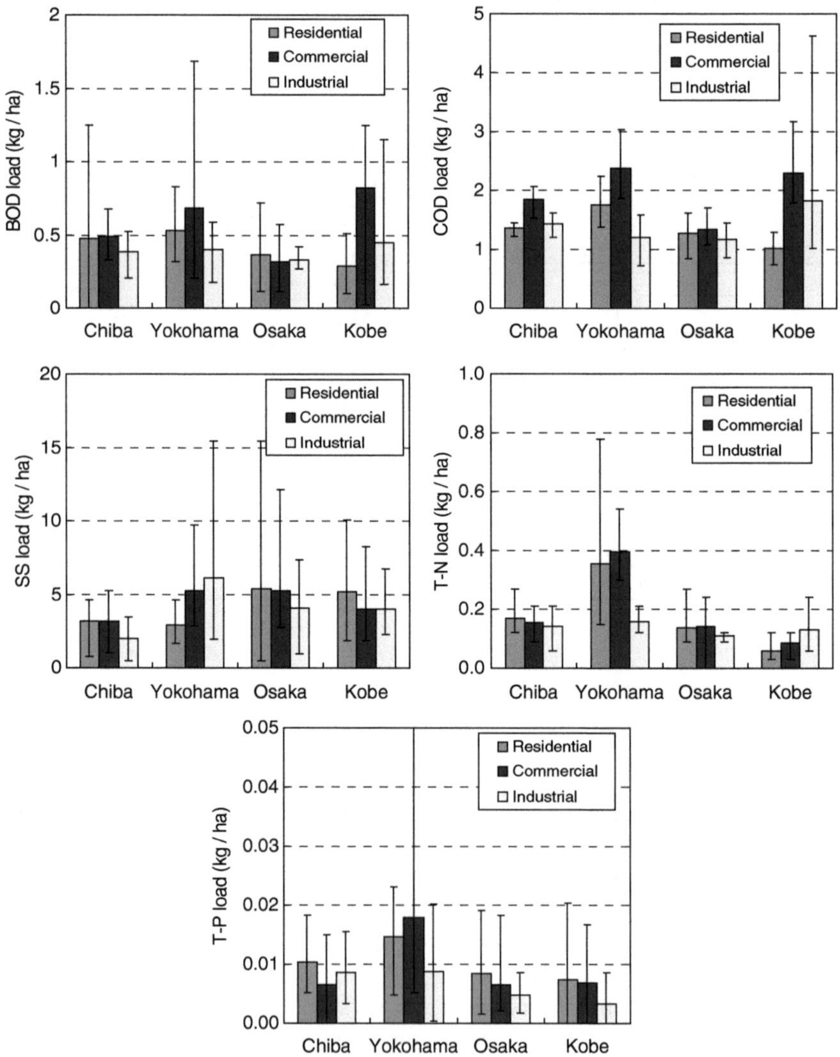

Fig. 3 Runoff loads in each city

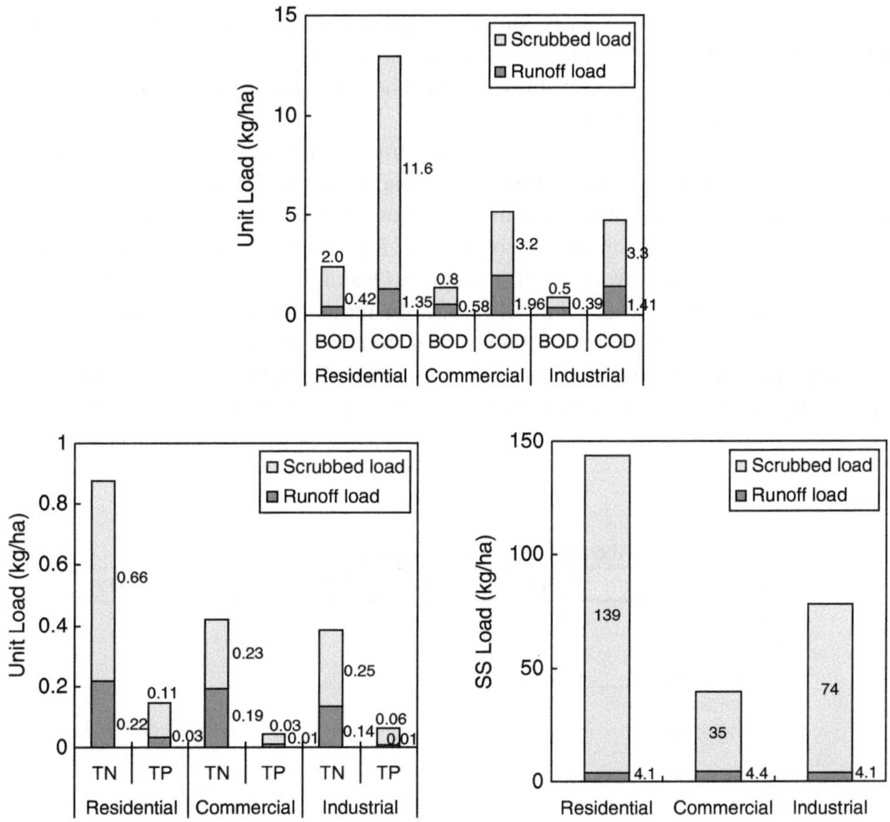

Fig. 4 Average unit loads (runoff and scrubbed loads)

loads larger than runoff loads. BOD, COD_{Mn}, and SS showed a trend of values in commercial district being larger than those in residential district, which were almost equal to those in industrial district. The mean runoff loads and scrubbed load shown in Fig. 2 were decided to be the load units obtained in this study.

Factors of Variation

A two-way layout analysis of variance was conducted on the collected 72 data sets of runoff loads using the nearby land use state and the city as the two factors of variation for each water quality constituent. The analysis showed that the factors affected the runoff loads of COD_{Mn} and TN at a level of significance of 5%. The factors were statistically suggested to little affect the runoff loads of the other constituents: BOD, SS and TP.

A two-way layout analysis of variance was also conducted on scrubbed loads. The results are shown in Table 3. For all water quality constituents, land use was judged to have affected the scrubbed loads. The effects of regional characteristics (city) were likely to be small on SS and TP.

To investigate the relationship between traffic volume and runoff load, the mean traffic volume was monitored at the road sections. In the survey, the numbers of pedestrians, automobiles and commercial vehicles, such as trucks and buses, per hour were counted. The results are shown in Fig. 5. In residential districts the traffic volume was small, except in Osaka. In commercial districts, the number of pedestrians was relatively large. In industrial districts, traffic volume especially that of commercial vehicles, was larger than in residential and commercial districts. The traffic volume in industrial districts was 100 times larger than in residential districts. However, the difference in runoff load between the

Table 3 Results of the analysis of variance (Scrubbed load)

Source of variation	BOD	COD	SS	TN	TP
Land use	**	**	**	**	**
City	**	**		**	
Land use × city					

*5% Significance.
**1% Significance.

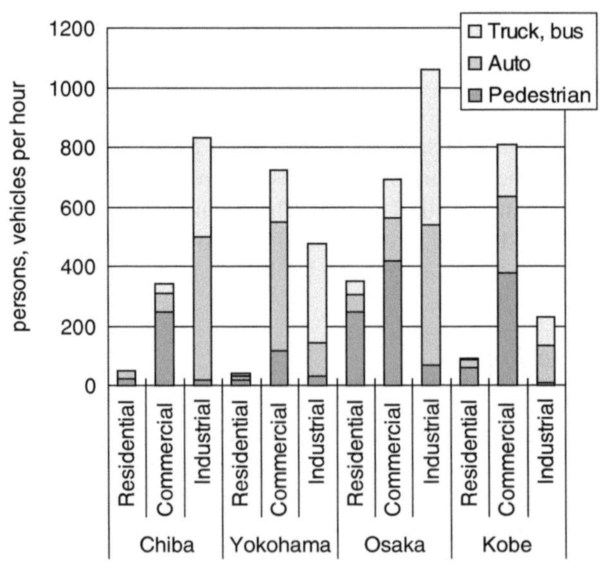

Fig. 5 Traffic volume at the sites surveyed

two land use states, shown in Fig. 4, was much smaller and was only several folds. The large pollution loads in commercial districts compared to those in residential and industrial districts suggest that pedestrians have large effects on runoff loads. Commercial activities and wastes produced by pedestrians are possibly a major source of pollutants.

Percentage of Runoff Load in Potential Load (Pollutant Runoff Coefficient)

The percentage of runoff load in potential load was defined as the "pollutant runoff coefficient" and is expressed as:

$$R = \frac{L_r}{L_s + L_r} \times 100 \tag{1}$$

where R, L_r and L_s are the pollutant runoff coefficient, runoff load, and scrubbed load, respectively. The mean pollutant runoff coefficient is shown in Fig. 6 for each city. In commercial districts, the coefficients were about 40% for BOD, COD_{Mn} and TN but were smaller than 20% for SS and TP. The results suggest that pollutants that exist in particle forms are scarcely discharged just by natural rain. The coefficient was smaller in residential districts than in the other districts.

The data was compared with the pollutant runoff coefficients determined by the Public Works Research Institute in 1975 [5]. The values calculated in this study were smaller than those of the Public Works Research Institute in general, but a similar trend was shown between the two when the ratio was calculated for each water quality constituent.

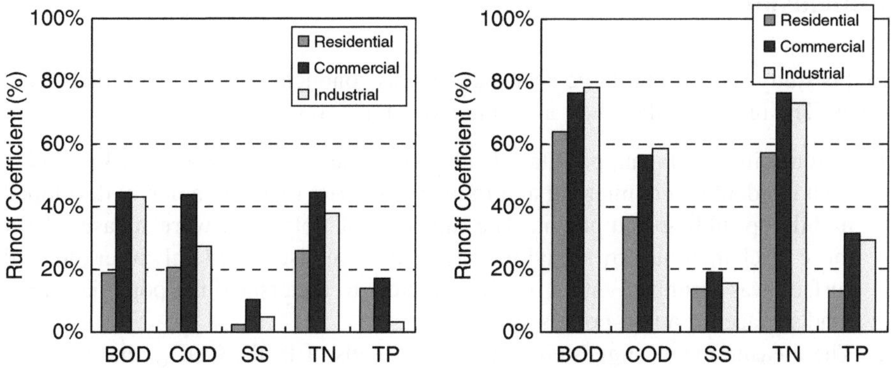

Fig. 6 Pollutant runoff coefficients. *Left*: this study, *right*: by the Public Works Research Institute (1975)

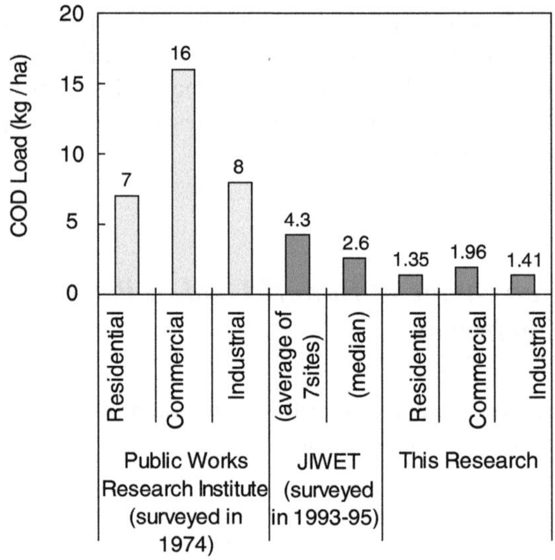

Fig. 7 Unit pollutant load from road surface

History of Unit Load in Japan

In Japan, national research institutes [5] and nonprofit foundations [6] have surveyed pollution loads from road surface. The data sets summarized in Fig. 7 cannot simply be compared with each other because the sites of the survey differed, but a decreasing trend over the years is shown. This was possibly attributable to reductions in dust fall from the atmosphere due to successful environmental policies, exhaust gas restrictions and road cleaning.

Conclusion

In this study, a basic and large-scale survey was conducted on pollutant loads from roads. The results of the study are summarized as follows.

1. In four cities of Japan, pollutant loads from road were surveyed at 24 points in each residential, commercial and industrial district by using a newly developed mobile sprinkling equipment. The amounts of pollutants were measured for those discharged just by spraying water and those discharged only after the road surface was scrubbed with a brush separately to understand the potential of the road surface for a non-point source.
2. The measurements suggested that the unit loads of BOD, COD_{Mn} and SS were the largest in commercial districts and were almost the same in industrial and residential districts, but with smaller values.

3. COD_{Mn} and TN loads were found to be likely affected by nearby land use state. On the other hand, the unit runoff load of SS was found to be little affected by regional characteristics.
4. As defined by Eq. 1, the pollutant runoff coefficients of SS and TP were smaller than those of the other water quality constituents.
5. A review of road surface load surveys conducted in Japan showed that the unit loads determined in this study were smaller than the values measured over 30 years ago and unit loads have decreased year after year.

Future works will include accumulation of precise data of expressways and highways and continuous updating of data by conducting monitoring. Understanding the percentages of loads from roads in the total load discharged from cities during rainfall is also indispensable for deciding the priority order of measures. Finally, the data collected in this study should be carefully handled because the study did not reproduce the friction by automobile tires and disturbance of surface runoff water during rainfalls. The unit loads from road surface determined in this study will be used as a standard unit load in sewerage projects in Japan.

Acknowledgements The authors express their gratitude to the cooperating staff of municipalities in the field survey for miscellaneous supports.

References

1. Ministry of Environment (2003) Quality of the Environment in Japan 2003, http://www.env.go.jp/policy/hakusyo_e/hakusyo.php3?kid = 221
2. Japan Sewerage Works Association (1999) Guideline and Commentary on Comprehensive Basin-wide Planning of Sewerage Works (in Japanese)
3. Lee Jun Ho, Bang Ki Woong (2000) Characterization of urban stormwater runoff. Water Research. 34:1773–1780
4. Kayhanian M, Stransky C, Bay S, Lau S-L, Stenstrom MK (2008) Toxicity of urban highway runoff with respect to storm duration. Science of the Total Environment. 39:386–406
5. Public Works Research Institute (1975) (in Japanese)
6. Ministry of Land, Infrastructure and Transport, Japan Institute of Wastewater Engineering Institute (2002) Tentative Guideline on Urban Nonpoint Pollution Control. (in Japanese)

Trace Metals in Stockholm Sediments

Sebastien Rauch, Malin Kylander, Maryam Khosravi, and Arne Jamtrot

Introduction

Trace elements typically occur at low concentrations (≤0.1%) in the environment. However, many trace elements are found at elevated concentrations as a result of human activities. Contaminated areas include cities where populations are exposed to contaminants. Stockholm is Sweden's largest city with a population of approximately 770,000 and a total of 1,200,000 inhabitants in the metropolitan area. As in most cities, anthropogenic activities are causing environmental contamination, including the contamination of aquatic systems [1, 2].

The project described here aims at determining the level of relevant trace metals in Stockholm sediments, that is Ag, Cd, Cu, Pb, Pt, Rh, Sb, Sn, W and Zn. Pt and Rh are emitted from automobile catalysts into the environment [3] and transport in stormwater has resulted in the contamination of aquatic systems [4]. Although little is known about the toxicity of these metals, uptake by benthic organisms has been reported [5], raising concern over potential chronic effects. Other relevant trace metals include W (used in tyre studs) and Sb (emitted from fossil fuel combustion, waste incineration and automotive parts, listed as priority pollutant by the European Union) as well as more commonly studied metals including Cd, Cu, Pb and Zn

S. Rauch (✉)
Department of Civil and Environmental Engineering, Water Environment Technology,
Sven Hultins gatan 8, 41296 Göteborg, Sweden
e-mail: Sebastien.Rauch@chalmers.se

M. Khosravi
Department of Civil and Environmental Engineering, Chalmers University of Technology,
412 96 Göteborg, Sweden

M. Kylander
Department of Geology and Geochemistry, Stockholm Univer sity, Stockholm, Sweden

A. Jamtrot
Environment and Health Administration, City of Stockholm, Stockholm, Sweden

S. Rauch et al. (eds.), *Highway and Urban Environment*,
Alliance for Global Sustainability Bookseries 17,
DOI 10.1007/978-90-481-3043-6_29, © Springer Science+Business Media B.V. 2010

Methodology

A set of 17 sediments samples was collected in the Stockholm area in December 2006. Collection was performed using piston corers. The top 2 cm was retained for analysis. The samples include 11 samples from the outflow of Lake Mälaren into the Baltic Sea (Lambarfjärden, Klubbensborg, Gröndal, Riddarfjärden, Kastellholmen, Ulvsundsjön, Barnhusviken, Saltsjön, Årstadal, Årstaviken, Lilla Värtan and in the Archipelago downstream from Stockholm) and six samples from lakes and ponds in the Stockholm area (Trekanten, Drevviken, Bornsjön). Lake Bornsjön is considered as a reference. Low concentrations are also expected at Lambarfjärden owing to its location upstream from the urban area.

Samples were prepared by microwave-assisted acid digestion (Mars5, CEM, USA) and analyzed by inductively coupled plasma-mass spectrometry (ICP-MS). Sediment samples were dried overnight at 105°C. Dry sediments were then sieved and the 0–75 μm fraction was retained for analysis. Approximately 0.5 g of dry sediments was placed in microwave digestion vessels. Digestion was performed in closed vessels after addition of 8 mL *Aqua regia* (6 mL HCl + 2 mL HNO_3) using a two-step temperature increase to 185°C and a maximum allowed pressure of 200 psi. The leachate was then slowly taken to dryness on a hot plate and redissolved in 5% HNO_3. The procedure does not provide a full digestion, but a strong leach suitable for the study of anthropogenic metals in sediments.

Prepared samples were analyzed by ICP-MS using a quadrupole instrument (Elan 6,000, Perkin Elmer, USA) and standard operating conditions. Calibration was performed by the analysis of multi-element standard solutions, except for Pt and Rh for which solutions of individual elements were analysed due to interference issues. Interference on Pt and Rh determination were corrected mathematically [5, 6].

Several precautions and quality control steps were implemented in the study. These include acid washing of all containers, the use of high purity acids (Suprapure grade, Promochem AB, Sweden) and water (18 MΩ cm, Milli-Q, Millipore, USA), the analysis of procedural blanks (prepared and analyzed randomly). In addition, reference material BCR-723 was analyzed to assess the accuracy of Pt and Rh analysis.

Trace Metal Concentrations in Surface Sediments

Copper, Zinc, Lead and Cadmium

The distribution of Cd, Cu, Pb and Zn in Stockholm sediments is presented in Fig. 1. Copper, zinc, lead and cadmium concentrations in surface sediments range from 27 to 475, 76 to 1,200, 15 to 772 and 0.4 to 4.0 μg/g, respectively.

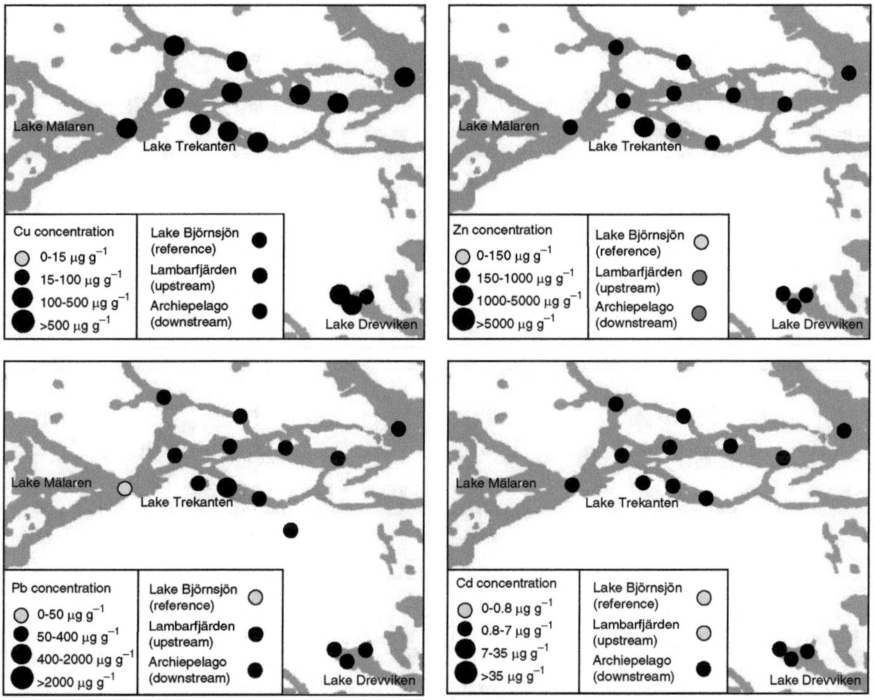

Fig. 1 Distribution of Cu, Zn, Pb and Cd in Stockholm sediments. The two highest concentrations represent high and very high concentrations in Swedish EPA sediment concentration guidelines [7]

The lowest concentrations are found at the background sites (Lake Bornsjön and Lambarfjärden), although Cu and Zn are considered to be moderately high at Lambarfjärden. The highest concentrations are found in the inner city area. Copper concentrations are considered to be high throughout the city, while high Zn concentrations are only found at Lake Trekanten. Urban Zn and Pb concentrations are in general moderately high, with high concentrations only observed at Lake Trekanten and Årstadal, respectively. Cadmium concentrations are in general low or moderately high.

Antimony, Tungsten, Tin and Silver

Antimony concentrations range from 0.1 to 8.9 μg/g. The lowest concentration was found at Lake Bornsjön. Concentrations do not exceed 2.5 μg/g, except for Årstadal, where an average concentration of 8.9 μg/g was measured (Fig. 2).

Tungsten concentrations range from 1.1 to 37 μg/g, with relatively low concentrations at most sampling sites and elevated concentrations only present at a limited

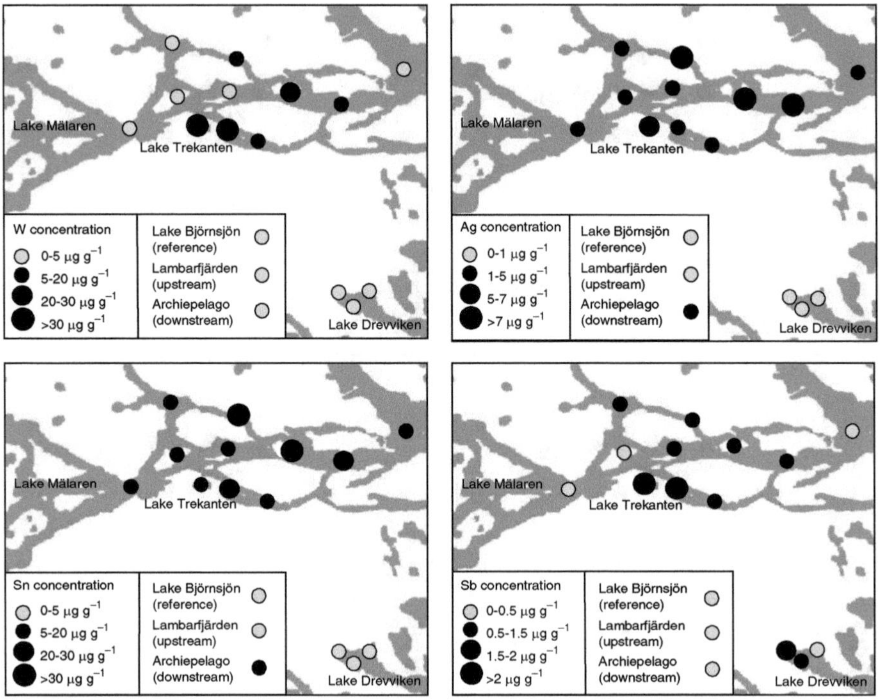

Fig. 2 Distribution of W, Sb, Sn and Ag in Stockholm sediments

number of sampling sites. Concentrations at Kastellholmen, Trekanten and Årstadal exceed 20 µg/g (Fig. 2).

Tin concentrations in surface sediments range from 1.2 to 38 µg/g (Fig. 2). The lowest concentration was found at Lake Bornsjön and maximum concentrations were measured at Barnhusviken. Average concentrations at Barnhusviken, Årstadal, Kastellholmen and Saltsjön exceed 25 µg/g.

Silver concentrations range from 0.2 to 8.5 µg/g with the lowest concentrations at background sites (Bornsjön and Lambarfjärden) and the highest concentrations in the inner city area (Barnhusviken, Kastellholmen, Saltsjön and Trekanten) (Fig. 2). The relatively high Ag concentration in the Archipelago downstream from Stockholm indicates an export of Ag into the Baltic Sea.

Platinum and Rhodium

Platinum and rhodium concentrations in sediments range from 2.4 to 25 ng/g and from 0.3 to 5 ng/g, respectively. The lowest concentrations are observed at background sites (Mälaren and Bornsjön), as well as at the downstream site (Archipelago), while the highest concentrations were measured in the inner city (Barnhusviken and Saltsjön) and in Lakes Trekanten and Drevviken (Fig. 3).

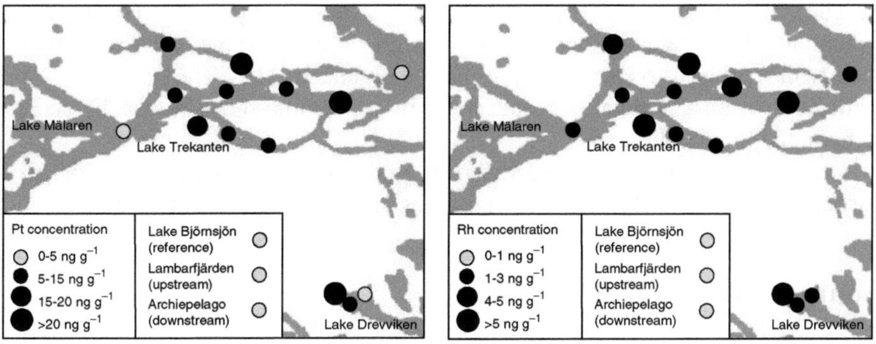

Fig. 3 Distribution of Pt and Rh in Stockholm sediments

Table 1 Sediment quality guidelines values

Metal	High/very high content [7]	Apparent effect threshold [8]	Probable effect concentration [9]
		μg/g	
Cd	>7	7.6	4.98
Cu	>100	840	149
Pb	>400	260	128
Zn	>1,000	520	459
Ag	–	4.5	–
Sb	–	3.0	–

Assessment of Potential Risks

Potential risks from the occurrence of metals in sediments were assessed by comparison of concentrations with Swedish EPA sediment quality guideline values [7], as well as reported apparent effect threshold concentrations (AET, concentration of a selected chemical above which statistically significant biological effects always occur [8]) and probable effect concentrations (PEC, concentration above which adverse effects are expected [9]) (Table 1).

Guideline concentrations are exceeded at most sites for Cu, with the exception of background sites (Lake Bornsjon and Lambarfjarden) and one site at Lake Drevviken. Zn and Pb concentrations are only high at Lake Trekanten and Cd is not found at high concentrations. In contrast, measured concentrations exceeded PEC at nine sites for Cu, at five sites for Zn and at nine sites for Pb. PEC concentrations were not exceeded for Cd.

The AET concentration is exceeded at seven sites for Ag and at one site for Sb. Potential risk for Ag is further supported by the guideline value of 3.7 μg/g associated with frequent adverse effects to marine organisms [10]. The results therefore indicate that present metal concentrations are a potential threat to aquatic ecosystems in the Stockholm area.

Other metals have no defined sediment guideline values and the potential risks associated with their elevated concentrations are therefore difficult to define.

Source Characterization

Source characterization was performed on the basis of observed spatial trends, correlations and knowledge of metal uses. Elevated concentrations in the urban area relative to reference and upstream sites indicate that most analyzed metals have urban sources.

Automobile traffic is believed to be the main source of Pt, Rh, W and Sb owing to a decreasing trend from a highway stormwater input at Lake Drevvikken. A smaller concentration decrease was also found for Cu, Pb, Zn and Sn, indicating that automobile traffic may contribute to elevated concentrations of these elements. Other sources of Cu, Pb, Zn and Sn metals include urban surfaces (e.g. roof, galvanized materials, coatings, asphalt).

Elevated W, Sb, Sn, Cu, Zn and Pb concentrations at Årstadal and/or Trekanten indicate the occurrence of location-specific sources. The sampling sites, which are within 2 km of each other, are located in the vicinity of an industrial area characterized as environmentally hazardous and comprising industries and a goods terminal.

A decreasing trend downstream from Stockholm was observed for all metals, except Ag, indicating that Ag is release into the Baltic Sea with a higher efficiency than other analyzed metals. This is possibly due to the use and emission of Ag as nanoparticles [11].

Conclusions

Elevated trace metal concentrations were found in sediments in Stockholm as a result of urban contamination. Copper is present at high concentrations at most sampling sites according to the Swedish EPA sediment concentration guideline, while Zn and Pb are present at high concentrations at specific sites. In addition, the apparent threshold concentrations are exceeded at several sites for Ag and at one site for Sb. Elevated concentrations were also found for Pt, Rh and W, but no guideline exist for these metals. This study indicated that metals are threat to aquatic ecosystems in the Stockholm area and emissions should be reduced to lower associated risks.

Acknowledgements This study was performed with financial support from the Environment and Health Administration, City of Stockholm. Stockholm Vatten is thanked for sampling.

References

1. Jonsson A (2000) The trace of metals: use, emissions and sediment load of urban heavy metals, Ph.D. thesis, Linköping University, Sweden
2. Lindström M, Jonsson A, Brolin AA, Håkansson L (2001) Heavy metal sediment load from the city of Stockholm. Water Air and Soil Pollution: Focus 1:103–118
3. Palacios MA, Gomez MM, Moldovan M, Morrison G, Rauch S, McLeod C, Ma R, Laserna J, Lucena P, Caroli S, Alimonti A, Petrucci F, Bocca B, Schramel P, Lustig S, Zischka M, Wass U, Stenbom B, Luna M, Saenz JC, Santamaria J (2000) Assessment of environmental contamination risk by Pt, Rh and Pd from automobile catalyst. Science of the Total Environment, 25:1–15
4. Rauch S, Hemond HF, Peucker-Ehrenbrink B (2004) Recent changes in platinum group element concentrations and osmium isotopic composition in sediments from an urban lake. Environmental Science and Technology, 38:396–402
5. Moldovan M, Rauch S, Gomez M, Palacios MA, Morrison GM (2001) Bioaccumulation of palladium, platinum and rhodium from urban particulates and sediments by the freshwater isopod *Asellus aquaticus*. Water Research, 35:4175–4183
6. Rauch S, Lu M, Morrison GM (2001) Heterogeneity of platinum group metals in airborne particles. Environmental Science and Technology, 35:595–599
7. Swedish EPA (2000) Environmental Quality Criteria – Lakes and Watercourses. Report 5050, ISBN: 91-620-5050-8
8. Cubbage J, Batts D, Breidenbach J (1997) Creation and Analysis of Freshwater Sediment Quality Values in Washington State, Washington State Department of Ecology, Olympia, WA
9. US EPA (2000) Prediction of sediment toxicity using consensus-based freshwater sediment guidelines, EPA 905/R-00-007
10. Bothner MH, ten Brink MB, Manheim FT (1998) Metal concentrations in surface sediments of Boston Harbor – changes with time. Marine Environmental Research, 45:127–155
11. Benn TM, Westerhoff P (2008) Nanoparticle silver released into water from commercially available sock fabrics. Environmental Science and Technology, 42:4133–4139

Environmental Impacts of the A6 Highway in La Silva Stream in the Region of Bierzo. (Leon). Spain

Lucas Vadillo, Virginia Rodriguez, Esther Alberruche, Osvaldo Aduvire, Margarita Lacal, Victorio Monteserin, and Ines Iribarren

Introduction

La Silva valley is affected by old coal mine spoil dumps and several mine adits with neutral drainages. When mining was in activity, the adits generated acid mine drainages (AMD). The natural attenuation availability of La Silva neutralized them and the stream supported a brown trout (*Salmo trutta*) population. The A6 highway construction has changed the physiognomy and the characteristics of 2 km of La Silva headwaters. Due to the blasting of the rock mass (Luarca slates), high acidity levels are being produced. ARD is generated when pyrite (FeS_2) contained in the slates is exposed to air and water in oxidizing and non-alkaline conditions. The reactions involved in this processes are well described [1, 2]. ARD causes the degradation of the water quality and impacts the aquatic biota [3–5]. A case of ARD generation has occurred in Pennsylvania due to the construction of the highway I99 [6]. Due to the high concentrations of dissolved metals in the stream (mineral acidity), mainly Fe, Al and Mn, the alkalinity of the tributaries is insufficient to attenuate the acidity. The pH in the stream reaches 7 after 8 km. The contamination spreads to 15 km downstream, where the river-bed is covered by iron and aluminum precipitates and colloids giving reddish, ochre-yellowish and whitish colors. Only a few benthic species can withstand this conditions and the presence of ichthyofauna is unviable. The purposes of this survey are: to determine hydrogeochemical characteristics and spatio-temporal evolution of the ARD draining to La Silva; to analyze the dilution and precipitation mechanisms of some metals

L. Vadillo (✉), V. Rodriguez, E. Alberruche, M. Lacal, V. Monteserin and I. Iribarren
Departamento de Investigación de Recursos Geológicos,
Instituto Geológico y Minero de España, Madrid, Spain
e-mail: l.vadillo@igme.es

O. Aduvire
Golder Associates Peru S.A, Peru

S. Rauch et al. (eds.), *Highway and Urban Environment*,
Alliance for Global Sustainability Bookseries 17,
DOI 10.1007/978-90-481-3043-6_30,

in La Silva; to evaluate the pollution extent of ARD and the impact on biota (benthic macroinvertebrates and fishes).

The study area is located on La Silva Valley, at the eastern part of the region of Bierzo, province of León (Spain). The ARD is generated in the first 2 km of the headwaters and go down together with other tributaries over La Silva. La Silva Stream flows into the Tremor River, which in turn flows into the Boeza River.

The geology of the region is composed of slates Cambro-Ordovician age, unconformable underlying Upper Stephanian of Carboniferous age. Pyrite is the dominant mineral in the slate. The stretch affected by the construction is composed of a 500 m in thickness, clay-rich and dark, not carbonated or sandy slates which is intensely folded and with a very deep schistosity. This makes bedding not distinguishable. The overlying strata are composed of clay-rich slates intercalated with more sandy stretches containing iron layers which originate topographical ridge. The upper strata consist of clay-rich slates of 200 m in thickness. The materials are affected by a contact metamorphism, hardly visible to the naked eye. In the belated stages three extensional joints systems, with monomineral paragenesis formed by pyrite with idiomorphic crystals weathering change into limolite in surface, can be recognized. La Silva Stream washes these materials accumulated in great volume spoil dumps and trails resulting from the works of the A6 Highway leading to the acidity of water.

Methodology

In order to determine the physicochemical and biological water quality, samples of water, sediments, macroinvertebrates, and fish were taken at 13 sampling points in La Silva Stream and several tributaries (Fig. 1). pH, Eh, electrical conductivity, dissolved oxygen, turbidity, dissolved Fe, acidity and alkalinity were measured in situ with portable kits. SO_4^{2-} and HCO_3^- were measured by Absorption Spectrophotometry. TSS was determined by filtering and drying at 105°C. Metal concentration was measured by ICP-AES. Analytical results from the Confederación Hidrográfica del Norte (Ministerio de Medio Ambiente) have also been considered. pH and Eh were measured in sediments in a 1:2.5 sediment:water suspension. Metals were extracted by digestion with HF, $HClO_4$ and HNO_3, followed by drying and dissolution in HCl and subsequently analyzed by ICP-AES. The Tessier method of sequential extraction for the speciation of particulate trace metals [7, 8] was performed, and metal concentration was analyzed by ICP-AES. Benthic macroinvertebrates were collected with a 300 μm mesh kick net and fish was collected by electric fishing. Family identification was performed for EPT [9] and IBMWP [10] indexes calculation.

Results and Discussion

Until the construction of the highway, La Silva was a mining valley. Mining began 2 km downstream from the headwaters. The evolution of the geochemical characteristics of La Silva Stream is shown in Fig. 2. The date of beginning of the

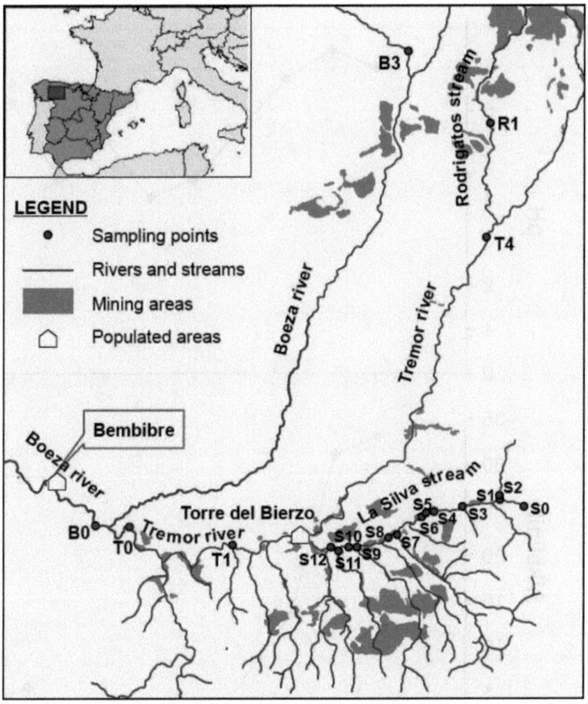

Fig. 1 Sampling points at the mining area of El Bierzo

construction is clearly noticeable, attending to the data of the Confederación Hidrográfica del Norte for the NO01590001 Station, located at 8,175 m from the headwaters and near to the S12 sampling point. These data also show the evolution of pH (from neutrality to nearly 4) and alkalinity (from 34 to 0 mg/L) in 2,000. Up to now these conditions have been maintained.

Prior to the construction, the natural attenuation processes and the dilution produced by the tributaries (mainly La Retuerta, Mostruelo and the adit) were sufficient to maintain a neutral pH in the stream and a population of brown trout. Some characteristics of the tributaries of La Silva can be seen in the Table 1.

Variation in Water Physicochemical Characteristics Along La Silva

In the headwaters of La Silva, the ARD mix with superficial runoffs during rainy periods and with small unpolluted streams which flow into the main bed. Both the runoffs and the discharges increase the flow of the stream, producing a dilution of the metallic content, a slight increase of pH and a precipitation that reduce especially Fe and Al contents. Main properties of the water samples of La Silva and its tributaries in winter 2005–2006 are shown in Table 2.

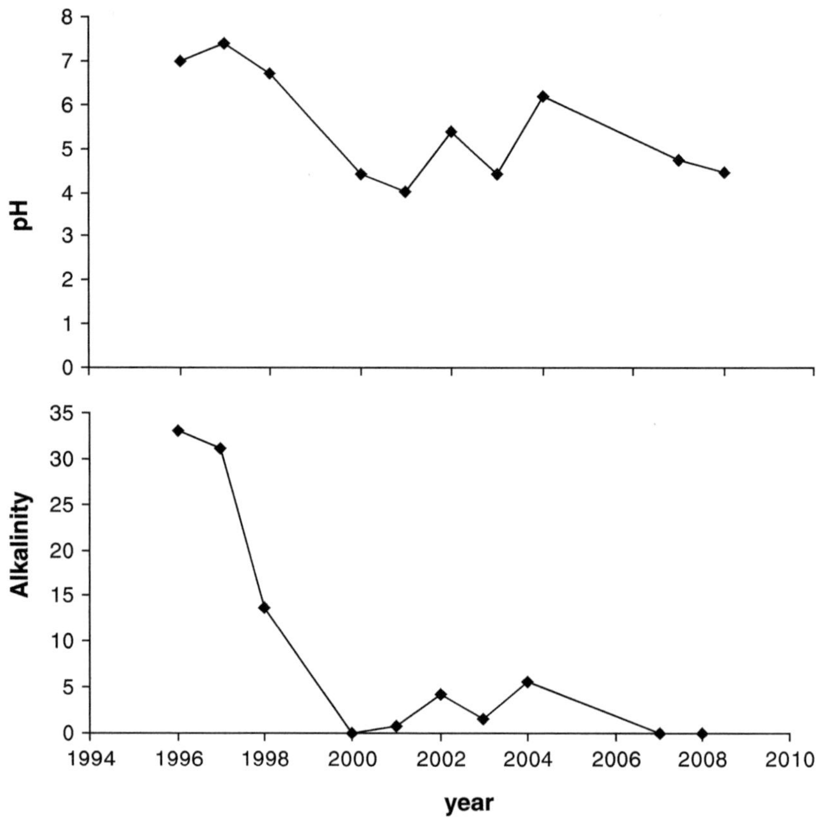

Fig. 2 Temporal evolution of the pH and the alkalinity (in mg/L) in La Silva from 1996 until 2008

Table 1 Characteristics of the tributaries of La Silva: flow, pH, alkalinity and distance from headwaters

Tribuary	Flow	pH	Alkalinity	Distance
Retuerta Stream	100	6.46	1	1,520
Adit	20	7	436	4,000
Mostruelo Stream	300	6.14	68	5,000
Fuente del Palomo Stream	200	4.78	2	7,150
Valdivieco Stream	60	5	1	7,500
Rinules Stream	20	4.45	0	8,200

Flow: unit in L/s; Alk.: alkalinity, unit in mg/L $CaCO_3$; Dist: distance, unit in m.

As can be seen in Fig. 2, these processes induce a decrease of the concentration of Fe from 93.7 mg/L at pH 3.2 to nearly 0 at pH 4, at 4,000 m from headwaters. This is probably due to the iron still dissolved in the water that ended to hydrolyze, resulting in the precipitation of iron oxyhydroxides, some of whom are transported downstream (Fig. 3).

Table 2 Characteristics of La Silva and its tributaries. Winter 2005–2006

Sampling point		Dist.	pH	SO_4^{2-}	Fe	Al	Cu	Ni	Mn	Zn	Alk.
S0	La Silva	0	3	730	81.3	68.31	0.42	0.83	6.02	0.7	0
S1	La Silva	1,500	3.2	746	93.7	234	2.6	4.35	7.91	0.22	0
S2	Retuerta	1,520	6.3	48	0.19	0.11	0.00	0.00	0.02	0.02	1
S3	La Silva	3,000	3.2	1,073	25.5	65.38	0.64	1.294	2.37	0.946	0
S4	La Silva	3,975	3.6	778	3.10	62.76	0.36	1.16	4.16	0.89	0
S5	Adit	4,000	7	689	8.29	0.09	0.00	0.00	7.10	0.00	436
S6	La Silva	4,025	4.5	836	3.83	42.44	0.29	0.87	4.24	0.77	0
S7	Mostruelo	5,000	6.8	99	0.00	0.01	0.00	0.01	0.00	0.14	68
S8	La Silva	5,025	4.7	556	1.06	20.10	0.17	0.51	2.89	0.44	0
S9	Palomo	7,150	6.8	423	0.72	0.197	0.00	0.118	3.11	0.14	2
S10	La Silva	7,175	4.9	425	1.08	7.98	0.086	0.28	2.49	0.26	0
S11	Valdivieco	7,500	6.4	513	3.16	0.19	0.00	0.09	3.71	0.11	1
S12	La Silva	7,550	5.2	418	1.15	4.54	0.07	0.25	2.32	0.27	0

Dist: distance, unit in m; concentration of SO_4^{2-} and metals, unit in mg/L; Alk.: alkalinity, unit in mg/L $CaCO_3$.

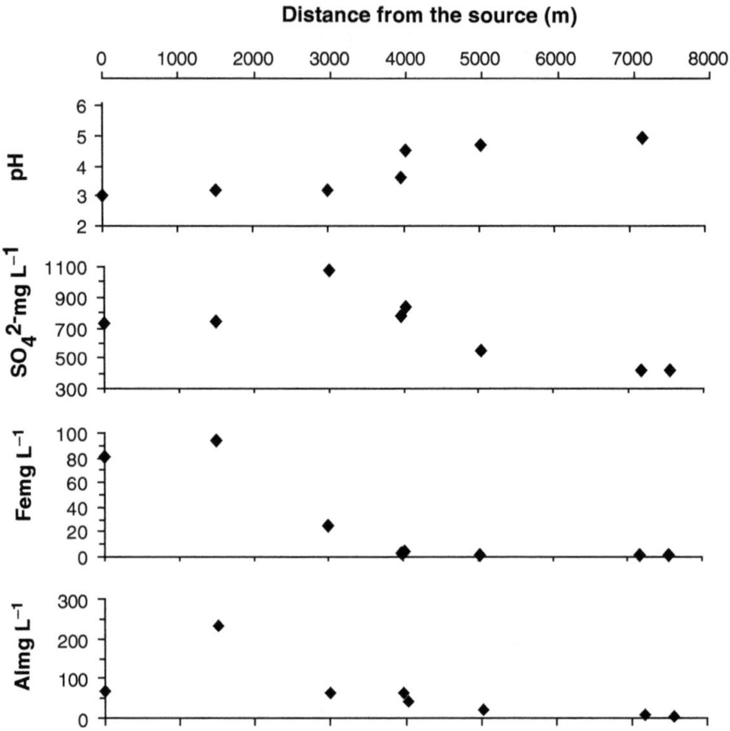

Fig. 3 Some parameters in La Silva Stream as a function of distance

Secondly, in the case of Al, a similar process as with iron occurs. The Al concentration decreases from 234 mg/L at pH 3.2 to 4.54 mg/L at pH 5.2 by means of hydrolysis and precipitation of Al over a stretch of nearly 8,000 m at the lower reach of the stream, coinciding with the final hydrolysis range of this element (pH from 3.5 to 5).

Thus, La Silva Stream has a slim layer of ochre precipitates all over its bed as a consequence of the oxidation of Fe^{2+} and the hydrolysis of Fe^{3+} and Al^{3+}. This layer includes amorphous (in its initial phase) precipitate like hydroxides (ferrihydrite and gibbsite) and oxyhydroxysulfates (schwertmannite, hydrobasaluminite and alumina) with significantly quantities of silica and sulphate, and even little adsorbed quantities of Cu, Zn, As, among others.

Acidity in La Silva Stream

The acidity produced by the ARD is due to the protons H^+ released to hydrolyzed metals in solution:

$$Fe^{2+} + 1/4O_2 + 3/2H_2O \rightarrow FeOOH + 2H^+$$
$$Fe^{3+} + 2H_2O \rightarrow FeOOH + 3H^+ \quad (1)$$
$$Al^{3+} + 3H2O \rightarrow Al(OH)_3 + 3\ H^+$$
$$Mn^{2+} + 1/4O_2 + 3/2H_2O \rightarrow MnOOH + 2H^+$$

These reactions are used to calculate the acidity produced by ARD using the expression: Acidity = $50[2Fe^{2+}/56 + 3Fe^{3+}/56 + 3Al^{3+}/27 + 2Mn^{2+}/55 + 1{,}000(10^{-pH})]$, where metal concentrations are in mg/L and 50 is the equivalent weight of $CaCO_3$ [8]. Water samples were taken in winter 2008 for analysis at S1 and an acidity curve was performed at the S1 station. This curve, in addition with another one corresponding to the winter 2006, is shown in Fig. 4. The main properties of the water at the S1 sampling point (La Silva) were: 2.75 pH, 56 mg/L Fe^{2+}, 108 mg/L Fe^{3+}, 504 mg/L Al^{3+}, 8.96 mg/L Mn^{2+}. Applying the above expression, the acidity was calculated to 3,289 mg/L $CaCO_3$. As can be seen in Fig. 4, the total acidity is close to the value obtained by the expression [8].

The acidity at S1 is very high (3,289 mg/L $CaCO_3$), and the alkalinity of the tributaries is very low. Thus they are not capable to neutralize the acidity, except for

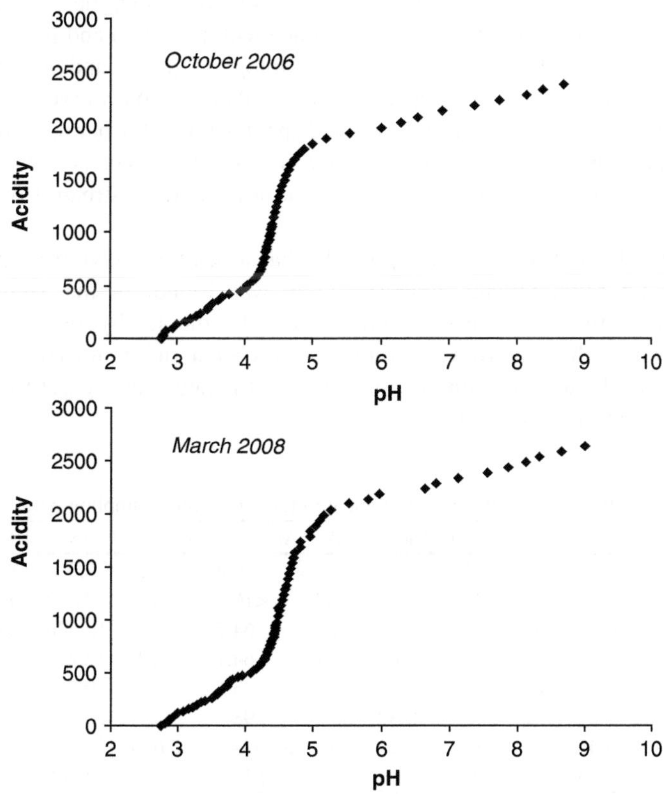

Fig. 4 Acidity curves in La Silva headwaters (acidity: units in mg/L $CaCO_3$)

the adit (S5) that shows 436 mg/L alkalinity and 20 L/s flow (Table 2). Even though this effluent is insufficient to neutralize the water of La Silva, it is capable to raise the pH from 3.6 to 4.5. Only after La Silva flows into the Tremor and having covered 8,000 m from the headwaters, the water reaches neutrality. In spite of this situation, the plume of contamination, noticeable due to the drag of Fe and Al precipitates, reaches Bembibre Village, 15,500 m downstream.

In the current environmental conditions, the biodiversity in La Silva is very poor. A comparison of the physicochemical and biological characteristics in differently polluted sampling points is showed in Table 3. B3 sampling point (Boeza River) is not affected by mining and shows a high environmental quality. B0 (Boeza River) is located downstream from the confluence of Tremor River. T0, T1 and T4 are located in the Tremor River, which is affected by old mining but have almost recovered. R1 sampling point is located in the Rodrigatos Stream, on the process of recovery. In both Tremor and Rodrigatos, the past mining activity is only noticeable by the high concentration of sulphates and of iron oxides. In order to compare results, S1 and S10 sampling points (affected by ARD) have been included. As can be seen in Table 3, La Silva Stream (S1 and S10) shows much lower IBMWP and EPT index values than the reference point (B3), being always lower than 5 (very poor biological quality). On the other hand, the sampling points T4 (Tremor River upstream from the confluence with La Silva) and R1 (Rodrigatos Stream) show values closer to B3 (very good biological quality and good biological quality, respectively). The negative effect of La Silva is also observed downstream from the confluence of the Tremor River and spreads much further up to the confluence of Boeza River. Thus, T1 and T0 show lower values than T4 (fair biological quality). The effect is still visible at B0, in Boeza River downstream from the confluence of Tremor (good biological quality).

Regarding the ichthyofauna, the point B3 has a healthy brown trout population whereas in La Silva Stream none specimen was found. In the Tremor River, upstream from the confluence of La Silva (T4), the ichthyofauna is present. However, as a result of the flowing water from La Silva, in T1 and T0 (downstream from La Silva) the ichthyofauna is nonexistent. It is necessary to get to the Boeza River (B0) to find brown trout.

Table 3 Physicochemical features and biological indexes in several sampling points

Sampling point		pH	SO_4^{2-}	Cond.	Turb.	Alk.	Al	Fe	Fe (S)	IBMWP	EPT
B0	Boeza	7.1	39	122	35	31.7	60.8	22	8,976	89	9
B3	Boeza	6.9	2	22	0.05	12.2	<10	<10	7,592	174	13
T0	Tremor	6.8	105	244	91	34	64.5	66	12,506	54	3
T1	Tremor	7.25	135	293	12.69	42.7	64.1	19	12,277	48	5
T4	Tremor	7.6	126	301	2.53	60	43.6	24	9,701	139	11
R1	Rodrigatos	7.2	304	575	5.52	65	98	83	10,037	86	6
S1	La Silva	3.45	438	781	0.21	0	52,644	10,640	175,178	0	0
S10	La Silva	4.73	289	507	25	0	8,548	808	12,665	5	2

Cond: conductivity, unit in μS/cm; Turb: turbidity, unit in FTU; Alk: alkalinity, unit in mg/L $CaCO_3$; Al, Fe: concentration in water samples, units in μg/L, Fe (S): concentration in Fraction III in sediments samples [7], units in μg/g.

Conclusion

The pollution by ARD in La Silva has resulted in high acidity levels as well as a high content of heavy metals. This is a consequence of the construction of the A6 Highway and the weathering of rocks with a high content of pyrite that have been exposed to the surface due to the blasting. Only 8,000 m downstream, the pH reaches neutrality and the concentration of metals diminishes. This high pollution inhibits the natural recovery ability of the stream. The drag in colloidal phase of aluminum salts and iron precipitates, as well as the precipitation of small amounts of iron at pH 6–8, and therefore causes the plume of contamination to be noticeable more than 15 km from the source. The result of the contamination is the absence of aquatic life in this stream and also along the stretch of the Tremor River where, La Silva flows into.

References

1. Nordstrom DK (1985) Rate of Ferrous Iron Oxidation in a Stream Receiving Acid Mine Affluent. USGS Water- Supply Paper 2270:113–119
2. Singer PC, Stumm W (1970) Acidic mine drainage: the rate-determining step. Science 167:1121–1123
3. Letterman RD, Mitsch WJ (1978) Impact of mine drainage on a mountain stream in Pennsylvania. Environ Pollut 17:53–73
4. Nichols LE, Bulow FL (1973) Effects of acid mine drainage on the stream ecosystem of the east fork of the Obey River, Tennessee. J. Tennessee Acad. Sci. 48:30–39
5. Scullion J, Edwards RW (1980) The effects of coal industry pollutants on the macroinvertebrate fauna of a small river in the South Wales coalfield. Freshwater Biol 10:141–162
6. Hammarstrom JM, Brady K, Cravotta CA (2004) Acid rock drainage at Skytop, Centre County, Pennsylvania. USGS Open-File Report 2005-1148
7. Tessier A, Campbell PGC, Bisson M (1979) Sequential extraction procedure for the speciation of particulate trace metals. Anal Chem 51:845–851
8. Hedin RS, Nairn RW, Kleinmann RLP (1994) Passive Treatment of Coal Mine Drainage. Bureau of Mines Information Circular, U.S. Dept. Interior, Bur. Mines Info. Circ. IC-9389:35
9. Klemm DJ, Lewis PA, Fulk F, Lazorchak JM (1990) Macroinvertebrate field and laboratory methods for evaluating the biological integrity of surface waters. EPA/600/4–90/030. U. S. Environmental Protection Agency. Environmental Monitoring Systems Laboratory, Cincinnati, Ohio 45268
10. Alba-Tercedor J, Sánchez-Ortega A (1988) Un método rápido y simple para evaluar la calidad biológica de las aguas corrientes basado en el de Hellawell (1978). Limnética 4:51–56Long term (€)

Part V
Contaminated Sites and Treatment

Arsenic Contamination After Wood Impregnation: Speciation, Sorption and Leaching

Emma Johansson, Kristine Ek, Malin Norin, and Ann-Margret Strömvall

Introduction

Arsenic (As) is a semi-metallic element, all forms are toxic and explain its use as a wood impregnation agent component. In the past, wood impregnation plants had no protection against dripping and leakage, resulting in a contaminated area around the impregnation equipment. In Sweden, the most commonly used impregnation agents containing arsenic are Boliden industry salt (BIS) and CCA fluids. BIS contains zinc (Zn) and As, and CCA fluids contain chromium (Cr), copper (Cu) and As. The most common CCA fluid is called Boliden K33. The As content of BIS is 18%, while in Boliden K33 it is 24% [1]. Since 2004, only professionals are allowed to use CCS-impregnated timber in Sweden. Several more environmentally friendly impregnation agents have been introduced in recent years. Remediation of arsenic contaminated sites will be common in the future and of the completed soil remediation projects in Sweden in 2006, 15% were former wood impregnation sites [2]. Research on suitable remediation techniques is in progress, and tests have demonstrated that arsenic in contaminated soil can be stabilized by addition of iron oxides [3].

In nature, arsenic is present in both inorganic and organic species. Inorganic As generally occurs in arsenite/As(III) form as $H_2AsO_3^-$ or $HAsO_3^{2-}$ or the form arsenate/As(V) as $H_3AsO_4^0$ or $H_2AsO_4^-$ [4]; the form depending on the redox potential and pH value [5]. Both forms often occur together, since As is reduced and oxidized rather slowly. As(V) is dominant in oxidizing environments and present as $HAsO_4^{2-}$ in alkaline and as $H_2AsO_4^-$ in acidic environments [6]. In reducing environments, As(III) is dominant in the neutral form $H_3AsO_3^0$ and only occurs in the

E. Johansson, K. Ek, and M. Norin
NCC Construction Sverige AB, SE-405 14, Göteborg, Sweden

A.-M. Strömvall (✉)
Water Environment Technology, Department of Civil and Environmental Engineering, Chalmers University of Technology, SE-412 96, Göteborg, Sweden

S. Rauch et al. (eds.), *Highway and Urban Environment*, Alliance for Global Sustainability Bookseries 17,
DOI 10.1007/978-90-481-3043-6_31, © Springer Science+Business Media B.V. 2010

ionized form $H_2AsO_3^-$ in basic environments at pH > 9.2. Arsenic is phytotoxic and hazardous to living organisms, the most toxic and mobile forms being inorganic As(III) species [7]. Inorganic As is transformed into organic As by the process of methylation, by which methylarsenic acid (MMA) and dimethylarsenic acid (DMA) are formed. The organic forms are less toxic and more stable compared to the inorganic forms and can bioaccumulate [8]. The organic forms are present at low concentrations in soil, water and sediment.

Arsenic binds strongly on soil colloids and is highly dependent on the content of clay, hydr/oxides surfaces of Al, Fe and Mn, calcium carbonates and/or organic matter [9]. It changes with pH, and the various forms have a different adsorption capacity; only smaller amount of As binds to soil particles at high pH since it forms anions in solution. It is therefore important to determine the form of As in order to understand leaching processes. In general, Fe hydr/oxides have the highest degree of involvement in the adsorption of As in both acidic and alkaline soils, but Al, Mn and organic matter may be important in the former. Arsenite binds more weakly and is less dependent on pH than As(V) [6]. The adsorption of As(V) on iron hydr/oxides is higher at low pH while that of As(III) is higher at higher pH. The formation of ferric hydr/oxides contributes to a natural reduction of the dispersal of As(V) in the environment [10]. In Sweden, Fe oxides are common in soil while pure Al oxides are less frequent [11]. In reducing conditions, iron hydr/oxides are dissolved, making the adsorbed As dissolve and become more mobile [6, 12]. Ferric hydr/oxides are metastable under most groundwater conditions and tend to dissolve in a discordant way, forming ferrous iron and releasing As into the solution [13]. In addition, the As form changes in reducing conditions; from As(V) to As(III) [6]. Furthermore, it is probable that salinity also affects the solubility of As, since both dissolved and particulate Fe concentrations decrease with increasing salinity [14].

The general leachability (K_d value) of As is estimated to 30 L/kg [15]. The K_d value is the ratio between the concentration of the substance in the solid and the dissolved phase:

$$K_d = \frac{solid\ phase\ concentration\,(mg\,/\,kg\ DS)}{dissolved\ phase\ concentration\,(mg\,/\,l)} \tag{1}$$

However, the concentration of As in soil affects the K_d value and more As is dissolved at high concentrations. A comparison of batch and column tests showed that higher concentrations yield up to three times higher K_d values [16]. For CCA-polluted soil, the K_d obtained in experimental conditions was lower compared to the calculated K_d.

The objectives of this study were to (i) determine concentrations of the different As species in arsenic contaminated soil and groundwater, (ii) determine the correlation of As concentration to the chemical soil characteristics, and (iii) study the leachability of As under different conditions and compare the laboratory results with As concentrations found in groundwater at the site. The results of the study will be used as a basis for decisions about further investigations and remediation actions at the site.

Experimental and Methods

The study site is situated less than 1 km from the centre of a city on the west coast of Sweden. Activities at the site include welding, engine and wood treatment and assembly. The CCA fluid Boliden K33 was used at the site from 1943 to 1979. As an example, 4,800 kg of Boliden K33 was purchased in 1978. The impregnation equipment, in the form of a barrel, was not kept in a protected area. After impregnation, the wood was dried in open areas close by. Sawdust formed during treatment of the dried wood was probably combusted in the area. In an environmental investigation carried out in 1998, the contaminants were judged highly hazardous and the dispersal potential was regarded as high. Due to the current use of the land area for industrial activities, the site was classified as risk class 2, which corresponds to "great risk" and not "high risk" as indicated by class 1 [17].

The property is flat with mostly hardened surfaces (asphalt) and has a total area of 80,000 m^2. The contaminated area around the former impregnation barrel is 1,800 m^2. The groundwater flow is generally towards the north where a stream is located. The groundwater surface is assumed to be at the same level as the water surface of the stream. The soil to the north of the former impregnation barrel consists of muddy sand and to the south of filling material comprising rocky gravely sand over lake sand on top of clay.

Samples

Soil samples were collected by means of an auger sample in September 2007, at 11 locations in the area of the former impregnation barrel. At each location, four to five samples were taken at different depths to a maximum of 2–3 m. The soil samples were analyzed for concentrations of total As and other metals. From the total of 52 soil samples collected in the area around the former impregnation barrel, 15 were selected on the basis of the As concentration levels, and new samples were mixed and classified as A-C with the following degrees of contamination:

- Soil A: 15–40 mg As/kg of dry substances (DS).
- Soil B: 40–400 mg As/kg DS
- Soil C: 400–3,000 mg As/kg DS

Five samples from each category were mixed in a mortar without crushing the material. Particle fractions >10 mm in diameter were removed by hand; however there were very few particles of that size. The mixed samples were then stored (+4°C) in diffusion-tight plastic bags designed for soil samples. The concentration of total As, As (III), As(V), DMA, MMA, other elements, water content and loss of ignition were then analyzed.

A set of batch tests was carried out aimed at studying the effects of mixing and aerobic/anaerobic conditions on the leachability of the different As species.

Soil was taken from sample C, with high concentrations of As and 100 g was added to each of the three beakers. Deionised and filtrated (Milli-Q) water was thereafter added to each beaker (L/S = 6). The first beaker was left undisturbed in an air/oxygen environment; the second was left undisturbed under anaerobic conditions by means of nitrogen gas gently bubbled into the water phase. In the third beaker, soil and water was vigorously stirred in an air/oxygen environment. After 20 h, water samples were taken from all beakers. The samples were filtrated through 0.45 μm cellulose acetate filters and the filtrates were sent for analysis of total As, As(III), As(V), DMA and MMA concentrations. The stirred sample was difficult to filtrate because the filter became clogged by particles and therefore this sample was pre-filtrated through a 1.2 μm micro fiberglass filter.

Leaching tests were carried out following a standardized procedure with two-step leaching (SS-EN 12457–3). In the first step, the soil was shaken with Milli-Q water (L/S = 2) for 6 h. The leachate water was filtrated and sent to the laboratory for chemical analysis. The sample was thereafter shaken again (L/S = 6) for 18 h and the leachate filtrated and sent for chemical analysis of total As.

Groundwater samples were collected on three occasions from the same well in the contaminated area around the former impregnation barrel. Before sampling, the groundwater well was turned over five times. On the first sampling occasion the groundwater was not filtrated in the field. On the second and third occasion the water was filtrated in the field and samples taken with a pump with an attached filter (0.45 μm) in the groundwater tube. All samples were collected in plastic flasks filled up to the top, sealed with a lid and stored cold.

The soil samples, that is the original 52, the new mixed samples A–C, and the groundwater samples were analyzed for concentrations of total As, other elements and dry substance by a certified environmental laboratory. For each sample, 1 g of dried soil was microwave digested in nitric acid and hydrogen peroxide. The samples were subsequently filtered or centrifuged (depending on particle content) before analysis. Total element concentrations were analyzed by inductively coupled plasma atomic emission spectroscopy (ICP-AES). Loss of ignition was determined at the Chemical Environmental Laboratory at Chalmers following a standardized procedure.

The mixed soil samples A–C, leachates from the laboratory experiments and the groundwater samples were analyzed for concentrations of the following As species: arsenite/As(III), arsenate/As(V), monomethylarsonic acid (MMA) and dimethylarsinic acid (DMA) by a certified environmental laboratory. The laboratory is not certified for the As speciation method, but the analyses were carried out by specialists in a research group at a Swedish university. One gram of the soil sample was treated with a phosphate buffer to leach the arsenic. The leachate was then centrifuged and filtered through a 0.45 μm filter. Immediately prior to analysis, the sample was acidified to 0.07 M with HNO3 and diluted. Samples were analyzed by ion chromatography inductively coupled plasma mass spectrometry (IC-ICP-MS). Leachates and groundwater samples were only filtered through a 0.45 μm filter and subsequently treated and analyzed in the same way as the soil samples.

Results and Discussion

The results revealed varying concentrations of the metals in the soil, extremely high concentrations, mg/kg of dry substances (DS), of total As up to 3,000, but also Pb ≤ 3,100, Cd ≤ 16, Cu ≤ 2,000, total Cr ≤ 1,200 and Zn ≤ 870 in close proximity to the former impregnation barrel. All these values significantly exceed the Swedish guideline values for contaminated soil; the guideline value for total As is 15 mg/kg DS. A mean As concentration in the soil in the highly contaminated area around the former impregnation barrel was calculated to 250 mg/kg DS. Extremely high concentrations of As in soil has also been reported at an industrial site in Belgium [18].

The dry matter content (75% in soils B and C, and 81% in soil A) indicates that soil A consists of larger grain sizes with a lower ability to bind water. The organic matter content was low (1.5%) in all soils. Total concentrations of several metals were analyzed in all soil samples and the results are presented in Fig. 1.

Soil C was highly contaminated with As, while soils A and B contained lower As concentrations. Soil C also had elevated levels of Cr, P and Al compared to soils A and B. Soil B generally exhibited low concentrations of all elements, although the As and Al levels were higher compared to soil A. Compared to soils B and C, soil A had higher concentrations of Fe, Mn, Cu, Ni, Pb and Zn.

The arsenic speciation showed that the soil samples had high levels of arsenate/As(V), see Fig. 2. This was expected, since the soil was unsaturated and therefore in an oxidizing environment; arsenate is formed under oxidizing conditions [6]. The arsenate/arsenite relation was around 80 in soil C, 70 in soil B and 20 in soil A, and similar relations were observed in a previous study [19]. The concentrations of DMA and MMA were low, which was expected because of the low organic content of the soils [12].

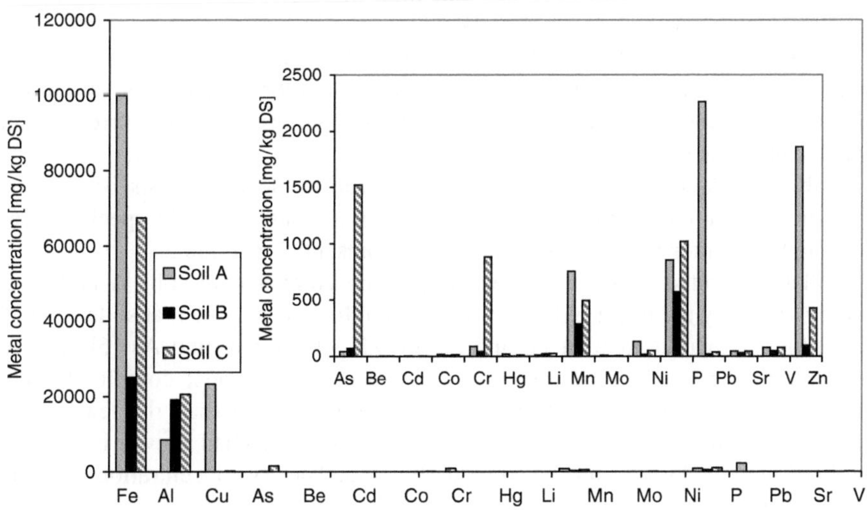

Fig. 1 Total metal and phosphorus concentrations in mixed soil samples A–C

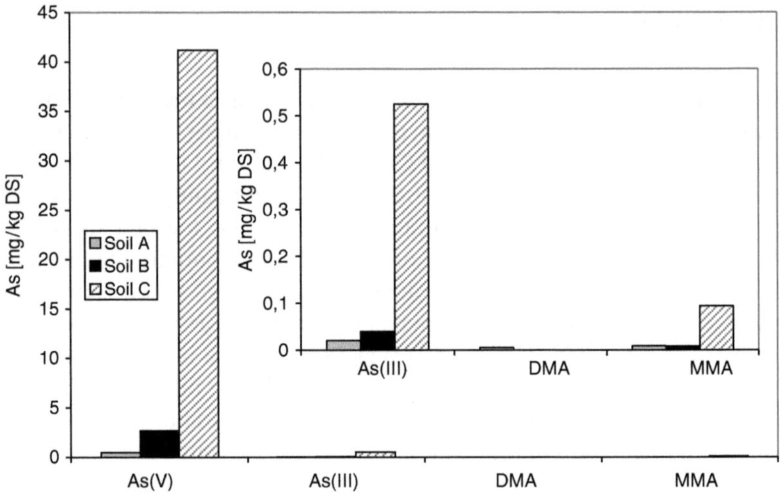

Fig. 2 Concentrations of As species in mixed soil samples A–C

Batch Tests – Leaching Aerobic/Anaerobic Conditions

The results of the As species' leachability from soil stored in beakers; one being stirred, one undisturbed and one containing an inert atmosphere, are compared in Table 1. The largest amount of leached As was found in the beaker with anaerobic conditions. This may be explained by the fact that As is more easily dissolved in reducing environments if bound as iron arsenate to the soil particles. In a reducing environment, iron oxides dissolve and thus the As in iron arsenate likewise dissolves and becomes more mobile [13].

The largest part (~90%) of all As leached in the samples was arsenate, As (V), which was expected due to the high concentrations in the soil. The lowest amount of leached As was found in the strongly oxidized environment in the stirred beaker, which causes As to bind to iron oxides.

Only total As concentrations were analyzed in the leaching tests because the shaking procedure oxygenates the samples and thus changes the As speciation [6]. Arsenic levels after 6 and 18 h of shaking are presented in Table 2.

Soil B exhibited a higher degree of leaching compared to soils A and C, as demonstrated by the lower Kd value. The reason for this may be that soils A and C had a higher total content of Fe and Al, metals which can bind As [6]. The As concentration in leachate from soils A and C decreased because complex binding of As with Fe increases over time. This binding is faster than the leaching of As into the water phase. On the other hand, in soil B, which had a lower Fe content, the leaching was faster than the complex binding. The As concentration therefore increased in leachate from soil B over time.

The K_d value was high in all soils; 800 (soil A), 77 (soil B) and 1,300 (soil C) times higher than the general K_d value of As of 30 L/kg [15]. These values indicate

Table 1 Total As, As species, Fe, Mn and Al concentrations in the water phase after anaerobic, aerobic (undisturbed) or aerobic (stirred) treatment of contaminated soil

	Stirred beaker-aerobic			Anaerobic			Undisturbed-aerobic		
Conc./Rel Conc./K_d	(µg/L)	(%)	(L/kg)	(µg/L)	(%)	(L/kg)	(µg/L)	(%)	(L/kg)
Total As	21			33			12		
Leached As[a]		0.0084			0.013			0.0048	
K_d value[b]			71,400			45,500			125,000
As(III) Arsenite	1.4	6		1.3	3		1.4	10	
As(V) Arsenate	20	89		38	94		13	89	
DMA	1.1	5		0.6	2		<0.1	<0.5	
MMA	<0.1	<0.5		0.4	<1		<0.1	<0.5	
ΣSpecies	23			40			15		
Fe	73	–		39	–		52	–	
Mn	82	–		96	–		110	–	

[a]Ratio between the As content in the water phase and in the soil (soil C; 1,500 mgAs/kg DS).
[b]As concentration in soil C (1,500 mg/kg DS) was used for the calculation of K_d.

Table 2 Total As, Fe and Al concentrations and K_d values from the leaching tests

	As in soil [mg/kg[a]]	As in leachate after 6 h (µg/L)	As in leachate after 18 h (µg/L)	K_d value after 6 h (L/kg)	K_d value after 18 h (L/kg)	Fe in soil (mg/kg[a])	Al in soil (mg/kg[a])
Soil A	41	1.7	1.5	24,000	27,000	100,000	8,500
Soil B	71	31	38	2,300	1,900	25,000	19,000
Soil C	1,500	34	28	45,000	54,000	68,000	21,000

[a]Of dry substances.

low leachability and lead to a classification of the soils as inert waste, which is incorrect because the As concentration in soil C is higher than the limit for hazardous waste (>1,000 mg/kg DS). The high K_d values in soils A and C may be explained by the binding of As to iron hydr/oxides [12] (Table 2). In addition, As(V) was the dominant form in the soils analysed, which was expected because As(V) adsorbs more strongly than As(III) to soil particles [1, 12]. Other studies also demonstrated equally high K_d values as those presented here [12]. In the present study, the K_d value was approximately three times higher in the batch test compared to the leaching test, which is supported by previous studies [15].

Groundwater

Table 3 shows As concentrations in groundwater from one sampling point on different sampling occasions and with different sample treatment. Groundwater samples revealed high concentrations of As, up to 42 mg/L but it varied greatly due

Table 3 Arsenic concentration in groundwater related to sampling occasion, groundwater table level, K_d value, Fe concentration and sample treatment

Sampling occasion	Gwt (m)	K_d value[a]	As conc (µg/L)	Fe conc (µg/L)	Sample treatment
Nov 21, 2007	0.9	410	610	23	Filtered on site and in laboratory
Nov 21, 2007	0.9	190	1,300	23	Unfiltered, acidified in laboratory
Nov 27, 2007	0.9	90	2,800	24000	Analysis redone March 3, 2008, acidified in laboratory
Dec 17, 2007	0.8	6	42,000	110	Filtered on site
Feb 07, 2008—	0.7	20	12,300	–	Filtered on site, acidified in laboratory
Feb 07, 2008	0.7	200	1,260	–	Filtered on site and in laboratory

[a]Calculated using a mean As concentration in the soil of 250 mg/kg DS. Concentration was calculated as a mean of all the analyzed soil samples from sampling points in the area.

to sample treatment and sampling occasion. Results also indicate that a fast complex formation occurs even after filtration of the sample.

The total As concentration is also dependent on the level of the groundwater table. Groundwater samples collected when the table was high contained more As compared to those collected when the table was lower (42 mg/L at 0.7 m and 0.61 mg/L at 0.9 m). This is probably due to the fact that, as it elevates, the groundwater reducing environment dissolves As that is bound to iron hydr/oxides in soil particles [12]. As the groundwater table fluctuates, As in soil can be dissolved into the groundwater, thus increasing the As concentration. Arsenic concentrations vary locally due to changes in the groundwater table level, inflow and precipitation [12]. At the site, the highly polluted area is mostly covered with asphalt, and stormwater is conducted through pipes to the stream. Direct infiltration of surface water is therefore small. It is considered that the seawater level has a large impact on the groundwater table level. As the groundwater is in direct contact with the seawater, a rise in the seawater level makes the As in the soil dissolve into the groundwater.

The calculated Kd values are significantly lower in the groundwater samples compared to those obtained in batch and leaching tests (Tables 1–3). Kd values are more than 300 times higher in the leaching tests and up to 7,500 times higher in the batch tests compared to the groundwater samples. This is because the rate of As dissolution from iron arsenate is higher in groundwater since it is a more reducing environment [13]. In addition, the higher salinity of groundwater probably causes more As to dissolve, since particulate Fe concentrations decrease with increasing salinity [14].

EU directive 2003/33/EG states that excavated material should be classified according to batch test results [20]. The present study demonstrate that the

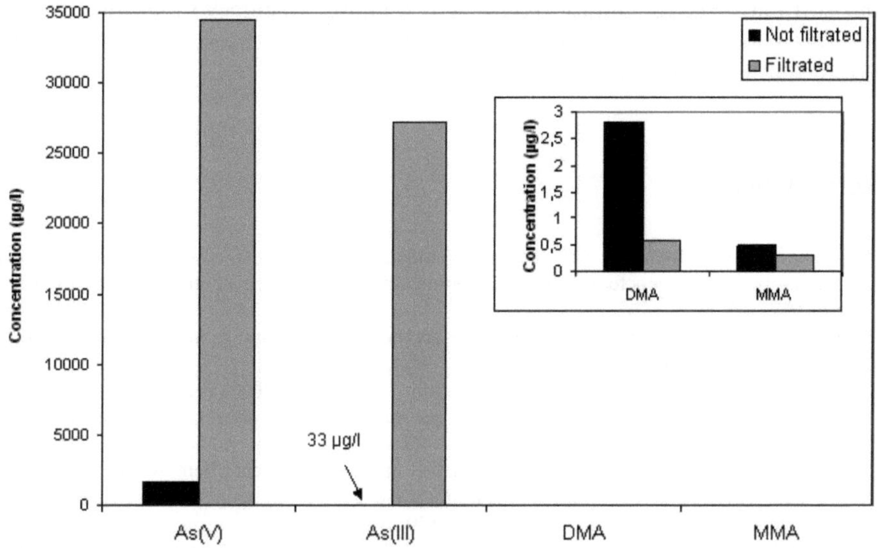

Fig. 3 A comparison of filtered and non-filtered samples in terms of As speciation of groundwater from the well in the highly contaminated area

leachability of As is underestimated in batch tests. Thus, excavated material can be classified as less hazardous than it is in reality.

On site filtration is crucial for the result of As speciation in water. If the water sample is not filtrated on site, iron oxides and soil particles containing oxides will remain. Arsenic(V) then reacts with these iron oxides and precipitates, while As(III) does not because it is not as easily adsorbed to metal oxides [12]. If the Fe content of the water sample is high, this effect can be significant. Thus, by not filtrating on site, the result for As(V) can be misleading. Figure 3 shows the As speciation in groundwater when the sample was filtrated compared to when it was not filtrated on site. DMA and MMA concentrations were low in all samples (0.050–0.55 µg/L). This was expected, because the organic content of the soils is low [12].

Conclusions

The investigated site is highly contaminated by As in both soil and groundwater and remedial actions are urgently needed. The groundwater concentrations of As(III) were high, despite the low amounts leached in the laboratory tests. Batch tests under aerobic conditions are not an appropriate method for determining the leachability of As from soil, because there is a risk that excavated material will be classified as less hazardous than it is in reality. Filtration of groundwater samples in the field is necessary. If the sample is not filtrated immediately, iron hydr/oxides and soil particles containing oxides will remain. Arsenic(V) then reacts with the iron oxides and precipitates, while As(III) does not because it is not as easily adsorbed to metal oxides.

References

1. Bhattacharya P, Mukherjee AB, Jacks G, Nordquist S (2002) Metal contamination at a wood preservation site: characterisation and experimental studies on remediation Sci. Tot. Environ. 290:165–180
2. Swedish EPA (2006a) Åtgärdslösningar - erfarenheter och tillgängliga metoder, Rapport 5637 ISSN 0282-7298, Stockholm (In Swedish)
3. Lidelöw S, Ragnvaldsson D, Leffler P, Solomon T, Mauric C (2007) Field trials to assess the use of iron-bearing industrial by-products for stabilisation of chromated copper arsenate-contaminated soil Sci. Total Environ. 387:68–78
4. Cullen WR, Reimer K (1989) Arsenic speciation in the environment Chem. Rev. 89:713–764
5. Vink BW (1996) Stability relations of antimony and arsenic compounds in the light of revised and extended Eh-pH diagrams Chem. Geol. 130:21–30
6. Sracek O, Bhattacharya P, Jacks G, Gustafsson J-P, Brömssen M (2004) Behaviour of arsenic and geochemical modelling of arsenic enrichment in aqueous environments Appl. Geochem. 19:169–180
7. Petänen T, Lyytikäinen M, Lappalainen J, Romantschuk M, Kukkonen JVK (2003) Assessing sediment toxicity and arsenite concentration with bacterial and traditional methods Environ. Pollut. 122:407–415
8. Leermarkers M, Baeyens W, De Gieter M, Smedts B, Meert C, De Bisschop, Morabito R. Quevauviller PH (2006) Toxic arsenic compounds in environmental samples: Speciation and validation Trends Anal. Chem. 25:1–10
9. Sadiq M (1997) Arsenic chemistry in soils: an overview of thermodynamic predictions and field observations Water Air Soil Pollut. 93:117–136
10. Sherman DM, Randall SR (2003) Surface Complexation of arsenic(V) to iron(III) (hydr) oxides: Structural mechanism from ab initio molecular geometries and EXAFS spectroscopy Geochim. Cosmochim. Acta 67:4223–4230
11. Swedish EPA (2007) Modeller för spridning av metaller från mark till vatten, Naturvårdsverkets rapport 5741, ISBN 91-620-5741-3, Stockholm (In Swedish)
12. Smedley PL, Kinniburgh DG (2002) A review of source, behaviour and distribution of arsenic in natural waters Appl. Geochem. 17:517–568
13. Krause E, Ettel VA (1989) Solubilities and stabilities of iron arsenate compounds Hydrometal. 22:311–337
14. Johansson E (2005) Partitioning of trace metals between the dissolved and particulate phase at the outlet of River Höjeå, Sweden: importance of salinity and organic matter nature conservation: Water Management. University of Lund
15. Swedish EPA (1997) Development of generic guideline values. Model and data used for generic guideline values for contaminated soils in Sweden. Rapport 4639, Naturvårdsverket, Stockholm
16. Swedish EPA (2006b) Laktester för riskbedömning av förorenade områden (Huvudrapport och underlagsrapport 1a), Naturvårdsverkets rapport 5535, ISBN 91-620-5535-6, Stockholm (In Swedish)
17. Swedish EPA (2002) Methods for inventories of contaminated sites: environmental quality criteria guidance for data collection, Naturvårdsverkets rapport 5053, ISBN 91-620-5053-2, Stockholm

18. Cappuyns V, Van Herreweghe S, Swennen R, Ottenburgs R, Deckers J (2002) Arsenic pollution at the industrial site of Reppel-Bocholt (north Belgium) Sci. Tot. Environ. 295:217–240
19. Georgiadis M, Yong C, Solo-Gabriele H (2005) Extraction of arsenate and arsenite species from sediment Environ. Pollut. 141:22–29
20. EU (2002) EU-directive on Landfill of Waste as of decision Dec. 19, 2002;"Criteria and procedures for the acceptance of waste at landfills" (2003/33/EG)

[illegible]

Effect of Heavy Metal on Complex Permittivity Characteristics of Saturated Sand

Muhammad F.M. Zain, Mohd R. Taha, Hashem A. Mattarneh, Rabah M.A. Ismail, and Maslina Jamil

Introduction

In the past decades, both terrestrial and aquatic environments have been polluted with metals [1]. Soil normally contains a low concentration of heavy metals such as iron (Fe), which is an essential micronutrient for the optimum growth of plants. A metal like lead (Pb) is usually not found in agricultural soil, and is toxic to plants. However, in areas where agricultural, industrial or municipal wastes are land-applied as fertilizer, concentrations may be much higher. Excessive levels of heavy metals can be hazardous to humans and ecosystems [2]. Although heavy metals are released in varying quantities into the soil from parent materials, increasing environmental contamination has been caused by human activities, such as mining, smelting, fossil fuel combustion, agricultural practices and waste disposal [3, 4]. In order to characterize subsurface contamination, the most commonly used methods involve collection of representative samples of soil and pore fluid and then analyzing them for targeted species in the laboratory. In general, this way of characterizing the contamination of a soil-fluid system is the only one acceptable to regulatory agencies [5]. However, sample collection and analysis in the laboratory pose significant problems: (1) sampling of the soils is extremely time consuming and expensive; (2) sampling is destructive in the case of removing the soil samples; (3) the sampling is not continuous with time; and (4) samples can be contaminated during sampling, transportation, and analysis in the laboratory.

Thus several geophysical methods have been developed which utilizes the contrast caused by the contaminant on physical properties of the soil [6, 7]. Electromagnetic and dielectric methods show high potential for characterization hydrocarbon contaminated soil and determination of the level of contaminant.

M.F.M. Zain, M.R. Taha, R.M.A. Ismail, and M. Jamil
Faculty of Engineering, Universiti Kebangsaan Malaysia, Malaysia

H.A. Mattarneh
Civil Engineering Dept, Universiti Tenaga National, Malaysia

S. Rauch et al. (eds.), *Highway and Urban Environment*,
Alliance for Global Sustainability Bookseries 17,
DOI 10.1007/978-90-481-3043-6_32,

In additional use of this method rely upon the availability of information about the dielectric properties of the contaminated soil. Development of such information has formed a focus of research by many authors around the world over several years [7–9].

This study focuses on the factors affecting the dielectric response of soil-water-heavy metal mixture at low frequency ranges. In this paper, the complex permittivity of soils was measured at the frequency range of 100 kHz–1 MHz by means of using LCR meters. The effects of measurement frequency, heavy metal type and concentration on the dielectric response were evaluated by analyzing the complex permittivity of specimen at low frequencies.

Complex Permittivity and Polarization

When a capacitor containing material is connected to a sinusoidal voltage source, all materials including soils have a loss current component which is in phase with the applied voltage [10]. The partially out-of-phase response of the materials due to the existence of the loss currents can be well described in terms of the complex permittivity (ε^*). The relative complex permittivity, which is normalized with respect to the permittivity of free space, can be represented by Eq. 1 [10]. In common usage, the word "relative" is frequently dropped (ASTMD150).

$$\varepsilon^* = \frac{\varepsilon}{\varepsilon_o} = \varepsilon' - i\varepsilon'' \tag{1}$$

In Eq. 1, ε_0 is the permittivity of the free space ($\varepsilon_0 = 8.85 \times 10^{-12}$ F/m) and i denotes an imaginary number. The real part of the complex permittivity (ε') is commonly called the permittivity or dielectric constant of a material and represents the capacitive behavior or polarizability of the material. The imaginary part (ε'') is called the loss factor and represents the energy losses due to polarization and conduction [11, 12]. In the permittivity measurement, losses due to polarization and conduction are measured together [13]. Hence, the effective imaginary permittivity (ε_{eff}'') becomes Eq. 2:

$$\varepsilon_{eff}'' = \varepsilon'' + \frac{\sigma}{\varepsilon_o \omega} \tag{2}$$

where σ is zero-frequency current (DC) conductivity, ω is the angular frequency ($\omega = 2\pi f$) and f is the frequency. When an electric field is applied to a medium, some charges are bound, yet these positive and negative charges can move locally relative to each other, which results in a polarized medium. The polarizability of the medium is represented by permittivity. In other words, permittivity is a measure of the extent to which the electrical charge distributed in a material can be polarized by the application of an electric field. The spectral response of permittivity with frequency captures the various polarization mechanisms [14, 15].

In single-phase and homogeneous materials, there are three types of polarization mechanisms: (1) electronic polarization due to the displacement of an electronic cloud around an atom, (2) ionic polarization due to the relative displacement of atoms in a molecule and (3) orientation polarization involving the rotation of dipolar or polarized molecules [13–15]. Polarization mechanisms display one of two characteristic spectra such as resonance or relaxation at characteristic frequencies. In multiphase materials, the predominant polarization mechanism is spatial polarization, which is also known as interfacial polarization or Maxwell-Wagner polarization. This mechanism results from the differences in the polarizability and conductivity among each component composing the material [15]. In such a condition, charges accumulate at the interfaces to maintain the same current density throughout the contiguous layers [13]. Accumulation of charges at the component inter face appears as an increase in the capacitance of the material. Spatial polarization manifests as relaxation spectra of real permittivity at radio frequencies because the accumulation of charges under an alternating current field is strictly time dependent. Compared with other polarization mechanisms in single-component materials, this polarization mechanism has a very large spatial scale [16].

Experimental Method

Materials and Sample Preparation

Sandy soil was used to investigate the permittivity of soil in this study. The sieve analysis of the soil in accordance with the ASTM Standard D 442, summarized in Fig. 1. The quartz sand with porosity (N = 0.4, according to the procedure described in AASHTO manual 1974) was dried and three samples of 500 g have been prepared

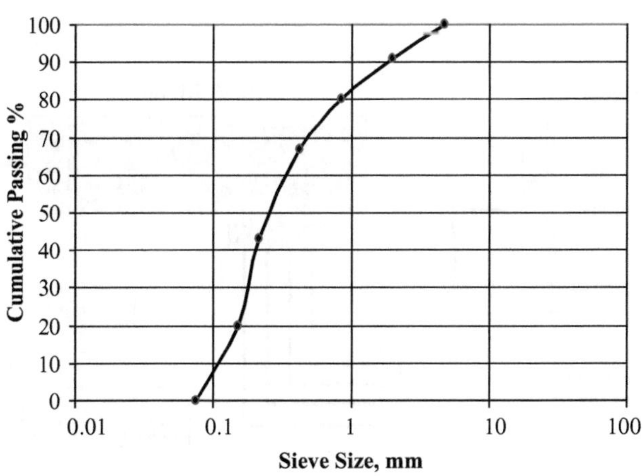

Fig. 1 The sieve analysis of the soil

from the dried sandy soil at certain amount of deionized water and two type of heavy metal (Fe and Pb). One soil sample was taken and recognized by the deionized water only (clean soil) while another two samples prepared to evaluate the effect of heavy metal contamination on saturated sand soil by adding 10 mg/L from the two type of heavy metals. The total liquid (water and heavy metal) was 40% by volume in order to saturate the sand soil with porosity 0.4. The soil samples were shacked for 15 min then kept for 3 h in sealed plastic container to avoid any evaporation. Three impedance measurements were obtained at each frequency and the average value was. The complex permittivity was deduced from the measured impedance of soil sample.

Electromagnetic Measurements and Test Cells

The test cell of internal dimensions 80 × 80 × 40 mm were used. Copper electrodes with dimensions 80 × 80 × 2 mm were attached to two opposite faces of the cell. Copper connections were passed through the cell walls and connected to the electrodes. All impedance measurements were acquired using the LCR meter, operating in voltage-drive mode, with the signal voltage being 1,000 mV. A linear sweep over the frequency range of 100–1,000 kHz was used with the data, recorded at 21 frequency points within this range.

The connection to the LCR meter was done by means of short, individually screened coaxial cables to the voltage (V), high/low and current (I) output/input terminals. Cable impedance, plate impedance and fringing impedance were determined using an appropriate model circuits. From the measurements, the impedances of soil sample were calculated and all data is logged by a PC. The complex permittivity (real part dielectric constant and imaginary part loss factor) of soil samples were deduced from the impedance of the soil. The setup of the system is shown in Fig. 2.

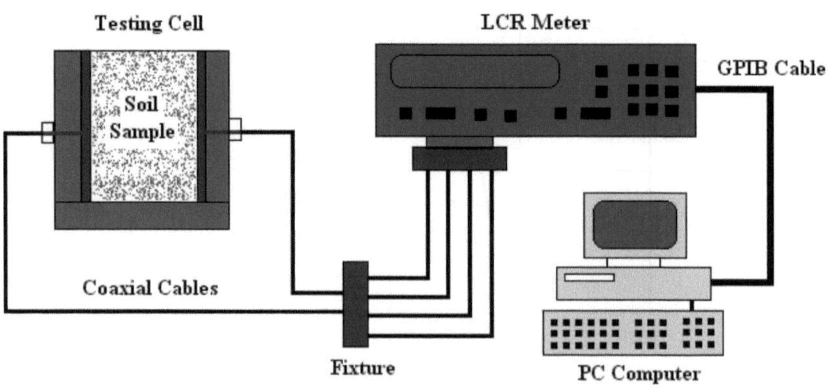

Fig. 2 Schematic diagram of the dielectric measurement system

Results and Discussion

The effect of heavy metal (Fe and Pb) on the complex permittivity was explored. The results of complex permittivity of saturated soil contaminated with Fe and Pb versus frequency are shown in Figs. 3 and 4 respectively. It is easy to note from Figs. 3 and 4 below that the dielectric constant (real part of permittivity) as well as the loss factor (imaginary part of permittivity) decrease with increasing frequency for both pure and contaminated soils. For contaminated soil the results show that the dielectric constant values are higher than the uncontaminated soil at frequencies less than 200 kHz, as shown in Figs. 3a and 4a, while at frequencies higher than 300 kHz the opposite trend is observed for both types of heavy metals.

The loss factor (imaginary permittivity) value for contaminated soil is higher than clean soil as shown in Figs. 3b and 4b. This increase is a result from increasing the ionic concentration. Therefore, the electric conduction is enhanced by free movement of charges whereby the ionic constituents act as charge carriers.

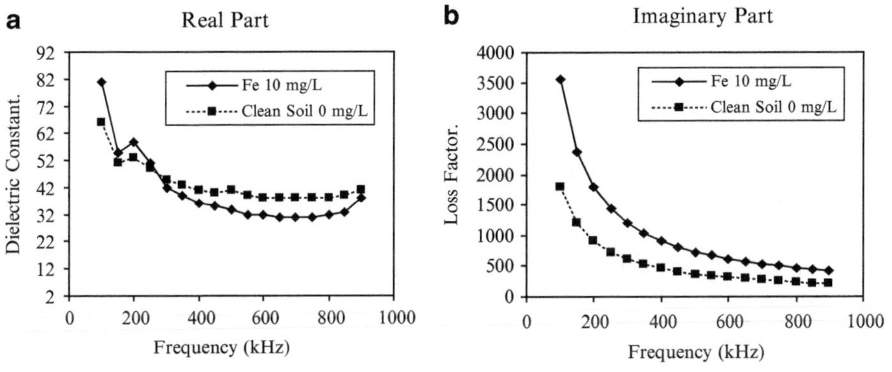

Fig. 3 The effect of iron on the complex permittivity. (**a**) Real part. (**b**) Imaginary part

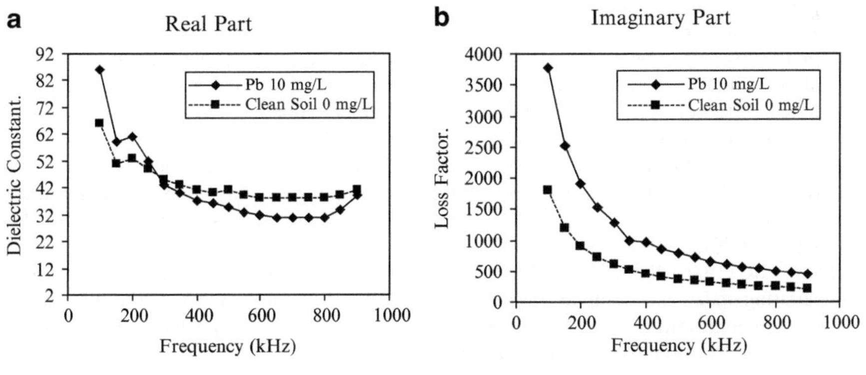

Fig. 4 The effect of lead on the complex permittivity. (**a**) Real part. (**b**) Imaginary part

Conclusion

The effect of iron (Fe) and lead (Pb) content on the measured electromagnetic properties of saturated soil material has been evaluated at a frequency range from 1 to 1,000 kHz. The dielectric constant and the loss factor were studied for soils contaminated by heavy metals at frequencies from 1 to 1,000 kHz. The research consummates that both the dielectric constant and the loss factor of soil samples decrease with increasing frequency.

The presence of heavy metals (Pb and Fe) cause an increase in loss factor of saturated soil over all frequency ranges from 1 to 1,000 kHz while the dielectric constant increase at low frequencies (less than 200 kHz) and decrease at high frequencies (higher than 300 kHz). This can be attributed to the electrode polarization at low frequencies. The results indicate that the measured complex permittivity of saturated soil may be used as a means of detecting and quantifying heavy metal contamination in soil.

Nomenclature

ε^*	Complex permittivity
ε''	Effective imaginary permittivity
σ	Zero-frequency current (DC) conductivity
ω	Angular frequency ($\omega = 2\pi f$)
f	Frequency
ε_o	Permittivity of the free space ($\varepsilon_o = 8.85 \times 10^{-12}$ F/m)
i	Imaginary number

References

1. Hopkin SP (1989) Ecophysiology of Metals in Terrestrial Invertebrates. Elsevier Applied Science, New York
2. United States Department of Agriculture (2000) Natural Resources Conservation Service. Soil Quality Institute 411 S. Donahue Dr. Auburn, AL 36832 334-844-4741 X-177, Urban Technical Note No. 3, Heavy Metal Soil Contamination, September 2000
3. Ross SM (1994) Toxic Metals in Soil-Plant Systems. Wiley, Chichester
4. Alloway BJ (1995) Heavy Metals in Soils. Blackie Academic & Professional, Glasgow
5. Okoye CN, Cotton RR, O'Meara D (1995) Application of resistivity cone penetration testing for qualitative delineation of creosote contamination in saturated soils. Geoenvironment 2000, ASCE, New York, NY, pp. 151–160
6. Olheft GR (1986) Direct detection of hydrocarbon and organic chemicals with ground penetrating radar and complex resistivity, Proceedings of the NWWA/API Conference on Petroleum Hydrocarbons and Organic Chemicals in Ground Water-Prevention, Detection, and Restoration, Houston, November 12–14, pp. 284–305

7. Chenaf D, Amara N (2001) Time domain reflectometry for the characterization of diesel contaminated soils, Proceedings of the second international symposium and workshop on time domain reflectometry for innovative geotechnical applications, Northwestern University
8. Arocena JM, Rutherford PM (2005) Properties of hydrocarbon-and salt-contaminated flare pit soils in northeastern British Columbia (Canada), Chemosphere 60:567–575
9. Abidin K (2002) Evaluation of soil porosity using a low MHz range dielectric constant, Turk J Eng Environ Sci, 26:301–307
10. Von Hippel AR (1954) Dielectric Material sand Applications. The Technology Press of Massachusetts Institute of Technology, Cambridge, MA
11. Kaya A, Fang HY (1997) Identification of contaminated soils by dielectric constant and electrical conductivity. J Environ Eng, 123:169–177
12. Shang JQ, Scholte JW, Rowe RK (2000) Multiple linear regression of complex permittivity of a till at frequency range from 200 MHz to 400 MHz. Subsurf Sens Technol Appl, 1:337–356
13. Santamarina JC, Klein KA, Fam MA (2001) Soils and Waves. Wiley, New York
14. Kingery WD (1963) Introduction to Ceramics. Wiley, New York
15. Santamarina JC, Fam M (1997) Dielectric permittivity of soils mixed with organic and inorganic fluids (0.2 GHz to 1.30 GHz). J Environ Eng Geophys, 2:37–51
16. Myounghak Oh, Yongsung Kim, Junboum Park (2007) Factors affecting the complex permittivity spectrum of soil at a low frequency range of 1 kHz –10 MHz. Environ Geol, 51:821–833

Blast-Furnace Sludge as Sorbent Material for Multi-metal Contaminated Water

Yuliya Kalmykova, Jesper Knutsson, Ann-Margret Strömvall, and Kristina Hargelius

Introduction

Cost effective alternative technologies and sorbent materials are in great demand for sustainable management and treatment of contaminated water. Natural materials and industrial waste, that is low-cost adsorbents, are available in large quantities and have the potential to be used for the treatment of metal contaminated water [1, 2]. Blast-furnace sludge is generated during the manufacture of pig iron [3]. Pig iron is produced in large blast furnaces, where preheated air is blown into the lower part of the furnace. A dusty gas is emitted from the top of the furnace and the gas is purified by means of wet treatment in flue gas scrubbers. The particles in the washing water settle through sedimentation, and the remaining muddy waste is termed blast-furnace sludge. At present this sludge is deposited in landfills and the steel industry in Europe produces ~500,000 t of the sludge yearly [4].

Environmental research and publications pertaining to the use of blast-furnace sludge are very limited [5]. The sorption capacity for metals in a single solution with furnace sludge has been investigated and found to be effective for metals such as Pb, Zn, Cd, Cu and Cr [6, 7] but the removal of Zn has been reported as low [8]. The possible release of and environmental risk posed by iron-cyanide complexes from deposited blast-furnace sludge has been found to be low [9].

The sludge has been chemically and mineralogically characterized [10]; the main elements found were C > Fe > > Ca > Si > > Al > Zn. Carbon was found in the form of coke and graphite, and the main minerals were calcite/CaCO3 > quartz/SiO2 > kaolinite/Al2Si2O5(OH)4 > hematite/Fe2O3 > magnetite/Fe3O4. Further stud-

Y. Kalmykova, J. Knutsson, and A.-M. Strömvall (✉)
Water Environment Technology, Department of Civil and Environmental Engineering, Chalmers University of Technology, SE-412 96 Göteborg, Sweden

K. Hargelius
Ramböll Sweden AB, Göteborg

S. Rauch et al. (eds.), *Highway and Urban Environment*, Alliance for Global Sustainability Bookseries 17,
DOI 10.1007/978-90-481-3043-6_33,

ies on the capacity of furnace sludge to remove metals from multi-metal solutions as well as real leachate and groundwater have been suggested.

The aim of this project was to investigate the potential of blast-furnace sludge as an adsorbent for metal contaminated water. The sludge was tested under extreme conditions that can occur in real environments by means of laboratory studies to determine the sorption capacity for metals (i) in multi solutions at different pH values and (ii) on an old sludge with multi-metals in real leachate.

Experimental and Methods

The blast-furnace sludge used in this study was from a Swedish steel producer and delivered by SSAB Merox AB in Oxelösund (a subsidiary of SSAB Svenskt Stål). The company's activities are concentrated on recycling and the sale of non-steel products from steel making processes.

Characterization of the Blast-Furnace Sludge

The crystalline components present in the blast-furnace sludge were studied qualitatively by x-ray powder diffraction (XRD) using a Siemens D5000 diffractometer and Cu Kα radiation. The compounds were identified by comparison with standards in the Joint Committee of Diffractions Standards database.

The buffer capacity of the furnace-sludge material was determined by 20 g of sludge added to 40 mL deionised and filtrated (Milli-Q) water, after which 0.1 M NaOH was added until the pH reached 12, after which 0.1 M HCl was added until the pH dropped to 6.0. The same procedure was used for the old sludge material with the addition of 8.7 g of material into 100 mL Milli-Q water and titration to pH 6.0. This titration was slow and continued for 14 days. The carbonate alkalinity of the old material was determined in a water solution after 5.0 g of the material had been in contact with 500 mL of Milli-Q water for 24 h. The carbonate concentration was calculated from the alkalinity by means of equilibrium calculations.

Sorption Beaker Tests

Sorption beaker tests were carried out to investigate the blast-furnace sludge sorption capacity at different pH values for multi-metal solutions in concentrations equal to extremely contaminated groundwater, see Table 1. The goal was to investigate the material's sorption capacity under extreme conditions that could occur in real-life environments. The sorption tests were carried out by means of batch beaker tests in which 5.00 g of dry sludge was added to each beaker. The sludge samples were mixed in 500 mL of Milli-Q water at pH values of 5.6; 7.0 and ~9 over a 24 h period.

Table 1 Selected metal concentrations, based on the assessment guidelines for contaminated groundwater in Sweden [11]

Metal ion	Conc. 1 mg/L extremely contaminated	Conc. 2 mg/L 5× extremely contaminated	Conc. 3 mg/L 10× extremel contaminated
Copper, Cu^{2+}	20/10 = 2.0[b]	10[b]	20 [b]
Zinc, Zn^{2+}[a]	20/10 = 2.0[b]	10[b]	20[b]
Chromium, Cr^{3+}	0.50	2.5	5.0
Nickel, Ni^{2+}	0.50	2.5	5.0
Molybdenum, Mo^{4+}[a]	0.50	2.5	5.0
Tin, Sn^{2+}[a]	0.50	2.5	5.0
Vanadium, V^{2+}[a]	0.50	2.5	5.0
Lead, Pb^{2+}	0.10	0.50	1.0
Cadmium, Cd^{2+}	0.050	0.25	0.50
Cesium, Cs^{+}[a]	0.050	0.25	0.50

[a]Swedish assessment guideline values missing.
[b]Due to the high assessment value, a ten time lower concentration was used.

In order to maintain a constant pH value, a 0.2 M acetate buffer was added to keep pH at 5.6 as well as a 0.02 M phosphate buffer for pH 7.0. In the beakers without pH adjustment and no buffer, the pH was initially 10.1–10.4, but dropped to 9.2 after the addition of the acidic multi-metal solution. Three metal concentrations were used for each pH level (Table 1). The concentration for each metal was selected to represent a real and extremely contaminated groundwater, five and ten times higher respectively, than the Swedish EPA assessment guidelines for contaminated groundwater [11].

Zero sample tests, with Milli-Q water and sludge being mixed for 24 h, were also done at the different pH values in order to determine the leaching of metals. Samples of the water solution were taken for metal analysis at 0 and 24 h.

A sorption test was also carried out on old sludge material and leachate from a landfill. The goal was to include ions from real lechate water and to test the material after ageing. The procedure was the same (zero samples included) as described above, but carried out at pH 9 with leachate taken from the landfill site. Depending on the low metal concentrations in the real leachate, it was spiked with the multi-metal solutions. The pH was initially 11.2, but decreased to 8.7 after the addition of the metal solutions.

All water samples were filtrated through 0.45 μm cellulose acetate filters and the filtrates acidified to pH < 2 before being sent to a certified commercial laboratory for metal analysis by means of inductively coupled plasma atomic emission spectrometry (ICP-AES). The samples were analyzed for concentrations of Cd, Cr, Cs, Cu, Mo, Ni, Pb, Sn, V and Zn.

The results of the sorption tests were presented by means of the calculation of the metal removal efficiency and sorption capacity. The Freundlich adsorption isotherms were also calculated and evaluated to identify whether or not the processes involved sorption.

Simulation of Species in the Multi-metal Solutions

Simulations of the metals' speciation in the test beaker solutions were carried out using CHEAQS Pro software [12]. The goal was to identify whether or not any of the metals were precipitated as salts in the test solutions. The program calculates the metals' species in a solution with equilibrium constants, drawn from various data bases and with input in the form of solution ion concentrations. The following beaker solutions were selected for the simulation: pH 7.0 with a phosphate buffer and pH ~9 without a buffer. The results were evaluated by comparison with the calculated removal efficiencies in the beaker tests with furnace sludge added. The XRD results revealed that the sludge contained several minerals comprising iron and carbonate. Therefore a second set of simulations was performed where the input concentration of ferric iron (Fe^{3+}) was 40 mg/L and the carbonate ($CO3^{2-}$) concentration 0.1, 10 or 1,000 mg/L. These simulations were only carried out for the multi-metal solutions at pH 7.0.

Result and Discussion

The results of the XRD qualitative analysis revealed that the sludge is mainly composed of different iron hydr/oxides, carbonates and silicates. The following minerals were found:

Iron hydroxide-oxide: $Fe_{1.833}(OH)_{0.5}O_{2.5}$
Hematite: Fe_2O_3
Magnetite: Fe_3O_4
Wuestite: FeO
Iron sulphate: $Fe(SO_4)(H_2O)$
Quarts: SiO_2
Calcium silicate: $CaSi_2O_5$
Calcite: $CaCO_3$

The results are in agreement with quantitative XRD analysis of blast-furnace slag [10]. The only difference was that kaolinite $Al_2Si_2O_5(OH)_4$ was one of the main minerals in the latter study. The active surfaces of the iron oxides in the sludge may explain the good sorption capacity for metals. The mineral goethite FeO(OH), closely related to the $Fe_{1.833}(OH)_{0.5}O_{2.5}$, has well documented sorption capacities for metals [13]. The other iron oxides detected could also be rearranged to active surface OH groups in contact with water, thereby acting as effective sorbents for metals through complex binding and surface precipitation [14]. Kaolinite and carbon, such as coke and graphite, may also enhance the sorption capacity of the material [10], but these components were not screened for in this study. The carbonates endow the material with a high buffer capacity, and the risk of a decrease in pH for water solutions in contact with the sludge material is low. The sludge had a buffer capacity of 0.32 mmol H +/g for decreasing pH to 6.0. The old furnace sludge had a higher buffer capacity of 1.2 mmol H +/g. The alkalinity of the old sludge was 64 mg/L, and the concentration of carbonate was calculated to 0.47 mg/L.

Blast-Furnace Sludge – Removal Capacities

Results from the sorption beaker tests showed good removal capacity, 93–100%, for Cd, Cr, Cu, Ni, Pb and Sn at all concentrations and pH ~9 (Fig 1a). The pH decreased slightly in the beakers from 9.2 to 8.1 during the experiment. The removal capacity for V and Zn was 46–98%, and for Cs and Mo in the range 19–71%. The removal capacity for Cd, Cr, Cu,

Sn and Zn was also good, 92–100% in the solutions at pH 7.0 (Fig. 1b); pH remained stable during the experiment. The capacity for Ni was 83–89%, for Mo 2–12%, and a leaching of V and Cs was indicated. At pH 5.6 (maintained stable) the removal capacity was lower, 32–99%, for Cr, Cu, Mo, Sn and V (Fig. 1c). The removal of Cd and Ni was only 6–14%, indicating a leaching of Zn, Cs and Pb. Due to the low capacity for several of the metals, blast-furnace sludge should not be used as a sorbent material in solutions at pH 5.6 or lower. On the other hand, the carbonates in the sludge effectively protect water solutions from acidification. A leaching of Zn and Pb from the sludge in pure water was indicated at pH 5.6, at pH 7.0 a leaching of V, and at pH ~9 a leaching of Zn, V and Pb.

Old Furnace Sludge – Removal Capacities

The results for the old furnace sludge stirred with real leachate and spiked with the multi-metal solutions and pH ~9 are presented in Fig. 2. The pH in the beakers decreased from 9.2 to 8.7 during the experiment. The removal capacity of this material was slightly lower than that of the 'new' sludge. The capacity for Cd, Cr, Cu, Ni, Pb, Sn and Zn was in the range 98–100%. As with the 'new' sludge, the capacity was lower for Mo and V, 8–96%, and even a leaching of Cs was indicated. A leaching of V and Zn was indicated from the old sludge in pure water at pH ~9. Ageing of the material and the use of real leachate reduce the sorption capacity and should be taken into consideration when laboratory results are applied to real-life situations.

Blast-Furnace Slag – Sorption Capacities

The sorption capacities in the solution with the highest metal concentration are presented in Table 2. The results show a trend of capacities increasing with pH for most of the metals. The material had a slightly higher capacity for Cu than for Zn, which could be explained by the differences in terms of electron configuration and the Jahn-Teller effect [15] that strengthens the complex binding of copper to sorption surfaces. For the metals in concentrations from 0.50–5.0 mg/L, the sorption capacity order was: Cr = Sn > Ni > V > Mo. Lead was the only metal in the concentration range 0.10–1.0 mg/L and it should be noted that furnace slag has been found to have

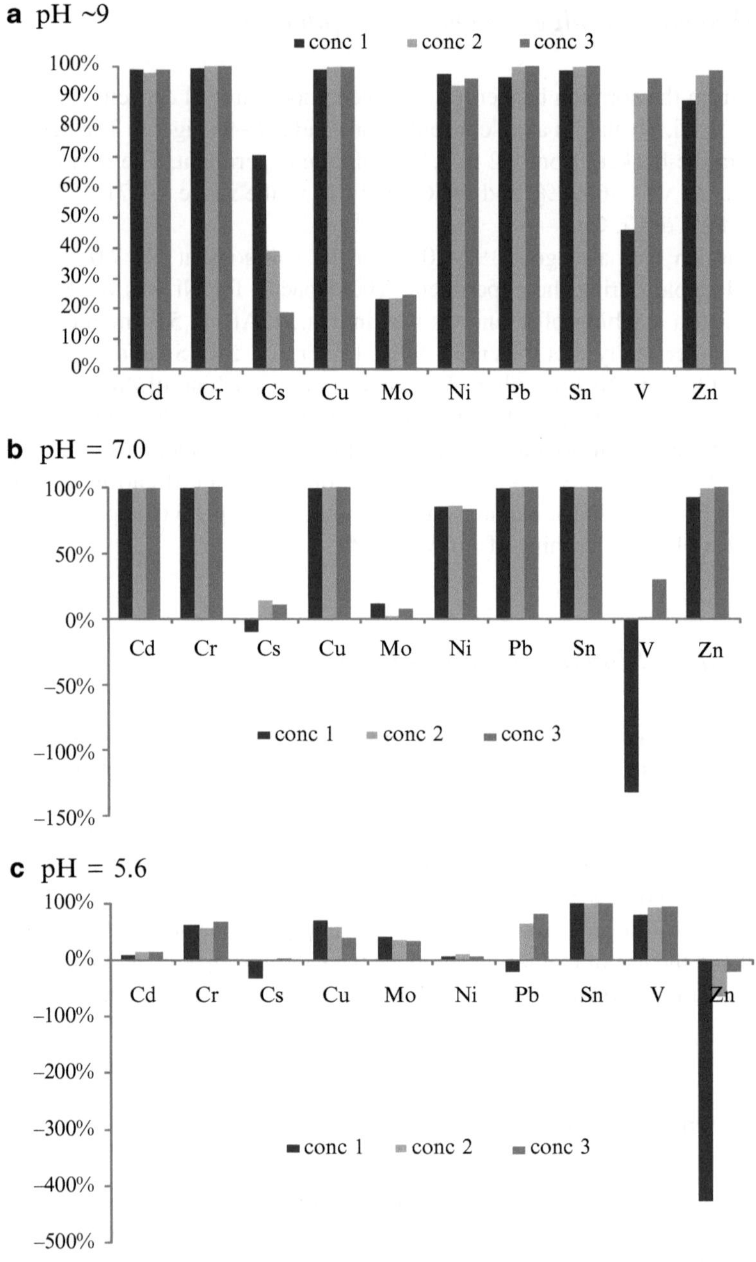

Fig. 1 Removal capacity of blast-furnace sludge for metals in different concentrations stirred in beakers (24 h) (**a**) pH ~9, (**b**) pH = 7.0 and (**c**) pH = 5.6

an extremely good capacity for sorption of lead, up to 80 mg/g [7]. The sorption capacity for Cadmium was slighter better than Cs for solutions in the 0.050–0.50 mg/L range. The figures in the table are not the maximum capacities of the material,

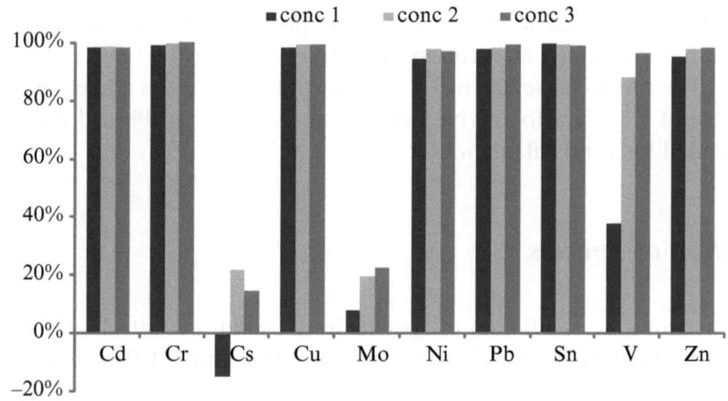

Fig. 2 Removal capacity of old furnace sludge at pH ~9 for metals in different concentrations in real leachate spiked with multi metals and stirred in beakers (24 h)

Table 2 Sorption capacity for metals in solutions at different pH with high initial concentrations and furnace sludge as sorbent

		Sorption capacity mg/g				
		Furnace sludge		Old		
Metals	C_0 mg/L	pH 5.6	pH 7.0	pH ~9	pH ~9	Maximum capacity[a] mg/g
Cu	21.6	1.7	4.1	4.1	4.3	16–24
Zn	21.8	–	4.0	3.8	4.1	4.2–9.7
Cr	4.4	0.59	1.0	0.97	1.0	9.6–16
Sn	3.8	0.76	0.96	0.66	0.75	–
Ni	5.0	0.062	0.82	1.0	1.0	–
V	4.1	0.78	0.29	1.1	1.0	–
Mo	4.7	0.31	–	0.25	0.23	–
Pb	0.84	0.14	0.19	–	0.20	64–80
Cd	0.46	0.013	0.086	0.094	0.090	6.7–10
Cs	0.45	–	–	0.018	0.014	–

[a] Maximum capacity found at 20–80°C [7]
– = No data

but only those for the concentrations used in this study. Initial concentrations of the metals have been found to heavily affect the capacities in batch tests [16, 17]. Higher capacities than presented here were found [7], due to the much higher metal concentrations used, up to 10 g/L. The capacity order was: Pb > > Cu > Cr > Cd > Zn; following the order of the metals' electronegativity and hydrated ionic radii. This order may also be explained by the metals' electron configuration and ligand field stabilization energy (LSFE), as found by [16].

The results of the batch tests, that is initial and final concentrations, were used to calculate the Freundlich adsorption isotherms for each metal. The calculations yielded the coefficients n and KF, and the regression coefficient R2, which show the probability of the occurrence of a sorption process, see Table 3. For a removal process to be determined as sorption, n should be in the 0.60–3.3 range, and R2 close to 1. The results presented in Table 4 show that sorption is a likely mechanism for removal of Cd, Cu, Mo and Ni. Sorption is indicated for Pb and Zn, but more data is needed for valid calculations.

Simulation of Species

The results from the simulation of species in the solutions with high multi-metal concentrations, at pH 7.0 with a phosphate buffer and pH 9 without a buffer,

Table 3 Calculated Freundlich adsorption isotherm constants for metals in multi solutions in contact with blast-furnace sludge as sorbent

	Furnace sludge						Old sludge					
	pH 5.6			pH 7.0			pH ~9			pH ~9		
	n	K_F	R^2	n	K_F	R^2	n	K_F	R^2	n	K_F	R^2
Cu	0.63	1.46	0.97	–	–	–	2.24	0.53	0.7	1.5	0.96	0.96
Zn	–	–	–	–	–	–	–	–	–	1.63	0.4	0.95
Cr	1.02	0.43	0.96	–	6.48	0.90	–	–	–	–	–	–
Sn	0.3	8.43	1.00	0.25	9.68	0.95	–	–	–	0.53	5.25	1.00
Ni	0.99	0.13	0.93	0.96	0.85	0.99	0.74	2.31	0.95	1.31	0.88	0.91
V	3.34	0.01	0.98	–	–	–	–	–	–	–	–	–
Mo	0.85	0.45	1.00	–	–	–	1.04	0.2	1.00	1.67	0.02	0.99
Pb	–	–	–	–	–	–	1.02	0.35	1.00	1.31	1.82	0.74
Cd	1.25	0.09	1.00	1.52	2.36	1.00	0.87	2.44	0.89	1.02	2.08	0.99
Cs	–	–	–	–	–	–	–	–	–	–	–	–

– = No data.

Table 4 Suggested removal mechanisms for metals in solutions at different pH values and furnace sludge as sorbent

	Furnace sludge		Old sludge	
Metal	pH 5.6	pH 7.0	pH ~9	pH ~ 9
Cu	Sorption	Sorption[a]	Sorption	Sorption
Zn	–	Other	Other	Sorption[a]
Cr	Sorption	Other	Other	Other
Sn	Other	Other	Other	Other
Ni	Sorption	Sorption	Sorption	Sorption
V	Other	–	Other	Other
Mo	Sorption	Sorption[a]	Sorption	Sorption
Pb	–	n.d.	Sorption	Sorption
Cd	Sorption	Sorption	Sorption	Sorption
Cs	–	–	Other	Other

[a] Data uncertain, could be sorption.
n.d = No data.
– = leaching > removal.

are correlated to the removal percentage determined in the experiments in Fig. 3. The results indicate that all metals in the beaker solutions were in the dissolved phase, but to a large extent appeared as metal-hydroxide or metal-phosphate complexes. It is thereby indicated that metals were not removed by means of precipitation as salts in the solutions. It should be noted that carbonates are of potential importance but they were not included in the first modeling. From the results it can be concluded that there is a strong correlation between the removal capacity and the occurrence of dissolved metal complexes at pH 7, but not at pH 9, which was the underestimated part of the complex bounded metals in the model.

Results from the second calculation of species in the beaker with furnace sludge at pH 7.0 and ferric (Fe^{3+}) concentrations of 40 mg/L and carbonate (CO_3^{2-}) of 0.1, 10 or 1,000 mg/L are presented in Fig. 4. The correlation of the precipitated salts and the observed removal capacity for Cd, Cr, Cs, Cu, Ni, Sn and Zn show a strong

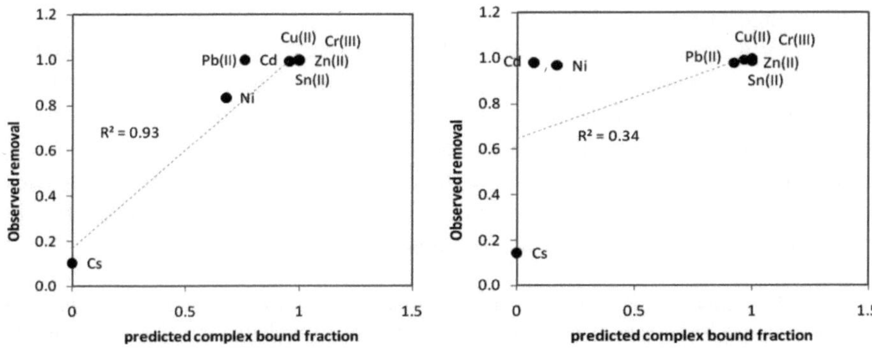

Fig. 3 Correlation of calculated percentage of metal complexes with the measured removal capacity in beaker tests with high metal concentrations and old furnace sludge: *left* - at pH 7 and with phosphate buffer; *right* - at pH 9 without buffer and old sludge

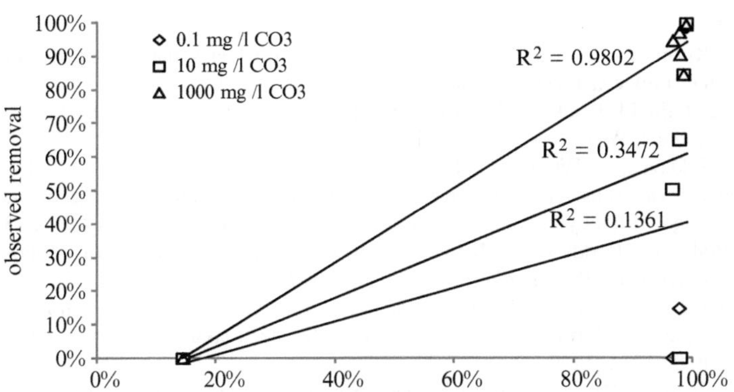

Fig. 4 Measured removal capacity as a function of metals precipitated at different carbonate concentrations, for high metal concentrations at pH 7

correlation (R2 = 0.98) at 1,000 mg CO_3^{2-}/L. At the lower carbonate concentrations, 0.1 and 10 mg/L, the correlation is less valid (R2 = 0.14 and 0.35 respectively).

It was concluded that the metals will only be precipitated as carbonates at extremely high carbonate concentrations. The carbonate concentration was 0.47 mg/L from the measurement of alkalinity for the old sludge in water solution. At the lower carbonate concentrations, more relevant for natural water in contact with the sludge, other mechanisms such as sorption are more likely to occur. The CHEAQS data base was unable to make the calculations for V and Mo, and in the second simulation Pb could not be dealt with by the software.

Conclusions

The results show that blast-furnace sludge has a good ability to remove the majority of the studied metals at pH 7–9. It is therefore recommended for use as a sorption material for water contaminated by metal, but the risk of the leaching of V and Cs should be investigated in each case. The slightly lower capacity of the old sludge and the use of real leachate should be taken into consideration when laboratory results are applied to real-life environments. The leaching of Cs and the lower capacity of the old sludge to remove Mo and V should be noted. For further studies, column experiments with real contaminated leachates are recommended for the investigation of the material's behavior under flowing conditions.

Acknowledgements SSAB Merox AB in Oxelösund is gratefully acknowledged for financial support. We also wish to thank Hanna Friberg and Therese Stark at Merox, as well as environmental consultant Kristina Hargelius who initiated the project, for their valuable help and support.

References

1. Babel S, Kurniawan TA (2003) Low-cost adsorbents for heavy metals uptake from contaminated water: a review. J. Hazard. Mater. 97:219–243
2. Bailey SE, Olin TJ, Bricka RM, Adrian DD (1999) A review of potentially low-cost sorbents for heavy metals. Water Res. 32:430–440
3. Peacey JG, Davenport WG (1979) The Iron Blast Furnace, Pergamon Press: Oxford
4. Van Herck P, Vandecasteele C, Swennen R, Mortier R (2000) Zinc and lead removal from blast furnace sludge with a hydrometallurgical process. Environ. Sci. Technol. 34:3802–3808
5. Das B, Prakash S, Reddy PSR, Misra VN (2007) An overview of utilization of slag and sludge from steel industries. Resour. Conserv. Recycl. 50:40–57
6. Martín MI, López FA, Pérez C, López-Delago A, Alguacil FJ (2005) Adsorption of heavy metals from aqueous solutions with by-products of the steelmaking industry J. Chem. Technol. Biotechnol. 80:1223–1229
7. López-Delgado A, Pérez C, López FA (1998) Sorption of heavy metals on blast furnace sludge. Water Res. 32:989–996
8. Jallan G, Pandey GS (1992) Blast furnace gas cleaning sludge as adsorbent of some toxic ions. Res. Ind. 37:143–145

9. Rennert T, Mansfeldt T (2002) Sorption and desorption of iron-cyanide complexes in deposited blast furnace sludge. Water Res. 36:4877–4883
10. Mansfeldt T, Dohrmann R (2004) Chemical and mineralogical characterization of blast-furnace sludge from an abandoned landfill Environ. Sci. Technol. 38:5977–5984
11. 11. Swedish Environmental Protection Agency (2002) Methods for Inventories of Contaminated Sites - Environmental quality criteria - Guidance for data collection Report 5053
12. Verweij, W (2005) CHEAQS Pro 2005. http://home.tiscali.nl/cheaqs/
13. Fischer L, Brümmer GW, and Barrow NJ (2007) Observations and modeling of the reactions of 10 metals with goethite: adsorption and diffusion processes. Europ. J. Soil Sci. 58:1304–1315
14. Farley KJ, Dzombak DA, Morel FMM (1985). A surface precipitation model for the sorption of cations on metal oxides. J. Colloid Interface Sci. 106:226–242
15. Bersuker IB (1984) The Jahn-Teller Effect - A Bibliographic Review, Plenum Press, New York
16. Kalmykova Y, Strömvall A-M, Steenari B-M (2008) Adsorption of Cd, Ni, Pb, Cu and Zn in low concentrations on Sphagnum peat. J. Hazard. Mater. 158:885–891
17. Benedetti MF, Milne CJ, Kinniburgh DG, Vanriemsdijk WH, Koopal LKM (1995) Metal-ion binding to humic substances - application of the nonideal competitive adsorption model. Environ. Sci. Technol. 29:446–457

Evaluation of Zn, Cu and Pb Sorption-Desorption Phenomena in Kaolinite-Sand Media Filtration Pilot Scale Installation

Paulo J. Ramísio and José M.P. Vieira

Introduction

A large number of heavy metals and organic micro pollutants in highway runoff appear as a result from automobile traffic. Cu, Pb, Zn and Cd - sometimes Ni and Cr - are considered the most important heavy metals associated with both mobile and stationary sources [1]. These metals are either dissolved in the stormwater or bound to particulates [2, 3]. Unlike organics and xenobiotic compounds, metal elements are not degraded in the environment. As a consequence metal elements exert both a short-term toxicity impact, characterized by concentration or activity, and a long-term toxicity impact, characterized by mass accumulation [4].

Previous studies of highway runoff have shown that Zn, Cd, Ni and Cu are predominately dissolved; Pb and Cr are equally distributed between dissolved and particulate-bound phases, while Fe and Al are predominately bound to particles [4].

Strategies for highways runoff contaminants control, such as infiltration methods, detention methods, vegetated filter strips and wetlands have been proposed.

Using infiltration ponds particles can be retained in pond bottom layers, but dissolved heavy metals might contaminate aquifers. Even with detention ponds, where particles could settle, small particles and dissolved heavy metals can contribute to groundwater contamination.

A decision with regard to the effectiveness of filtration or sedimentation is balanced by particle size gradation and density characteristics as well as loading characteristics and specific site constraints.

Since filtration media are mainly designed for mechanical retention removal, they only contribute to the removal of heavy metals in the particulate form. The use of a reactive bed for the filter media will enhance the dissolved heavy metals retention processes due to sorption and precipitation, transferring pollutants

P.J. Ramísio and J.M.P. Vieira
University of Minho, Portugal

S. Rauch et al. (eds.), *Highway and Urban Environment*, Alliance for Global Sustainability Bookseries 17,
DOI 10.1007/978-90-481-3043-6_34, © Springer Science+Business Media B.V. 2010

from the dissolved form to the media matrix. This will contribute to the control of receiving waters (surface and groundwater) pollution due to highway runoff. Therefore, the immobilization of dissolved metal complexes and particulate-bound metal elements through reactive filtration mechanisms can induce the development of an innovative "in situ" retention technique, characterized by high heavy metal retention efficiency and simple operational procedures.

In developing an effective treatment practice, feasibility must be evaluated based on characteristics of the runoff loading and the physicochemical mechanisms that can be utilized given the site [5]. The development of an "in situ" retention technique, applying current construction materials with low variability proprieties, could help to develop a prototype technical solution for the reduction of heavy metal pollution from highway runoff.

Experimental studies with different filter bed types have been developed at a laboratorial scale: iron-oxide-coated sand, natural zeolites and granular activated carbon [3].

The use of minerals in the constitution of reactive filter bed may enhance the feasibility of this technical effluent at a real scale installation. However, the behavior of the immobilized heavy metals by desorption must be known, since by this way the retained heavy metals can return to the dissolved form. Therefore, the heavy metals desorption behavior is relevant.

Laboratory experiments are necessary in order to evaluate infiltration data and sorption-desorption phenomena in a reactive-filtration pilot-scale installation. A research program was developed to study those data and related processes and phenomena. Previous work has reported the heavy metals partitioning, the hydraulic behavior, physicochemical control parameters and the efficiency of dissolved heavy metal removal involved in the dual-media reactive filter, with a sand and kaolinite filter bed, over a 1 year experimental study [6].

This paper reports Zn, Cu and Pb mass retention in the filter media layers and the immobilized heavy metals behavior by desorption procedures.

Experimental Details

The experimental installation consists of two reservoirs and three equal diameter cylindrical columns of transparent polyethylene. The layout of the installation is depicted in Fig. 1.

The inner diameter of the columns is 172 mm. The bottom structure was constructed in order to allow a sliding movement at the end of the experiment.

Downstream head control was established by means of a weir in order to easy measure the head loss through the filter media.

In each column, different adsorptive filtration beds with 70 cm height where installed: a single sand medium in column C1; a dual media of 10% of kaolinite and 90% of sand in column C2; a dual media of 20% kaolinite and 80% sand in column

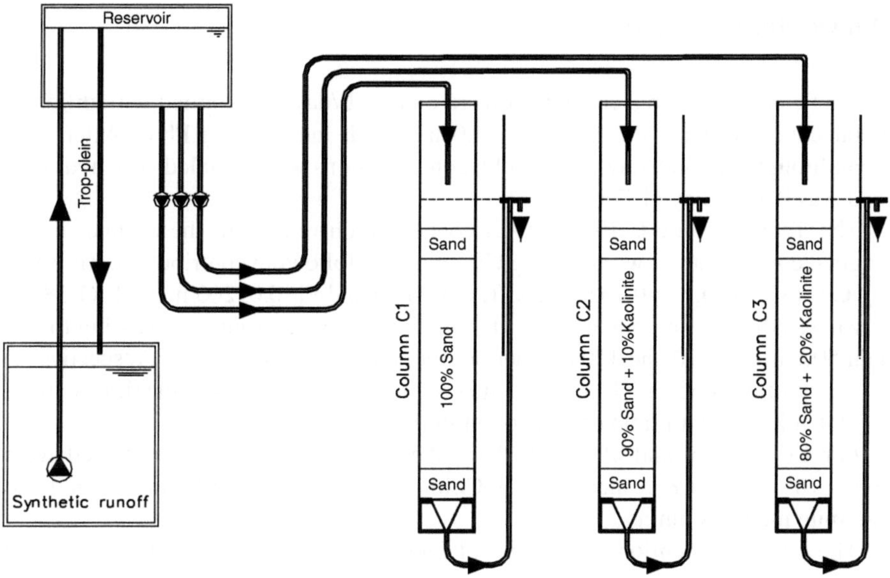

Fig. 1 Experimental layout

C3. For mechanical protection, a 10 cm sand layer was applied to the top and to the bottom of the filter bed. The filter media composition was defined in previous pilot scale studies based on their hydraulic conductivity behavior.

Each column was fed with a synthetic highway runoff by regulated flow pumps. This synthetic solution was obtained from the dilution of Zinc acetate dehydrate $(CH_3COO)_2 Zn \times 2H_2O$), Copper (II) acetate monohydrate ($(OOCCH_3)_2 Cu \times H_2O$) and Lead (II) acetate trihydrate ($(CH_3COO)_2 Pb \times 3H_2O$). The maximum solubility values for these reagents, at 20°C, are: 430, 72, and 410 g/L, respectively.

Laboratory analyses were carried out according to Standard Methods [7]. Reagent grade chemical products and ultra-pure water were used. All glassware coming in contact with the samples was washed with nitric acid and rinsed with de-ionized water.

After the samples were collected, the physicochemical parameters (pH, conductivity and temperature) as well as the hydraulic head were measured immediately. The sample was then filtered on a 0.45 μm porosity membrane, and acidified to pH < 2 with HNO_3. All the samples where then preserved at 4°C. Pb was analyzed by Graphite Furnace Atomic Absorption Spectrometry (GFAAS), while Zn was analyzed by Inductive Coupled Plasma Atomic Emission Spectrometry (ICP-AES). Cu was analyzed by GFAAS or ICP. Detection limits were of 0.001 mg/L (GFAAS) and 0.01 mg/L (ICP-AES) for Cu, 0.002 mg/L (GFAAS) for Pb and 0.01 mg/L (ICP-AES) for Zn. Materials and experimental details are presented in [6].

Operational Procedures

Synthetic effluent with a constant flow rate of 10 L/day was applied with different metal concentrations: 8.0 mg Zn/L, 1.0 mg Cu/L and 0.4 mg Pb/L. After Zinc breakthrough (at 156th day), Cu and Pb concentrations were doubled. The hydraulic load was 0.43 m/day (157 m/year).

The Total Organic Carbon (TOC) concentration related to the acetate, in the original concentration, was 6.72 mg C/L, with the following contribution: Zn-5.870 mg C/L (87.38%); Cu-0.7553 mg C/L (11.24%) and Pb-0.09266 mg C/L (1.38%).

Zn was the most mobile metal with retention efficiency values decreasing to less than 50% in a 15, 70 and 110 day period, for columns C1, C2 and C3, respectively.

The evolution of dissolved heavy metal removal efficiencies in the three experimental columns with time can be found in [6].

For Cu and Pb, and after 260 days, the retention efficiencies obtained were above 70% and 40%, respectively, in Column C1, and above 90% in the sand and kaolinite media columns.

The different retention efficiencies of Zn observed in the three columns suggest a direct relation to the filter bed characteristics, since the breakthrough period in column C2, with a 10% kaolinite filter bed, was almost half the breakthrough period in column C3, with a 20% kaolinite filter bed. Although other processes may occur, these results indicate that the cation exchange capacity is relevant and therefore sorption processes are the major phenomena.

Desorption Procedures

At the end of the experiment, and for each column, the filtration media were separated in six layers and numbered from the bottom to the top. Representative samples of each layer where submitted to chemical desorption. The desorption experiment was carried out in a 24 h period, using a solution of: 2.5 mL of 0.1 M NaNO3, 2.5 g air dried filter media and 247.5 mL deionised water.

For each layer, two desorption procedures were applied. The first resulting from a 24 h mixing at 55 rpm of the effluent (labeled pH6), and the second with the same procedure but acidified at pH = 2 with HNO3, (labeled pH2). The pH evolution with time was monitored.

Although other methods are available, in this work the total metals retained in each layer will be evaluated by the values obtained by desorption at pH2.

The samples obtained after a 24 h mixing were filtered in a 0.45 μm membrane, acidified with HNO3 to pH < 2 and kept at 4°C before the analytical analysis. An electronic microscope observation was carried out in order to evaluate the particulate heavy metal retention in the filtration membrane.

Results for desorption at pH2 and pH6 for columns C1, C2 and C3, respectively, are presented in Tables 1 through 3.

Table 1 Column C1: Zn, Cu and Pb concentrations in the desorption experiment

	Zn (mg/L)		Cu (mg/L)		Pb (mg/L)	
Layer	pH2	pH6	pH2	pH6	pH2	pH6
C1–1	0.86	0.40	0.52	0.21	0.252	0.064
C1–2	0.65	0.49	0.58	0.36	0.199	0.087
C1–3	0.65	0.58	0.54	0.45	0.192	0.111
C1–4	0.68	0.63	0.64	0.62	0.199	0.184
C1–5	0.73	0.64	0.76	0.69	0.249	0.191
C1–6	0.61	0.60	1.28	1.26	0.287	0.291

Table 2 Column C2: Zn, Cu and Pb concentrations in the desorption experiment

	Zn (mg/L)		Cu (mg/L)		Pb (mg/L)	
Layer	pH2	pH6	pH2	pH6	pH2	pH6
C2–1	1.15	0.62	0.39	0.091	0.327	0.063
C2–2	4.49	0.59	0.23	0.006	0.128	0.004
C2–3	4.47	0.59	0.32	0.008	0.189	0.004
C2–4	4.87	0.64	1.19	0.036	0.703	0.021
C2–5	5.87	0.69	5.06	0.051	1.37	0.058
C2–6	2.00	0.70	3.38	0.23	0.876	0.078

Table 3 Column C3: Zn, Cu and Pb concentrations in the desorption experiment

	Zn (mg/L)		Cu (mg/L)		Pb (mg/L)	
Layer	pH2	pH6	pH2	pH6	pH2	pH6
C3–1	0.86	0.40	0.52	0.21	0.252	0.064
C3–2	0.65	0.49	0.58	0.36	0.1999	0.087
C3–3	0.65	0.58	0.54	0.45	0.192	0.111
C3–4	0.68	0.63	0.64	0.62	0.1999	0.184
C3–5	0.73	0.64	0.76	0.69	0.249	0.191
C3–6	0.61	0.60	1.28	1.26	0.287	0.291

Retention efficiencies and desorption results obtained can be used for estimating the fraction of heavy metals that can be desorbed. Hereafter, the values obtained for Zn, Cu and Pb, in the three columns and at the different layers are compared and discussed.

Column C1

The desorption results obtained for column C1 with a sand filter bed demonstrated small variations of the results from pH2 and pH6. Zn values have small differences through the different layers, while the results obtained for Cu and Pb denote higher values at the bottom layer, but increasing as the layer depth decreases. These results suggest that although heavy metals are retained in the sand layers, a very high

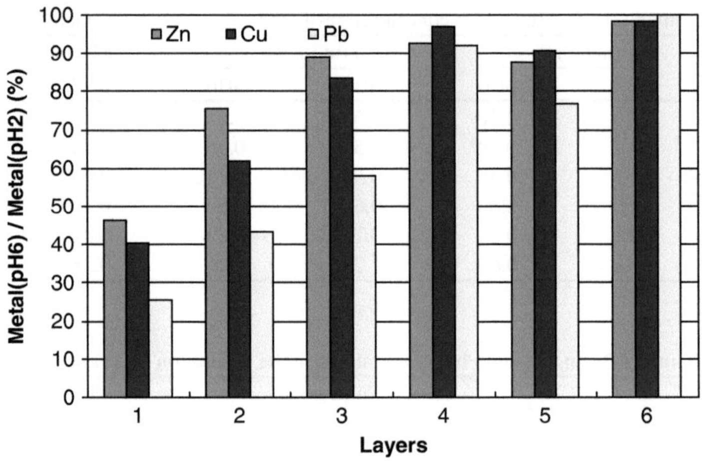

Fig. 2 Column C1: metal mass ratio

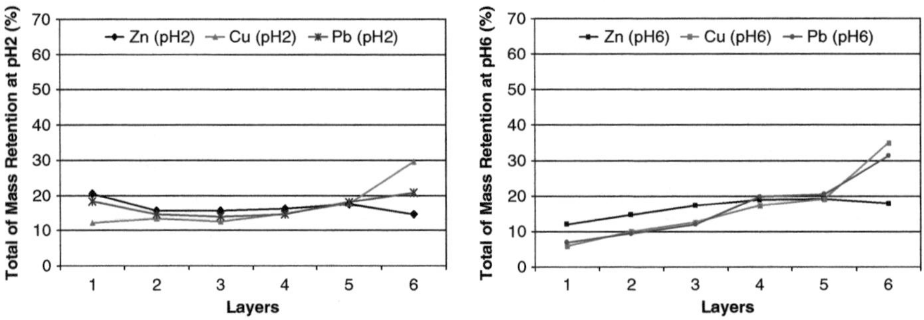

Fig. 3 Layer relevance in overall heavy metal retention at pH2 and pH6

percentage of these metals are release from the filter media at pH6, as presented in Fig. 2. The relevance of each layer in the overall retention of Zn, Cu and Pb, at pH2 and pH6, for column C1 is depicted in Fig. 3.

At pH2 all the reactive layers are equally relevant. At pH6 the retained heavy metals decrease for higher depths. High retention of Cu and Pb in the top layer suggests the occurrence of precipitation and particulate retention, favorable conditions in terms of groundwater protection.

Column C2

In this column, with the reactive layers with 5% of kaolinite and 95% of sand, the heavy metals retained were considerably higher than those in column C1. In the reactive layers heavy metals retention, at pH6, vary from 12% for Zn to <5% for

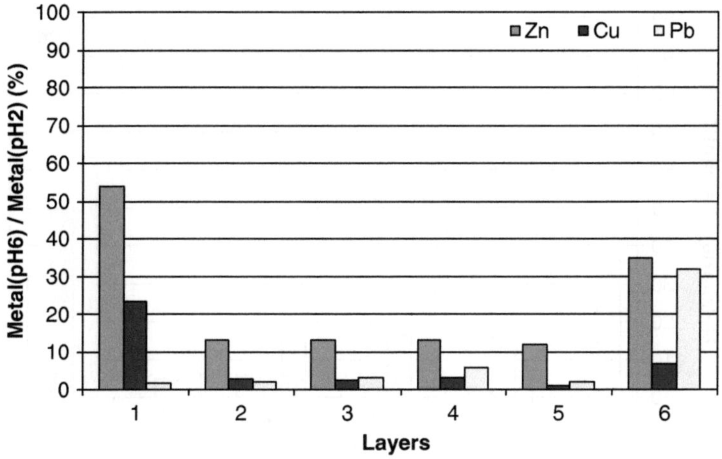

Fig. 4 Column C2: metal mass ratio

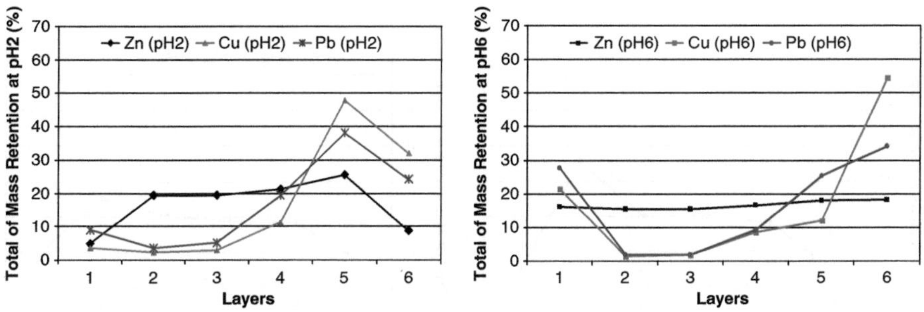

Fig. 5 Layer relevance in overall retention of Cu at pH2 and pH6

Cu and Pb, when compared with the corresponding values at pH2. In the top sand layer, higher heavy metals retention values of Zn and Pb were obtained (Fig. 4).

For Zn, the layer relevance in the overall retention is similar at pH2 and pH6, which is possibly due to the verified Zn breakthrough. For Cu and Pb, higher retention relevance is verified in the upper reactive layers as shown in Fig. 5.

Column C3

The results obtained in Column C3 are similar to those obtained in Column C2. The metals released at pH6 were slightly higher for Zn (about 15%) than those verified in column C2, and almost the same values for Cu and Pb, near 5% (Fig. 6).

The top layer of the reactive filter demonstrates higher retention than the lower layers, possibly due to the absence of a breakthrough occurrence. Higher resistance to desorption of heavy metals present in the reactive layers is verified.

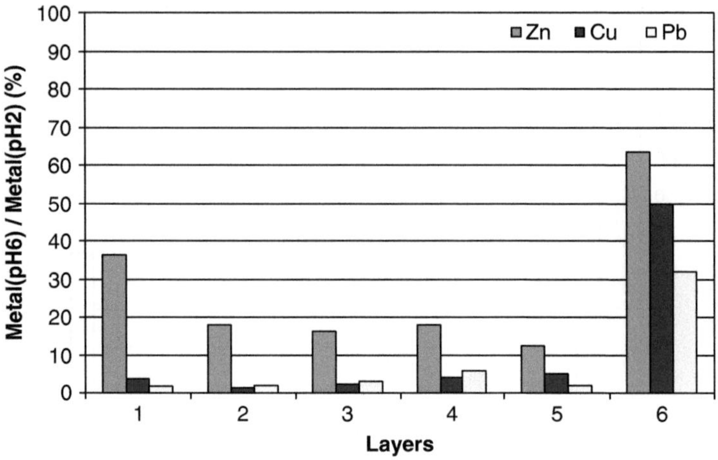

Fig. 6 Column C3: metal mass ratio

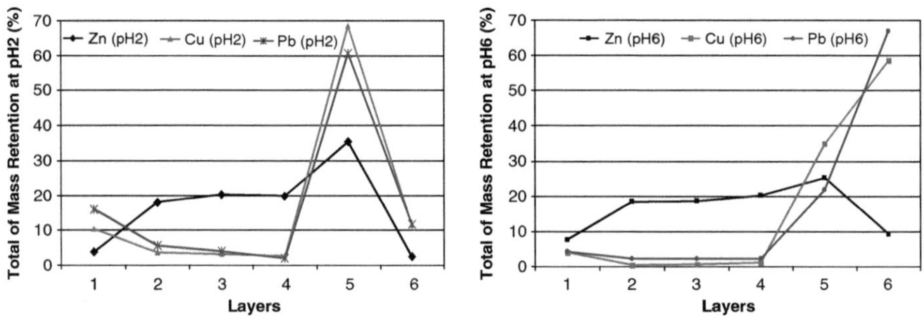

Fig. 7 Layer relevance in overall retention of Pb at pH2 and pH6

Heavy metals released at pH6 are lower in the bottom layers, which is a favorable result when looking for groundwater protection (Fig. 7).

Conclusion

Results obtained for chemical desorption allow evaluation of the heavy metals retention in each layer and their environmental availability at pH6.

In column C1, with a sand filter medium, a significant amount of retained metals is released at pH6.

In columns C2 and C3 with reactive filter media, high retention efficiencies were found. The environmental availability results obtained were 12–18% for Zn and lower than 5% for Cu and Pb.

After a 1 year experimental period, the retention efficiencies verified in column C1 for the dissolved Cu and Pb were 70% and 40%, respectively, and above 90% in the columns C2 and C3. This demonstrates the high longevity of filter media in these experimental conditions. A direct link of reactive layers and a high retention efficiency of dissolved heavy metals can be established. Moreover a strong resistance to chemical desorption of previously retained heavy metals in those reactive layers was also found.

The results obtained in this research project encourage the development of innovative engineering solutions for retention of heavy metals in highway runoff, and thus contributing for surface and groundwater protection.

References

1. Hvitved-Jacobsen T, Vollertsen J (2005) Urban Storm Drainage. Pollution – Concepts and Engineering, Ph.D. Course: Process Engineering of Urban and Highway Runoff, Aalborg University, Aalborg
2. Barbosa AE (1999) Highway runoff pollution and design of infiltration ponds for pollutant retention in semi-arid climates, Ph.D. thesis, Aalborg University, Aalborg, 52 pp
3. Farm C (2002) Metal sorption to natural filter substrates for storm water treatment – column studies. Sci Total Environ 298:17–24
4. Sansalone JJ, Buchberger SG, Al-Abed SR (1996) Fractionation of heavy metals in pavement runoff. Sci Tot Environ 189/190:371–378
5. Sansalone JJ (1999) Adsorptive infiltration of metals in urban drainage - media characteristics. Sci Tot Environ 253:179–188
6. Ramísio PJ, Vieira JMP (2007) Proceedings of the 8th Highway and Urban Environment Symposium. Amsterdam: Springer
7. Standard Methods for the Examination of Water and Wastewater, 19th Ed. (1995) American Public Health Association, American Water Works Association, Water Environment Association

Diagnosing the Impact of Traffic on Roadside Soils Through Chemometric Analysis on the Concentrations of More Than 60 Metals Measured by ICP/MS

Jose Antonio Carrero, Naiara Goienaga, Oihana Barrutia, Unai Artetxe, Gorka Arana, Antonio Hernández, José María Becerril, and Juan Manuel Madariaga

Introduction

Road traffic is one of the most important environmental problems in many cities. Its contribution to the global emission of environmental pollutants is increasing every year [1]. However, the dispersion of those pollutants is affected by the climatic conditions, placing finally in other areas such as the surrounding soil and water [2]. There are many studies showing the contamination of roadside soils as a consequence of traffic activity [2–5]. Among all possible pollutants originated by road traffic, metals and polycyclic aromatic hydrocarbons (PAHs) are the most dangerous ones. While organic compounds can be transformed or degraded by means of the microbial action of the soil in others innocuous compounds such as water and CO_2, metals can remain in the environment for a long time [6].

According to the bibliography [2, 3, 5, 7] roadside soils have long been known to contain high levels of heavy metals, Pb, Ba, Zn, Cd and Cu mainly. Well-known road traffic related metal emission sources of concern are brake linings, tyres, road pavement and exhaust fumes [8]. The metals of concern for emissions from brake linings are Cd, Cr, Cu, Ni, Pb, Sb and Zn. Wear from vehicle tyres is another major metal source, especially Zn and Ba. Tyres have traceable amounts of several other metals, for example Cd, Co, Cr, Cu, Hg, Mn, Mo, Ni and Pb. Metal emissions from exhaust fumes derives from fossil fuels and the aging processes of engines and catalysts. Combustion of leaded petrol was previously the major source. Since January 2000, Pb has been banned as an anti-knocking additive in petrol in the EU. However, there are still some emissions of Cd, Cr, Cu, Ni, Pb and V from fossil fuels [9].

J.A. Carrero (✉), N. Goienaga, G. Arana, and J.M. Madariaga
Dpt. Analytical Chemistry, Univ. Basque Country (UPV/EHU),
P.O. Box 644, 48080, Bilbao, Spain
e-mail: joseantonio.carrero@ehu.es

O. Barrutia, U. Artetxe, A. Hernández, and J.M. Becerril
Lab. Plant Physiology, Dpt. Plant Biology and Ecology, Univ. Basque Country (UPV/EHU), P.O. Box 644, 48080, Bilbao, Spain

S. Rauch et al. (eds.), *Highway and Urban Environment*,
Alliance for Global Sustainability Bookseries 17,
DOI 10.1007/978-90-481-3043-6_35, © Springer Science+Business Media B.V. 2010

There is general agreement that these metals decrease in concentration with depth and with distance from the roadway [2].

Roadside soils may be screened for all constituents of significance as a first step in establishing soil quality criteria for protection of soil ecosystems. In this sense, ICP/MS offers the capability to screen road soil samples for most elements in the periodic table. The aim of this study was to identify traffic-related elements (TRE) in the vicinity of both an old secondary road and a relatively new highway in the north of Spain, in order to perform a preliminary study to know the traffic impact on roadside soils and monitor the contribution of traffic to environmental pollution and risk.

Experimental – Method

The study area is a countryside located in Biscay, northern Spain (43°19′35.53′ N and 2°52′20.39′ W) near the city of Bilbao (Fig. 1). The first study area is a secondary road which has borne a high traffic density until the new highway was built two decades ago. Nowadays the medium traffic density of this older road is about 1,800 vehicles per day and 5.9% of these are of high tonnage [10]. The second one is a modern highway which is parallel to the old road all along the course of the studied area at 50 m of distance, separated by a stream. It bears most of the traffic in the area and has a traffic density of about 28,200 vehicles per day, and 5.7% of them are of high tonnage.

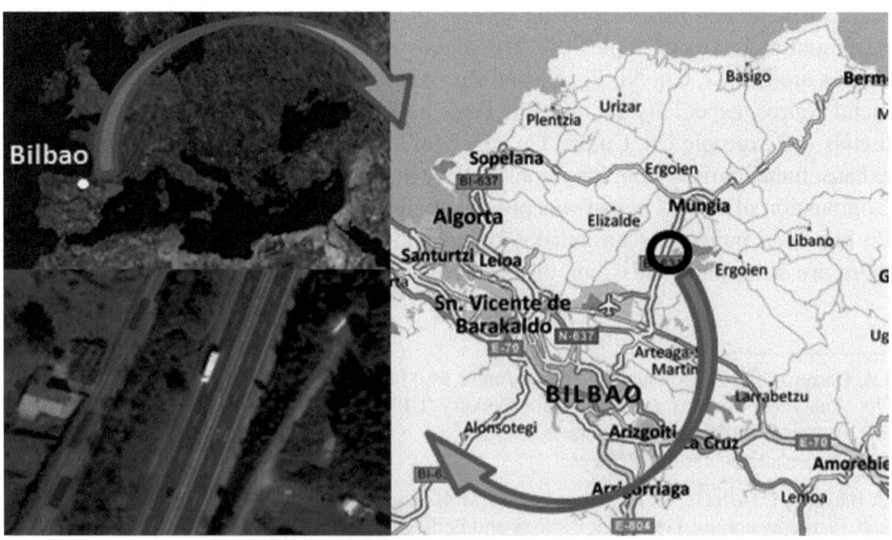

Fig. 1 Location of the study area in North Spain, next to the city of Bilbao (Biscay)

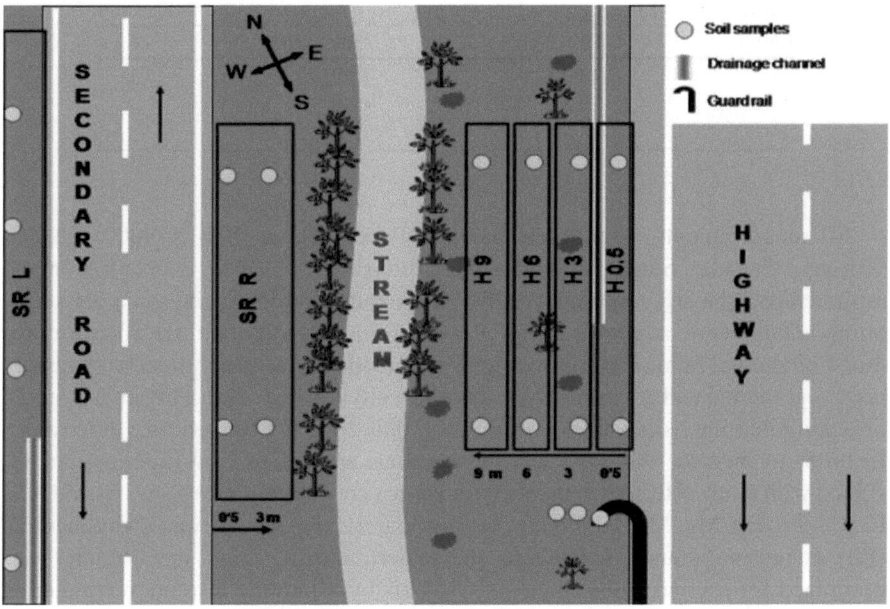

Fig. 2 Experimental design of the sampling area

Roadside soil samples were collected at different sampling points along both roads during May of 2007. Sampling in the highway was carried out at a distance of 0.5, 3, 6 and 9 m from the highway (H 0.5, H 3, H 6 and H 9 sampling areas) (Fig. 2). In the secondary road samples were collected at both roadsides (SR L and SR R sampling areas) but the sampling was only completed at 0.5 m in the west roadside and at 3 m in the east roadside because of the existence of a road cut and a stream that avoided sampling at a further distance.

Soil samples under the guard rail were also collected (GR 1–GR 3). Control soils were sampled 700 m away from both traffic roads in the north-west direction. Soil samples were collected from the upper 0–20 cm of soil, subsequently air dried in a fume hood during 24 h and then milled and sieved to <2 mm.

The analytical determination of heavy metals in soil samples were carried out following the US Environmental Pollution Agency EPA 3051A method which describes the "microwave assisted acid digestion of sediments, sludge, soils, and oils" [11]. The method consists in the digestion of 0.5 g of soil sample with a mixture of concentrated nitric acid and concentrated hydrochloric acid (9:3) in a fluorocarbon polymer (PTFE) microwave vessel at the temperature of 170°C (±10°C) during 4.5 min (Table 1). A first ramp of 1,000 W was applied during 7 min to reach 170°C as established in the method, and then it was maintained by means of a second ramp of 800 W during 4.5 min. The extracts were filtered through a 0.45 μm PVDF filter in a 50 mL polypropylene tube. The residual soil and the vessels were washed several times with Milli-Q water and the extracts were mixed.

Table 1 Power program used for the microwave assisted acid soil digestion

Ramp	Power (W)	Time (min)	Fun
1	1,000	7:00	1
2	800	4:30	1
3	0	15:00	3

All measurements were made using a Perkin Elmer SCIEX 9000 ICP/MS (Toronto, Canada). Sample solutions were diluted to 1% HNO_3 concentration and aspirated into the argon plasma (99.995%, Praxair, Madrid, Spain) via a peristaltic pump. Data were acquired using Perkin Elmer TotalQuant III software for multi-elemental semi-quantitative analysis. Common isobaric interferences are pre-programmed and corrections are automatically applied. The software has stored pre-calibrated intensities per concentration unit covering the required m/z range. In order to increase the accuracy, these values are updated by running samples spiked with a few selected elements with known concentration (Ag, Al, As, Cd, Co, Cr, Cu, Fe, Hg, Mg, Mn, Ni, Pb, Sb, Se, Sn, V and Zn at 20 ng/g) and using rhenium (Re) as internal standard. Prior to any experiment the mass spectrometer was optimized for routine multi-elemental analysis by aspirating a solution containing Mg, Rh and Pb (10 ng/mL of each). The operating conditions for sample introduction were as follows: plasma Ar flow rate, 15 L/min; auxiliary Ar flow rate, 1 L/min; nebulizer flow rate, 0.9 L/min, 1 replicate and one sweep per replicate.

Results and Discussion

Fifty elements (Ag, As, B, Ba, Be, Bi, Cd, Ce, Co, Cr, Cs, Cu, Dy, Er, Eu, Ga, Gd, Ge, Hf, Ho, In, K, La, Li, Lu, Mg, Mo, Na, Nb, Nd, Ni, Pb, Pr, Rb, Sb, Se, Sm, Sn, Sr, Tb, Te, Th, Tl, Tm, U, V, Y, Yb, Zn, Zr) were determined above the detection limit in soil extracts. These results were statistically treated with "The Unscrambler 9.2" software.

Principal component analysis (PCA) of all available data revealed two main components accounting for 64% of the total variance (PC1, 48%; PC2, 18%). The projection of the scores and loadings on the bi-dimensional space defined by these two principal components is shown in Fig. 3. Samples can be grouped according to sampling point in three groups. Samples from the west roadside of the secondary road (SR L) clearly differentiate from the rest, and are gathered in the left side of the x axis (PC1). Samples from the highway (H 0.5, H 3 and H 6) form another group in the upper side of the graphic. Finally, samples from the east roadside of the secondary road (SR R), further samples from the highway (H 9) and control soils put together in the left side.

Overlapping the scores and loading plots shows that the concentrations of Cd, Cr, Cu, Hf, Mo, Nb, Ni, Pb, Sb, Sn, Sr, Zn and Zr, are the variables which explain most of the variance of the SR L sampling point.

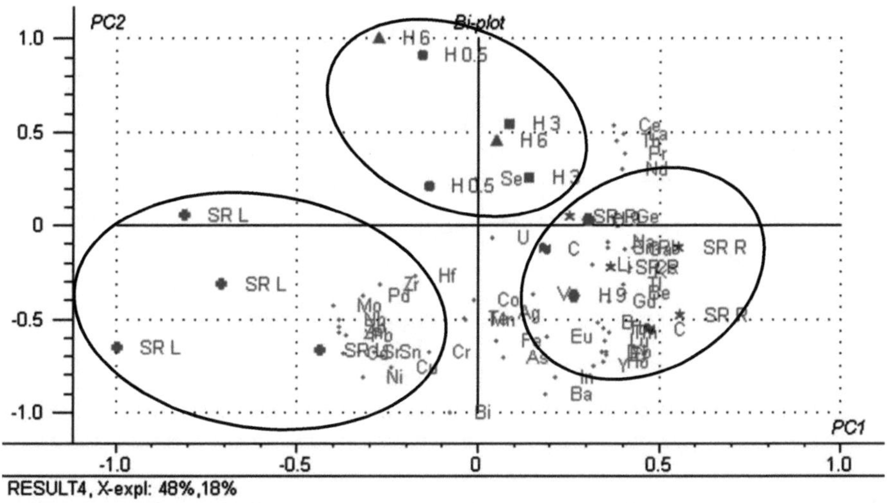

Fig. 3 Scores and loadings plot for the first two PCs resulting from the PCA of all elements determined in roadside soils. C: control; SR L: west roadside of the secondary road; SR R: east roadside of the secondary road; H 0.5, H 3, H 6 and H 9: highway soil samples at 0.5, 3, 6 and 9 m distance from highway

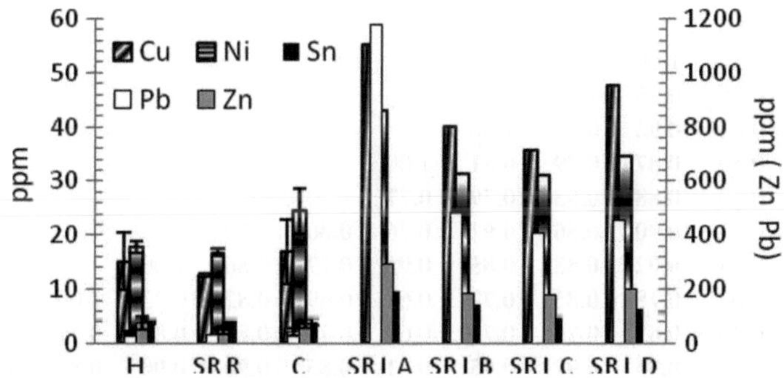

Fig. 4 Concentration of some TRE by sampling area. SR L A-SR L D: metal concentration in soils from the west roadside of the secondary road. H, SR R and C: average value of metal concentration in highway, east roadside of the secondary road and control respectively

In fact, the concentration of these metals was considerably higher in these soil samples as can be seen in Fig. 4, where concentration of some of these metals has been represented for samples from the west roadside of the secondary road (SR L A-SR L D) against the average value of the concentrations in the other sampling points (H: Highway; SR R: east roadside of the secondary road; and C: control). The rest of above mentioned metals (data not shown) have the same behavior.

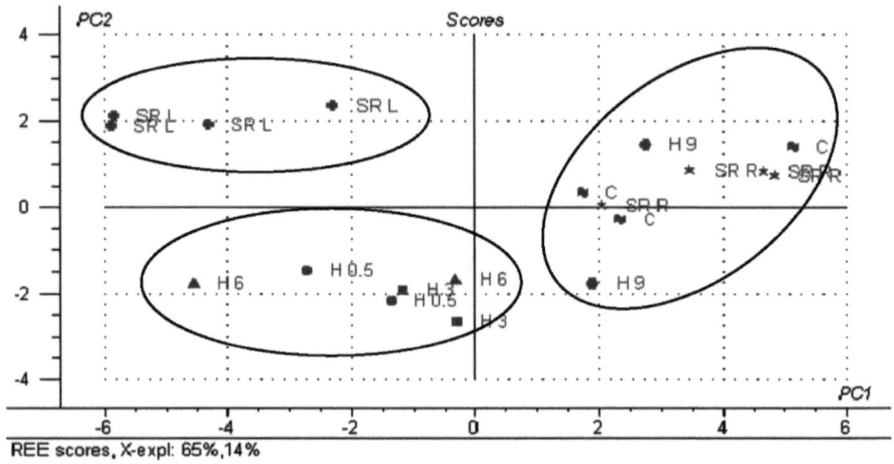

Fig. 5 Scores plot for the PCA of the rare earth elements and alkaline metals. C: control; SR L: west roadside of the secondary road; SR R: east roadside of the secondary road; H 0.5, H 3, H 6 and H 9: highway soil samples at 0.5, 3, 6 and 9 m distance from highway

Table 2 Correlations between traffic related metals

	Hf	Zr	Cd	Cu	Mo	Nb	Ni	Pb	Sb	Sn	Sr	Zn
Hf	1.00											
Zr	**0.94**	1.00										
Cd	0.34	0.56	1.00									
Cu	0.32	0.55	**0.95**	1.00								
Mo	0.40	0.64	**0.92**	**0.91**	1.00							
Nb	0.26	0.50	**0.87**	**0.79**	**0.81**	1.00						
Ni	0.31	0.50	**0.88**	**0.83**	**0.79**	**0.77**	1.00					
Pb	0.28	0.51	**0.90**	**0.86**	**0.92**	**0.76**	**0.80**	1.00				
Sb	0.21	0.44	**0.92**	**0.88**	**0.85**	**0.93**	**0.79**	**0.86**	1.00			
Sn	0.45	0.62	**0.78**	**0.85**	**0.77**	0.61	0.69	**0.82**	0.73	1.00		
Sr	0.28	0.46	**0.82**	**0.75**	**0.79**	0.69	0.70	**0.87**	**0.80**	0.71	1.00	
Zn	0.34	0.58	**0.96**	**0.97**	**0.95**	**0.81**	**0.85**	**0.90**	**0.90**	**0.83**	**0.78**	1.00

Following with the PCA of the soil samples, Fig. 5 represents the scores of the two first principal components which explain 79% of the total variance (PC1, 65%; PC2, 14%). In this case, concentrations of only rare earth elements and alkaline metals have been taken into consideration. Samples from the west roadside of the secondary road (SR L) and from the highway (H 0.5, H 3 and H 6) forms two different groups. A third group is formed by samples from the east roadside of the secondary road (SR R), further samples from the highway (H 9) and control soils.

Correlation analysis of the metal concentrations reveals two groups of metals showing high correlations among them (Table 2). A first group is formed by Hf and Zr, meanwhile the rest of metals (Cd, Cu, Mo, Nb, Ni, Pb, Sb, Sn, Sr and Zn) remains in another group.

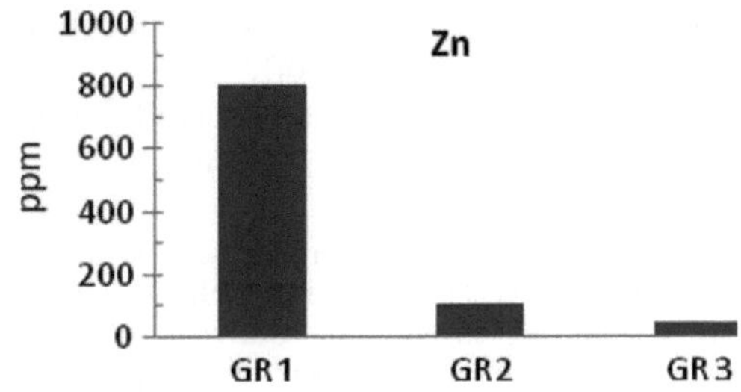

Fig. 6 Graphical representation of Zn concentration values in surroundings of the guard rail. GR 1: soil sample just under the guard rail. GR 2 and GR 3: soil samples moving away the guard rail

Finally, soil samples under the guard rail contain very high concentration of Zn values. Three different samples were collected nearby it. The one just under the guard rail (GR 1) points out the highest value for Zn concentration among all the soil samples analyzed, and shows a very strong decreasing gradient as they move away from it in very few meters (GR 2 and GR 3) (Fig. 6).

Conclusions

ICP/MS semi-quantitative analysis of the soil samples allows the fast determination of most soil elements as a first tool for evaluating the impacted area. A first list of metals that could be related to road traffic was identified by means of the chemometric analysis of the obtained semi-quantitative results ICP/MS: Cd, Cr, Cu, Hf, Mo, Nb, Ni, Pb, Sb, Sn, Sr, Zn and Zr. The study area was reduced. The secondary road appears to be much more polluted by the traffic emissions than the highway. TRE in the west roadside appear at higher concentrations than the rest of sampling areas, included the control soils, probably due to a longer-term accumulation in the first one as well as its topographical characteristics that would lead to the concentration of atmospheric particles in a smaller area (because of the presence of a road cut). The east roadside in the secondary road as well as soils further away from the highway appears not to be affected by the road traffic, and it shows similar metal concentration values to control samples.

Soil samples from diverse origins could be differentiated in basis of their rare earth elements and alkaline metal concentrations. It was concluded that soils from both roadsides in the secondary road have different source, as well as highway soils. West roadside seems to be original land, whereas the highway land is possible to be a filling, as for soil samples further away, at a distance of 9 m from the road seem to be original land too. Finally, east roadside has another different source, as the presence of a road cut suggests it.

Two groups of metals appeared in the correlation analysis suggest a different emission source for both of them. Hf and Zr are associated to catalytic converters in automobiles and they enter in the environment by atmospheric deposition of the exhaust fumes, whereas Cd, Cr, Cu, Mo, Ni, Pb, Sb, Sn and Zn have different pollution sources, like tyre wear, brake linings or mechanical abrasion and reach roadside soils by road run off.

Finally, it has also been observed that guard rails surrounding the roads are a very important input of Zn to soils under them.

Acknowledgements This work has been financially supported by the ETORTEK Program of the Basque Government through the BERRILUR II Project (ref. IE06-179). J.A. Carrero and O. Barrutia are grateful respectively to the University of the Basque Country and the Spanish Ministry of Science and Technology for their fellowships.

References

1. Hjortenkrans DST, Bergbäck BG, Häggerud AV (2007) Metal emissions from brake linings and tires: case studies of Stockholm, Sweden 1995/1998 and 2005. Environ. Sci. Technol. 41:5224–5230
2. Viard B, Pihan F, Promeyrat S, Pihan J-C (2004) Integrated assessment of heavy metal (Pb, Zn, Cd) highway pollution: bioaccumulation in soil, Graminaceae and land snails. Chemosphere 55:1349–1359
3. Garcia R, Millán E (1998) Assessment of Cd, Pb and Zn contamination in roadside soils and grasses from Gipuzkoa (Spain). Chemosphere 37:1615–1625
4. Carlosena A, Andrade JM, Prada D (1998) Searching for heavy metals grouping roadside soils as a function of motorized traffic influence. Talanta 47:753–767
5. Rauch S, Hemond HF, Barbante C, Owari M, Morrison GM, Peucker-Ehrenbrink B, Wass U (2005) Importance of automobile exhaust catalyst emissions for the deposition of platinum, palladium, and rhodium in the Northern Hemisphere. Environ. Sci. Technol. 39:8156–8162
6. Wu Y, Luo Y, Zou D, Ni J, Liu W, Teng Y, Li Z (2008) Bioremediation of polycyclic aromatic hydrocarbons contaminated soil with Monilinia sp.: degradation and microbial community analysis. Biodegradation 19:247–257
7. Bretzel F, Calderisi M (2006) Metal contamination in urban soils of coastal Tuscany (Italy). Environ. Monitor. Assess. 118:319–335
8. Falahi-Ardakani A (1984) Contamination of environment with heavy metals emitted from automotives. Ecotoxicol. Environ. Safety 8:152–161
9. Hjortenkrans D, Bergbäck B, Häggerud A (2006) New metal emission patterns in road traffic environments. Environ. Monitor. Assess. 117:85–98
10. Evaluación del tráfico de las carreteras de Vizcaya (2004) Gobierno Vasco
11. Microwave assisted acid digestion of sediments, sludges, soils and oils (2007) Test Methods for Evaluating Solid Waste, Physical/Chemical Methods. Environmental Pollution Agency

Influence of De-Icing Salt on K Supply and Street Trees Ecological Status in Riga, Latvia

Gunta Cekstere, Anita Osvalde, and Olgerts Nikodemus

Introduction

The street greenery as a very significant landscape's design element with important ecological value in Riga's central part is subjected to a multiple stresses, e.g. chemical injury. In urban area, safe driving conditions during winter season are of primary concern. The most common de-icing material used in Riga is sodium chloride (NaCl). The total area of streets and roads in Riga is 24.64 km^2 [1] and the average rate of NaCl application is about 10,000 t/year which leads to ca. 4.06 kg spread salt per m^2.

While Na^+ and Cl^- in small amounts are thought to be beneficial for some plants or even essential for halophytes [2–4], in general these ions are toxic for plants. It was shown that high concentrations of Na^+ and Cl^- in cells can inhibit membrane functions, induce ion deficiency, decrease chlorophyll concentration and the activity of several enzymes, producing disturbance of various metabolic processes, such as gas exchange and photosynthesis [5–10]. Plant injury can also be caused by salt-induced water stress [2]. One of the physiological mechanisms of salt toxicity in plants is closely related to unfavorable K/Na ratio. K as plant macronutrient has important functions as osmoregulation, maintenance of electrochemical equilibria, effects on protein conformation, regulation of enzyme activity etc. [3]. The most widespread tree species in the street greenery of Riga is *Tilia × vulgaris H.* In deciduous trees visual symptoms of Na^+ and Cl^- damage usually appear as post-flushing dieback and foliage discoloration [5], burning of leaves [7], as well as chlorosis and necrosis of leaves [9, 11].

The first studies in Riga to assess the impact of chloride on Riga street trees were carried out in 1966/1967 by [12]. Our previous work on chestnut and lime in Riga

G. Cekstere and O. Nikodemus
Faculty of Geography and Earth Sciences, University of Latvia

G. Cekstere and A. Osvalde
Laboratory of Plant Mineral Nutrition, Institute of Biology University of Latvia

S. Rauch et al. (eds.), *Highway and Urban Environment*, Alliance for Global Sustainability Bookseries 17,
DOI 10.1007/978-90-481-3043-6_36, © Springer Science+Business Media B.V. 2010

showed a severe imbalance in the mineral nutrition of urban trees [13]. Investigation in 2004/2005 revealed high concentrations of Na^+ and Cl^- in the street snow melt water, urban soil and *T.* × *vulgaris* leaves [10]. It is a topical problem as the street greenery currently covers 39% of the total street length in the centre of Riga and in general the status of trees could be characterized as damaged.

Our study was conducted to evaluate the toxic effect of soil and plant pollution level with sodium and chlorine on potassium supply and ecological condition of the street trees (*Tilia* × *vulgaris*) in Riga.

Materials and Methods

Riga, the capital of the Republic of Latvia, is situated along the Baltic Sea at the southern part of the Gulf of Riga. The central part of the city is located on the right bank of the Daugava river. The climate is moderately warm and humid. Summers are relatively cloudy and cool. The average temperature in July is +16.9°C. Winters are relatively warm with frequent thaws, the average temperature in January is –4.7°C. The average amount of precipitation is 700–720 mm/year [1].

Sampling

During the time period of March 2005 to August 2007, soil and leaf samples were collected from 14 sites (45 trees) in the central part of Riga, and from a control site (three lime trees) located in a park area – Viestura Garden (Fig. 1, Table 1). Soil samples were collected from the tree-rooting zone (road side) with a soil probe to a depth of 35 cm in March (the end of winter), June and July 2005, and June and August 2007.

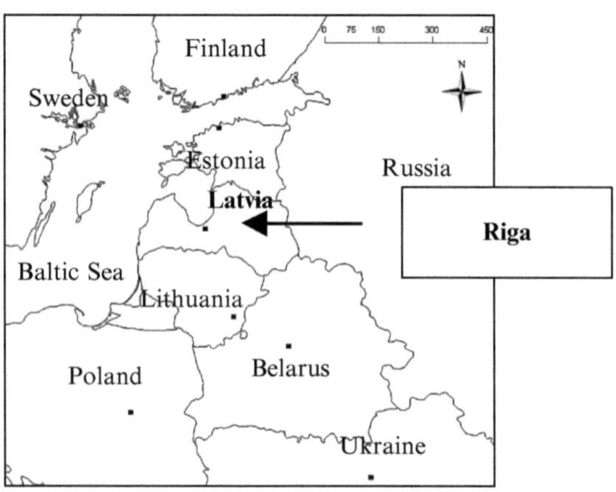

Fig. 1 Location of Riga in Europe

Table 1 Description of studied sites in Riga

Site	Distance of tree's stems to street (m)	Status of trees in August 2005 and 2007
1 – Park (three trees)	>50	2005, 2007: Healthy[a]
2 – Hanzas 1 (three trees)	~2	2005: Damaged; 2007: damaged
3 – Hanzas 2 (three trees)	~2.15	2005: Medium damaged: 2007: medium damaged
4 – Elizabetes (five trees)	~0.7	2005: Three-damaged, two-healthy; 2007: one-dead in 2006, two-damaged, two-healthy
5 – Kr.Valdemara (three trees)	~0.7	2005: Damaged; 2007: medium damaged
6 – Stabu 1 (three trees)	~0.6	2005: Damaged; 2007: damaged
7 – Stabu 2 (four trees)	~0.6	2005: Three-damaged, one-healthy; 2007: three-damaged, one-healthy
8 – Basteja 1 (three trees)	~3.5	2005, 2007: Healthy
9 – Basteja 2 (three trees)	1.0–1.5	2005: Damaged; 2007: damaged; one tree dead in winter 2008
10 – Raina 1 (three trees)	~7	2007: Healthy
11 – Raina 2 (three trees)	~7	2007: Healthy
12 – Brivibas 1 (three trees)	~1.5	2007: Medium damaged
13 – Brivibas 2 (three trees)	~7	2007: Healthy
14 – Brivibas 3 (three trees)	~1.5	2007: Medium damaged
15 – Brivibas 4 (three trees)	~7	2007: Healthy

[a]Healthy: 0–15% crown necrosis; medium damaged: 20–70%; damaged: 75–100%.

For each soil sample three sub-samples were obtained and thoroughly mixed, to form one sample. Soil samples in the control site (Viestura Garden) were collected under the perimeter of the crown of trees and leaves from several branches of the crown.

Leaf samples were taken with telescopic scissors from leaves just reaching maturity and full size. For each sample 50 leaves were collected from different branches of trees along roadsides during the vegetation season of 2005 (June, July and August) and 2007 (June and August).

Along with leaf sampling bioindication research was done to diagnose the ecological status of *Tilia* × *vulgaris*. Leaf necrosis as percentage of damaged leaf area/ total leaf area was estimated in June and at the end of August. Intensity of crown necrosis in percent was estimated in August.

Laboratory Analysis

Sample preparation. The soil samples were stored at +4°C, then dried at +35°C for 2 days and sieved <2 mm. Leaves were washed with distilled water, dried at +60°C and ground.

Soil extraction and plant tissue ashing. Na and K in the soil samples were extracted with 1 M HCl solution (soil-extractant mixture 1:5). Cl^- content was determined in distilled water extract (soil-distilled water mixture 1:5). Leaf samples were dry-ashed in concentrated HNO_3 vapor and re-dissolved in HCl solution (HCl-distilled water mixture 3:100) to determine Na and K [14]. To determine Cl^- leaf sampled were dry-ashed, then re-dissolved in distilled water.

Chemical analysis. Na^+ and K concentration in soil and plant samples was estimated by flame photometer *JENWAY PFPJ*. Cl^- concentration in all samples was determined by $AgNO_3$ titration [14]. Analytical replication was three times.

Results and Discussion

Results revealed that the use of de-icing material NaCl caused a considerably increased Na^+ and Cl^- accumulation in the street greenery soil and lime leaves located close to the carriageway compared with the park (Figs. 2 and 3, Table 1).

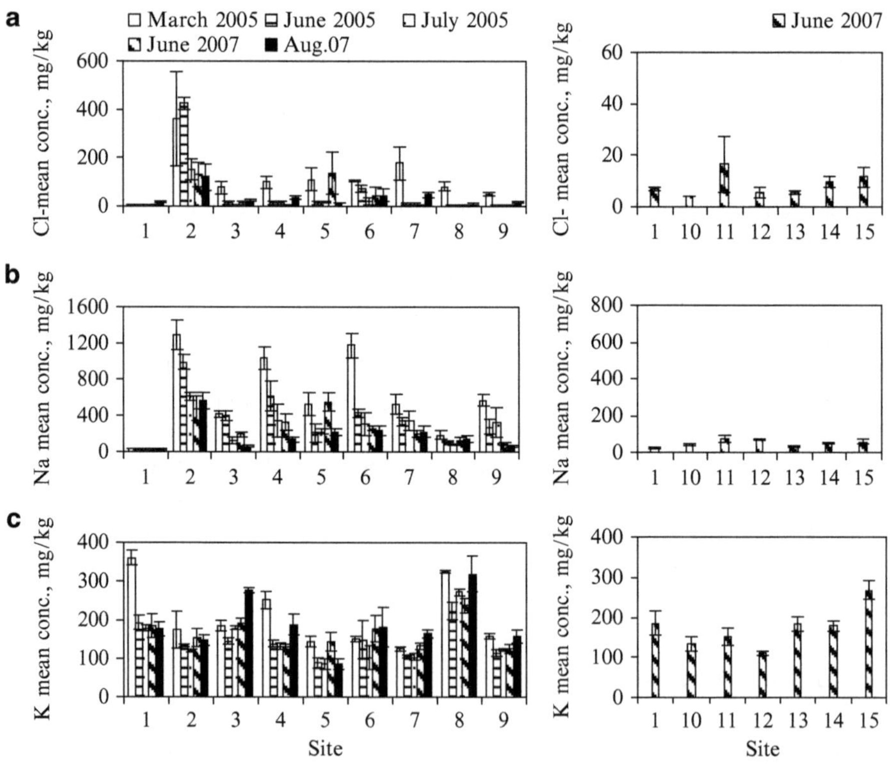

Fig. 2 Mean concentrations from 3–5 samples ± SE of Cl^- (**a**), Na (**b**) and K (**c**) (mg/kg) in soil samples from the central part of Riga (spring, summer 2005, 2007). Sites: 1 – park; 2 – Hanzas 1; 3 – Hanzas 2; 4 – Elizabetes; 5 – Kr.Valdemara; 6 – Stabu 1; 7 – Stabu 2; 8 – Basteja 1; 9 – Basteja 2; 10 – Raina 1; 11 – Raina 2; 12 – Brivibas 1; 13 – Brivibas 2; 14 – Brivibas 3; 15 – Brivibas 4

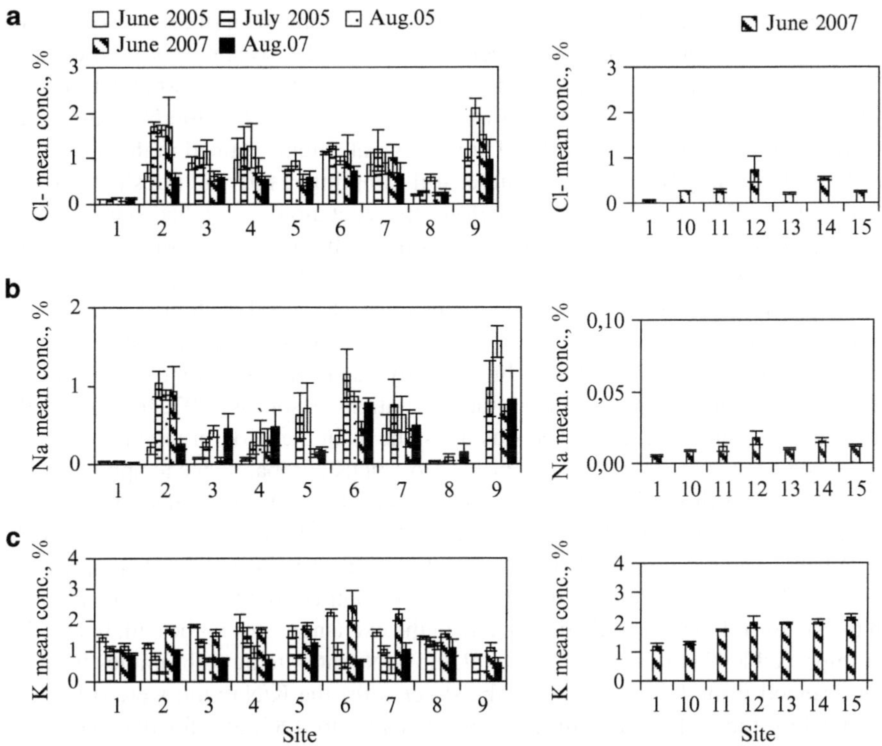

Fig. 3 Mean concentrations from 3–5 samples ± SE of Cl^- (**a**), Na (**b**) and K (**c**) (%) in leaf samples from the central part of Riga (spring, summer 2005, 2007). Sites: 1 – park; 2 – Hanzas 1; 3 – Hanzas 2; 4 – Elizabetes; 5 – Kr.Valdemara; 6 – Stabu 1; 7 -Stabu 2; 8 – Basteja 1; 9 -Basteja 2; 10 - Raina 1; 11 – Raina 2; 12 – Brivibas 1; 13 – Brivibas 2; 14 – Brivibas 3; 15 – Brivibas 4

Correlation coefficients and the level of statistical significance ($p < 0.05$) were found using SPSS 14.0 version. The coefficients were classified as follows: $r < 0.5$ week, $0.5 < r < 0.8$ medium close, and $r > 0.8$ close correlation.

Soil. The highest concentrations of Na and Cl^- in soil samples of street greenery were found at the end of March 2005 - beginning of spring. Levels exceeding 1,000 mg/kg Na were found at sites 2 (max. 1,568.1 mg/kg), 4 and 6, which exceeded the background values by up to 51 times. A marked decrease in Na and especially Cl^- concentrations occurred in spring and summer due to leaching from the root zone by precipitation and uptake by the plants. Compared to Na, the Cl^- concentrations in 2005 and 2007 were considerably lower, ranging from 4.27 ± 0.08 mg/kg (Site 10) (similar to park results) to 359.5 ± 194.2 mg/kg (Site 2) or on average more than 60 times higher than in the park.

The obtained concentrations of Cl^- in street soil samples collected in June 2005 and 2007 on average were not significantly different, while results of Na were significant higher in 2005 to compare with 2007. Such situation could be explained by the winter season of 2006/2007, which was warmer in comparison to 2004/2005 [15].

As a result, decreased amounts of de-icing salts were applied on streets and decreased accumulation of ions occurred in street soil. Besides, Cl^- as anions are more leachable from the soil and a more rapid decrease compared to Na^+ occurred. Cl^- usually follow the water and do not take part in chemical reactions, while Na^+ participate in chemical processes in the soil and are to a great extent retained in the upper part of the soil profile [16]. Thereby summer soil sampling results did not show the real situation with Na and Cl^- pollution before the vegetation season had started.

A different situation was stated with K (Fig. 2). In the vast majority of sampling sites in 2005 the concentrations of K in the soil were lower compared to the park. It was consistent with other studies in urban areas where soil contained construction and demolition waste [13, 17]. The decrease of K concentrations was stated almost for all studied sites from March to July, mainly due to K uptake by plants. Better K status in soil samples was found in the vegetation season of 2007. In most of the studied street sites, the level of K was the same as in the park, or slightly elevated. It showed a tendency to increase from June to August, probably due to application of K containing fertilizers. In general, the concentrations of K in soil samples during 2005 and 2007 ranged from 85.10 ± 13.38 mg/kg (Site 5) to 318.90 ± 46.22 mg/kg (Site 8).

When comparing results from 2 years, the level of K in street soil samples was statistically significant higher in June 2007, in contrast to Na. Less amounts of Na in the street soil, which could displace K in the exchange sites in the soil, resulted in an improved K/Na ratio in 2007. In total, the K/Na ratio in the soil samples from the park ranged from 4.15 to 13.48. In the street soils, the K/Na ratio were 0.13–6.45 for healthy limes, for medium damaged limes up to 4.82, and for damaged trees 0.07–2.66. It means that the small K/Na ratio in street soil samples had additive harmful impact on the tree status.

Plants. In general, the concentrations of Na and Cl^- in lime leaves increased during the summer of 2005 and 2007, showing opposite trends as those in the soil. The K content in lime leaves decreased (Fig. 3), thereby harmfully affected the tree status. The analysis of plant material showed that the soil salinity induced increased Na and Cl^- concentrations in lime leaves in sites with higher soil salinity during 2005 and 2007. During the vegetation seasons, Na ranged from 0.01% in June to 1.93% in August, which was up to 48 times higher than values from the park. The Cl^- concentrations in the leaves from the vast majority of street sites were significantly higher than those of the park site at all sampling times, as well as higher than the Na concentrations in lime leaves. The highest Cl^- concentrations, found in Site 4 in July (3%) and August (2.90%), were more than 25 times above the background level. As the lime tree decayed in autumn, this concentration could be characterized as especially toxic to lime trees. In general, concentrations of Cl^- in leaves higher than 2% were harmful to lime trees, causing tree decay.

The lowest concentrations of Na and Cl^- in lime leaves were found in sites situated about 3.5 m from the carriageway (sites: 8, 10, 11, 13, 15), as well as sites 12 and 14. These limes were characterized as healthy.

There were no significant correlation between the element concentrations in soil and leaves, except the concentrations of Na in July 2005 ($r = 0.39$; $n = 30$). The results showed negative medium close correlation between the level of K in soil of

March 2005 and Na in leaves during summer 2005 ($r_{June} = -0.66$, n = 24; $r_{July} = -0.65$, n = 30; $r_{Aug.} = -0.66$, n = 30), which demonstrated the importance of soil chemical composition to plants just after winter.

As the external salt concentration rises, the concentration of Na^+ increases and that of K^+ decreases in the roots and leaves. It appears that Na^+ out-compete K^+ for specific binding sites on the K^+ transport proteins, causing K^+ deficiency [3]. Besides, Na^+ in the soil can replace not only K^+, but also Ca^{2+}, Mg^{2+}, NH_4^+ and other cations on the soil exchange complex. This could lead to nutrient deficiencies [2, 18] causing injury.

A wide dispersion of K concentrations and supply levels were stated in lime leaves during 2005 and 2007 (Fig. 3). The smallest K result (0.15%) was even up to 6.33 times lower compared with the minimum K concentration in lime leaves in the park. At the same time the highest concentrations of K (max. 3.40%) were also stated in lime leaves with necrosis, which mean that sufficient supply with K could not prevent the appearance of leaf necrosis. In Riga, K concentration in healthy lime leaves without necrosis ranged from 0.76% to 2.42%, which was in good agreement with results generalized by [19]. For normal plant growth, K concentration in plant tissues should be 0.5–1.2%. The decrease in K content in leaves during the vegetation season could be explained by K reutilization in plant [2], as well as an increase of Na content ($r_{Na,K\ aug.2005} = -0.76$, n = 30; $r_{Na,K\ aug.2007} = -0.43$; n = 29). As a result, the decrease of K/Na ratio from June to August was stated. The K/Na ratio for healthy lime leaves without necrosis in Riga's street greenery ranged from 2.92 to 237 (in park: 24.12–270.83), but for severe necrosis (>30% leaf necrosis) ranged even from 0.15 to 72.00. In general, the stated K/Na ratios diapasons in Riga were substantially wider compared with results reported by [9] in Opola for *Tilia cordata* L. (1.76–15.65). It can be concluded that sufficient K/Na ratio in lime leaves did not have a determinant role, if the stated leaf necrosis were caused by Cl^- toxicity, or if the Na concentration in leaves reached toxic level.

The leaf chemical results revealed significant higher levels of Na and Cl^- in August 2005 ($p > 0.05$) compared with results from August 2007, whereas the level of K was significant higher during the vegetation season of 2007. This fact could be the main reason for improved ecological status of street trees and decreased leaf necrosis in 2007 compared to 2005 (Table 2). The current research revealed significant close correlation between the intensity of the leaf necrosis and the concentration of Na and Cl^- in leaves (maximum in August 2005, $r_{Na} = 0.92$; $r_{Cl} = 0.85$; n = 30), as well as negative correlation with concentrations of K in leaves (maximum in August 2007, r = –0.50).

The leaf necroses were observed in cases when the Cl^- content in leaves reached at least 6,600 mg/kg of dry weight. These concentrations were lower in comparison with other investigations on leaf injury caused by Cl^- toxicity [9, 12, 20]. These differences might be explained in the scope of factors such as the different tree species studied, methods used, a misbalanced supply of mineral nutrients, insufficient supply of water during the vegetation period, and too high concentrations of de-icing salt in the soil solution causing 'physiological drought' which promotes leaf necrosis development, as well as specific ion toxicity.

Table 2 Intensity of lime leaf necrosis (%) in Riga, summer 2005, 2007

	Range (mean ± SE)			
Site	June 2005	August 2005	June 2007	August 2007
1 (n = 3)	0 (0.0)	0 (0.0)	0 (0.0)	0 (0.0)
2 (n = 3)	5 (5.0 ± 0.0)	45–65 (56.7 ± 6.0)	5 (5.0 ± 0.0)	20–70 (36.7 ± 16.7)
3 (n = 3)	0–5 (3.3 ± 1.7)	10–30 (18.3 ± 6.0)	0–3 (1.0 ± 1.0)	10–15 (11.7 ± 1.7)
4 (n = 5; n = 4)	0–30 (7.0 ± 5.8)	0–85 (33.0 ± 14.7)	0–5 (3.3 ± 1.81)	0–70 (29.3 ± 16.1)
5 (n = 3)	0 (0.0)	10–60 (36.7 ± 14.5)	0–3 (1.0 ± 1.0)	0–5 (1.7 ± 1.7)
6 (n = 3)	5 (5.0 ± 0.0)	15–50 (28.3 ± 10.9)	3 (3.0 ± 0.0)	5–90 (43.3 ± 24.9)
7 (n = 4)	0–5 (3.6 ± 1.3)	5–65 (38.8 ± 12.5)	0–5 (2.8 ± 1.0)	0–90 (30.0 ± 20.7)
8 (n = 3)	0 (0.0)	0 (0.0)	0 (0.0)	0 (0.0)
9 (n = 3)	5 (5.0 ± 0.0)	70–85 (78.3 ± 4.4)	5–10 (6.8 ± 1.7)	15–25 (18.3 ± 3.3)
10–15 (n = 15)	0 (0.0)	0 (0.0)	0 (0.0)	–

The occurrence of the first toxic symptoms in the form of leaf necrosis caused by Na toxicity was detected when the Na content in most of the cases ranged from 0.14% to 0.26%. The obtained relationships between the Na content in the leaf samples and visual injury symptoms of the leaves are in good agreement with those reported by [7, 21].

Conclusions

The results revealed that the ecological status of street trees in Riga could be characterized as seriously injured due to high concentrations of Na and Cl^- in soil (maxNa 1,568.1 mg/kg; maxCl 744.9 mg/kg) and lime leaves (maxNa 1.93%; maxCl 3.00%). The concentrations exceeding 2.0% Cl^- in lime leaves were revealed as extremely toxic for *Tilia* × *vulgaris*, leading to tree decay. The concentrations of Na and Cl^- in lime leaves demonstrated a marked tendency to increase during summer, while the concentrations of K in leaves decreased (minimum 0.15%). The high variability in Na and K concentrations resulted in a wide range of K/Na ratios in Riga's greenery. Close correlations were found between the concentration of Na and Cl^- in lime leaves and the extent of leaf necrosis, although unfavorable K/Na ratio has additive negative impact.

Acknowledgement This work was supported by the European Social Fund.

References

1. Anonymous (2005) Riga Development Program 2006–2012 (in Latvian). Council of Riga, Riga
2. Marschner H (1995) Mineral Nutrition of Higher Plants. 2nd edn. Academic Press, London

3. Denny H (2002) Plant mineral nutrition. In: Ridge I (ed) Plants. Oxford University Press, Oxford, pp. 167–219
4. Taiz L, Zeiger E (2002) Plant Physiology, 3rd edn. Sinauer Associate, Sunderland
5. Gibbs JN, Palmer CA (1994) A survey of damage to roadside trees in London caused by the application of de-icing salt during the 1990/1991 winter. Arboricult J 18:321–343
6. Pedersen LB, Raundrup TB, Ingerslev B (2000) Effects of road distance and protective measures on deicing salt. J Arboricult 26:238–245
7. Bryson MG, Barker AV (2002) Sodium accumulation in soils and plants along Massachusetts roadsides. Comm. Soil Sci Plant Anal 33:67–78
8. Paludan-Müller G, Saxe H, Pedersen LB, Randrup TB (2002) Differences in salt sensitivity of four deciduous tree species to soil or airborne salt. Physiol Plant 114:223–230
9. Czerniawska-Kusza I, Kusza G, Duzynski M (2004) Effect of deicing salts on urban soils and health status of roadside trees in the Opole region. Environ Toxicol 19:296–301
10. Cekstere G, Nikodemus O, Osvalde A (2008) Toxic impact of the de-icing material to street greenery in Riga, Latvia. Urban forestry & urban greening (ISSn 1618–8667) vol 7, pp 207–217
11. Tester M, Davenport R (2003) Na^+ tolerance and Na^+ transport in higher plants. Ann Bot 91:503–527
12. Ripa A, Petersons E (1968) Let's save city greenery trees (in Latvian). D rzs un Drava, 10(11) 20–22, 19–21
13. Cekstere G, Osvalde A, Karlsons A, Nollendorfs V, Paegle G (2005) The effect of urban environment on the mineral nutrition status of street trees in Riga, the problems and possible solution. Acta Universitatis Latviensis. Earth Environ Sci 685:7–20
14. Rinkis GJ, Ramane HK, Kunickaya TA (1987) Methods of Soil and Plant Analysis (in Russian). Zinatne, Riga
15. Data of "State agency Latvian Environment, geology and meteorology agency" (2008) http://www.meteo.lv/public
16. Lundmark A, Olofsson B (2007) Chloride deposition and distribution in soils along a deiced highway- assessment using different methods of measurement. Water Air Soil Pollut 182:173–185
17. Meyer FH (1978) Trees in the City. Verlag, Eugen Ulmer, Stuttgart
18. Bergmann W (1988) Nutrient Flow in Crop Plants (in German). VEB Gustav Fischer Verlag, Jena
19. Polevoi VV (1989) Plant Physiology (in Russian). Visshaya shkola, Moscow
20. Supłat S (1996) Disturbances in tree leaf development as a reaction to an elevate chlorine content of tissues (in Polish). III Krajowe Sympozjum Kornik, 23–26 maja 1994, 2:560–567
21. Leh HO (1973) Investigations on the impact of NaCl as de-icing agent on street trees in Berlin (in German). Nachrichtenbl Deut Pflanzenschutzd 25:163–170

Heavy Metals Soil Pollution in Some Urban Location from Romania

Radu Lacatusu and Anca-Rovena Lacatusu

Introduction

Urban soils derive from those natural pursuant to anthropogenic influence caused by human settlement construction and their development. The transformation level is directly proportional with urbanization process intensity, with its nature and civilization degree reached by inhabitants. On the strength of these reasons, the urban space will be separated in zones. In some zones the natural soil has been strongly modified, sometimes removed as a consequence of civilian and industrial buildings, urban equipment, and communication routes systems constructions. In other zones the soil has been less modified, but has received and still is, receiving negative impacts, caused by daily household or industrial activities. The soils belonging to this category are soils from green spaces less modified by urbanism and soils from gardens of peripheral and suburban areas denizens. At last, a final category is represented by soils from urban space integral keeping its natural character. These are soils existing under the woods included in the urban space as parks along the time, and soils that have been suffered an anthropogenic modification.

It is indubitable that the abundance of chemical elements in urban soils is a result of geogenic abundance and anthropogenic influence. This is decisive for urban soils properties modification, mainly those of chemicals, and can explain the large amplitude of heavy metal content variations in urban soils [1, 2].

Urban traffic of different vehicle types has a significant role not merely to air pollution but also near the roads with heavy and congested automobile traffic soils pollution. Many times the input of additional material utilized for streets construction has a major contribution to soil chemical composition change. The permanent contribution on roadside soil chemical transformation is represented by emissions during combustion of fuels, predominantly leaded gasoline, in vehicle engines.

R. Lacatusu
Al. I. Cuza University, Iassy

R. Lacatusu and A.-R. Lacatusu
National R&D Institute for Soil Science, Agrochemistry and Environment Protection Bucharest

S. Rauch et al. (eds.), *Highway and Urban Environment*,
Alliance for Global Sustainability Bookseries 17,
DOI 10.1007/978-90-481-3043-6_37,

From the metal chemical element groups, the most aggressive impact belongs to lead. This is due to the use as an antiknock additive in some gasoline compositions [3]. The lead originating from gasoline combustion has been polluting all environment components, including inhabitants [4, 5]. Today, despite of the unleaded gasoline preponderant use, the historical impact of gasoline with lead additives during the last decades is still observable on soil. Other heavy metals from the gasoline, diesel and tire compositions have gotten into the soil because of fuels burning and wear of auto tires. Industrial emissions have a considerable contribution to urban soils pollution, including those located along roads. This paper presents the heavy metals abundance in the soils located along the main roads from three large Romanian cities, which are different in both dimension and industrial capacity.

Experimental Method

Soil samples from superior horizon (0–20 cm depth) of urban soils situated lengthways the main streets were sampled. Soil samples have been analyzed from a chemical point of view: soil reaction (pH in aqueous solution), carbonates content, argyle <2 μm content, organic carbon for soil type and subtype establishing. The results are presented in [6]. Heavy metals content determinations (Cd, Co, Cr, Cu, Mn, Ni, Pb and Zn) were performed by means of absorption spectrometry in an air-acetylene flame. After soil samples mineralization, using a mixture of HNO3 and HClO4, the total heavy metals content was determined in a hydrochloric solution. Soil mobile heavy metals were extracted with an EDTA-CH3COONH4 solution at pH 7.0. The heavy metals were fractioned by soil components, using a method developed by [7]. Results were analyzed statistically, for grouping centre parameters (arithmetic mean, geometric mean, median), and scattering parameters (standard deviation, minimum and maximum values). Moreover, correlative links were established between some of the chemical elements. Pedogeochemical abundance index (PAI) and geochemical abundance index (GAI) values have been calculated using a methodology elaborated by [8].

Results and Discussion

The Nature of Urban Soils from Bucharest, Iassy and Baia Mare

The Romanian metropolis Bucharest, with a surface area of about 228 km^2 and over a billion inhabitants, is located in a plane zone in the southeastern part of the country. The development of the city has occurred on natural soils belonging to different types: that is Chernozems and Brown Reddish types. By anthropogenic action, these natural soils have become urban soils as Urbic Regosols, Regosols and Humic Regosols.

Iassy is located in the extreme eastern part of the country, in a hilly zone, having 340,000 inhabitants and 93.9 km^2 of surface area. The urban soils, belonging to Urbic and Mixic Entiantrosols, have been derived from natural soils from Chernozems or alluvial soils types, some of them saline.

Baia Mare, located in the northeastern part of Romania, with over 140,000 inhabitants and a surface area of about 35.73 km^2, is a city well known for its complex sulphide ore extraction and processing industry. The natural soils developed on the settlement are Albic Luvisols and Brown Acid type, both being acid soils. Those strongly modified zones have resulted in Urbic Mixic and Urbic-Alluvic Entiantrosols.

Abundance of Heavy Metals

Statistical parameters of the eights studied heavy metals from urban soils situated lengthways the main streets of Bucharest, Iassy and Baia Mare cities (Tables 1 through 3) has revealed contiguous values for some chemical elements (Cd, Co, Cr, Mn and Ni) and contrasting values for other (Cu, Pb, Zn). Marked differences between the chemical elements Cd, Cu, Pb and Zn from Baia Mare soils, has been recorded. Because of the presence of nonferrous ore processing industries, the concentration of these heavy metals in Baia Mare soils are much higher as compared with those from Bucharest and Iassy. Thus, only arithmetic mean values of the Cu, Pb and Zn contents in Baia Mare urban soils exceed the maximum allowable limits for these elements by 1.6, 7.9 and almost 2.0 times, respectively.

Analysis of the medium content of the same chemical elements from Bucharest and Iassy urban soils showed that the values, except those for Pb, are ranging below

Table 1 Statistical parameters of total heavy metal content (mg/kg) in urban soils of Bucharest city (n = 87) situated lengthways of the main streets as compared to normal content (NC), maximum allowable limits (MAL) and alarm threshold (AT) for a less sensible

	Cd	Co	Cr	Cu	Mn	Ni	Pb	Zn
Parameter[a]	mg/kg							
xmin	1.1	15	18	27	394	14	27	70
xmax	4.6	31	41	86	730	50	359	404
x	2.2	23	29	51	507	37	119	165
σ	0.8	4	6	18	137	8	94	84
xg	2.1	23	27	50	501	38	108	143
Me	2.2	24	29	46	493	39	72	150
NC[b]	0.3	5	30	20	500	20	15	50
MAL[c]	3.0	50	100	100		50	100	300
AT[d]	5.0	100	300	250	2,000	200	250	700

[a] xmin - minimum value; xmax - maximum value; x - arithmetic mean; σ - standard deviation; xg – geometric mean.

[b]After [9].

[c]After [10].

[c]After Romanian order of Environment Ministry (no 756/1997) for soil pollution.

Table 2 Statistical parameters of total heavy metal content (mg/kg) in urban soils of Iassy city (n = 64) situated lengthways of the main streets as compared to normal content (NC), maximum allowable limits (MAL) and alarm threshold (AT) for a less sensible

	Cd	Co	Cr	Cu	Mn	Ni	Pb	Zn
Parameter[a]	mg/kg							
xmin	0.1	10	15	8	340	23	47	73
xmax	3.9	37	63	42	693	74	432	215
x	1.7	26	31	23	521	42	152	112
σ	0.6	9	10	9	164	12	61	52
xg	1.5	21	30	20	509	40	143	109
Me	1.4	22	28	21	517	37	138	101
NC[b]	0.3	5	30	20	500	20	15	50
MAL[c]	3.0	50	100	100		50	100	300
AT[d]	5.0	100	300	250	2,000	200	250	700

[a] idem Table 1.
[b] idem Table 1.
[c] idem Table 1.
[d] idem Table 1.

Table 3 Statistical parameters of total heavy metal content (mg/kg) in urban soils of Baia Mare city (n = 73) situated lengthways of the main streets as compared to normal content (NC), maximum allowable limits (MAL) and alarm threshold (AT) for a less sensible

	Cd	Co	Cr	Cu	Mn	Ni	Pb	Zn
Parameter[a]	mg/kg							
xmin	0.3	11	10	22	449	16	131	139
xmax	3.9	107	30	329	880	46	3510	2115
x	1.6	23	20	157	520	27	794	589
σ	0.5	16	5	93	163	8	611	517
xg	1.4	22	19	146	497	25	729	507
Me	1.3	20	18	139	501	23	707	483
NC[b]	0.3	5	30	20	500	20	15	50
MAL[c]	3.0	50	100	100		50	100	300
AT[d]	5.0	100	300	250	2,000	200	250	700

[a] idem Table 1.
[b] idem Table 1.
[c] idem Table 1.
[d] idem Table 1.

the maximum allowable limits (MAL), with some variation at the half interval between the normal content and MAL value, more precisely (Tables 1 and 2).

The medium lead content exceeds by 1.5 and 1.2 times, respectively, the MAL value for Iassy and Bucharest urban soils. The medium lead content in Baia Mare urban soils (Table 3) exceed by 3.2 times the alert threshold value. This high content is caused by the cumulative pollutant effect of traffic and industrial emissions of the nonferrous ore processing industry. Analyses of maximum values for the eight chemical elements in Baia Mare soils have revealed that Co, Cu, Pb and Zn exceed the alert threshold for a sensitive use of land.

For the other two cities, the maximum values exceed the alert threshold for Pb, by 1.4 times in Bucharest soils and by 1.7 times for Iassy soils. Accordingly, from an abundance point of view, lead is the dominant chemical element and is strongly influenced both by traffic and industrial emissions.

In addition to absolute values and statistical parameters of the heavy metal contents, the values of two abundance indexes: pedogeochemical index (PAI) and geochemical index (GAI) were also calculated. They are representing the ratio between the metal content medium value in analyzed soils and the normal content value in soils, clark value. Clark is representing the medium content of a chemical element in the lithosphere. Analysis of Fig. 1 clearly reveals that the two indicator values are small and very small, ranging round the one value for Cr, Cu, Mn and Ni.

This fact proves the predominant geogenic origin of these chemical elements. Higher values of the pedogeochemical index were recorded for Co, attesting concentrations of this chemical element in the soil. On the contrary, the IAG and IAP values for Cd, Zn, and especially Pb, have been striking the wrong note. Very high values of these two indicators denote the major contribution as anthropogenic, which enhance very much these chemical element contents, as compared to lithological and pedologic background (Fig. 1). The heavy metal abundance in urban soils situated lengthways the main streets in the three investigated Romanian cities is superior to

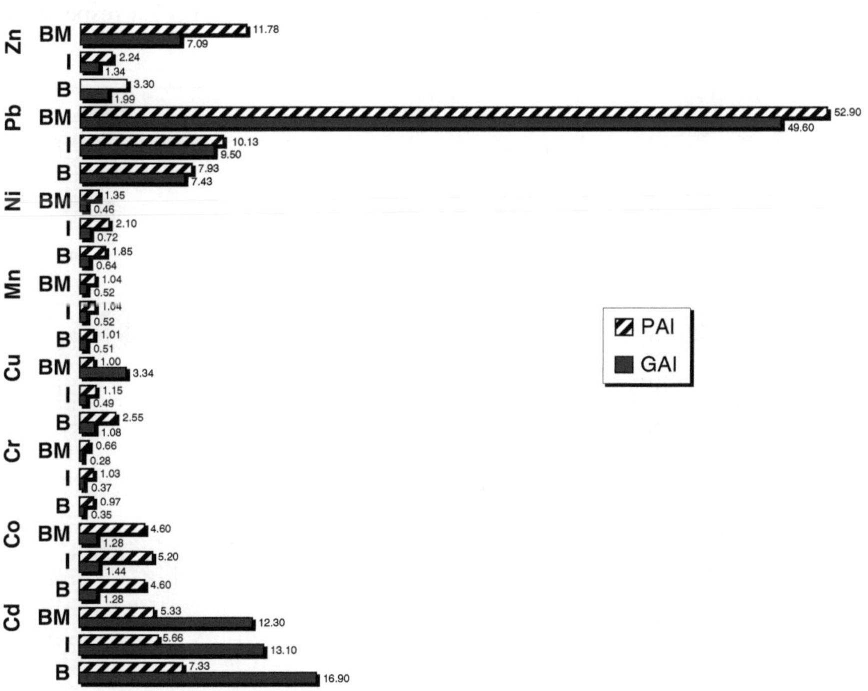

Fig. 1 Values of pedogeochemical (PAI) and geochemical abundance index (GAI) of heavy metals in urban soils situated lengthways the main streets of Bucharest (B), Iassy (I) and Baia Mare (BM) cities

heavy metal abundance global values, recorded in urban soils of these localities, including heavy metals from parks soils and those from industrial enclosures [1].

Cd, Pb and Zn Total Content Fractions

Although sequential extraction with different conventional reagents of those three heavy metals have resulted in percentage values closer from one soil to another, these percentages represent parts of very different total heavy metal contents, expressed in mg/kg. Thus, for the lead case, 29% of lead is associated with exchangeable components of a 34.5 mg/kg total lead content in Bucharest urban soils, while the same lead exchangeable fraction from Baia Mare soils, about 34% from total content, correspond to 270 mg/kg lead (Fig. 2). A very similar situation was found in the zinc case, when closer percentage values of the exchangeable content (34% in Bucharest soil and 36% in Baia Mare soil), in fact correspond to very contrasting values expressed in 56.1 and 212.0 mg/kg, respectively.

Analysis of the heavy metal concentrations in Baia Mare soil solution showed that it contained up to 12% Cd, 7% Pb and 9% Zn. These percentage values correspond to heavy metal contents expressed in mg/kg about: 0.19 Cd, 55.6 Pb and 53.0 Zn. In urban soils situated lengthways the main streets in Bucharest and Iassy, similar calculations (mg/kg) result in 0.17 Cd, 7.1 Pb, and 11.5 Zn respectively.

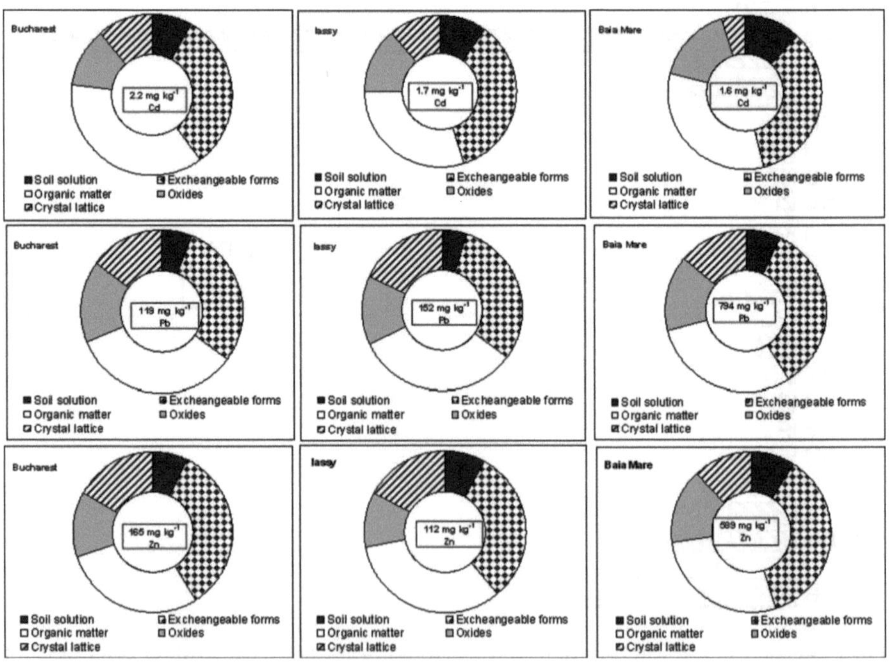

Fig. 2 Fraction percentage of Cd, Pb and Zn total contents from superior horizons of urban soils situated lengthways of the main streets of Bucharest, Iassy and Baia Mare

Compared to urban soils from Bucharest and Iassy, the Baia Mare soil solution contains about 7.6 times more Pb and 5.3 times more Zn.

Besides, there is a similar situation both for the other potential mobile fraction (exchangeable fraction and organic matter bound fraction), and for those harder mobile (fraction bound by oxides and crystal lattice of primary and secondary soil minerals). If predominant acid reactions of these soils are taken into account, the above conclusions explain the heavy metals abundance in mobile and potential mobile fractions of the Baia Mare soil.

Heavy Metals Contamination/Pollution Level

The contamination/pollution index, counted using a method proposed by [11], show a large palette of values (Table 4), the majority being spread below, and only three values over the unit that define the contamination domain.

Those latter values were recorded only for Cd, Pb and Zn in the urban soils lengthways the main streets of Baia Mare city. For other chemical elements, the index values reflect moderate and sever contamination levels. In this series values for Cd, Co, Cr and Ni are included. Very severe contamination levels were recorded for Cd in Baia Mare soils and for Pb and Zn in Bucharest and Iassy soils.

Table 4 The values of heavy metal contamination/pollution indexex of urban soils situated lengthways the main streets in Bucharest (1), Iassy (2), and Baia Mare (3)

Chemical element	Location	Values of contamination/ pollution index	Signification
Cd	(1)	0.43	Moderate contamination
	(2)	0.32	Moderate contamination
	(3)	0.94	Very severe contamination
Co	(1)	0.25	Slight contamination
	(2)	0.28	Moderate contamination
	(3)	0.49	Moderate contamination
Cr	(1)	0.43	Moderate contamination
	(2)	0.39	Moderate contamination
	(3)	0.50	Moderate contamination
Cu	(1)	0.54	Severe contamination
	(2)	0.61	Severe contamination
	(3)	1.04	Slight pollution
Ni	(1)	0.37	Moderate contamination
	(2)	0.49	Moderate contamination
	(3)	0.50	Moderate contamination
Pb	(1)	0.83	Very severe contamination
	(2)	0.95	Very severe contamination
	(3)	9.50	Very severe pollution
Zn	(1)	0.83	Very severe contamination
	(2)	0.74	Very severe contamination
	(3)	4.20	Very severe pollution

Consequently, we could affirm that urban soils situated lengthways the main streets from Baia Mare city are polluted with Pb, Zn, Cu and Cd, while the same situated urban soils from Bucharest and Iassy are contaminated very severe with Pb and Zn, severe contaminated with Cu and moderate contaminated with Cd.

Heavy Metals Correlations

Direct proportionality relationships; linear type, represented through a correlation coefficient (r), or parabolic type, represented by a correlation ratio (η), were established between some of the investigated heavy metals. The values, statistically assured by a 5% confident level, are presented in Table 5.

Clearly differentiation of correlations between Cd with Co and Mn, between Cr with Mn, Ni, Pb and Zn, between Cu and Mn with Zn, between Ni with Mn and between Pb with Cu and Zn, could be observed.

The strongest correlation between these chemical elements were recorded in Baia Mare soil, where concentrated anthropogenic influences, caused by nonferrous ore processing industry, generated a close dependence between component elements of complex sulphide ores.

The smaller and diffused heavy metal concentrations in the other two cities soils also showed a higher spreading degree. For this reason, the recorded values or their correlation coefficients have been smaller.

Table 5 Values[a] of correlation coefficients (r) and ratio (η) between heavy metals from urban soils situated lengthways the main streets in Bucharest (1), Iassy (2), and Baia Mare (3) cities

	Cd	Co	Cr	Cu	Mn	Ni	Pb	Zn
Cd								
Co	(1)r = 0.394 (2)r = 0.415 (3)η = 0.523							
Cr								
Cu							(1)η = 0.412 (2)η = 0.470 (3)η = 0.520	
Mn	(1)r = 0.204 (2)r = 0.192 (3)r = 0.232		(1)η = 0.412 (2)η = 0.470 (3)η = 0.520			(3)r = 0.420		
Ni			(1)η = 0.412 (2)η = 0.470 (3)η = 0.520					
Pb			(3)η = 0.583					
Zn			(3)η = 0.620	(1)r = 0.420 (2)r = 0.567 (3)η = 0.624	(1)r = 0.374 (2)r = 0.429 (3)η = 0.703		(1)r = 0.660 (2)r = 0.520 (3)η = 0.710	

[a]Filled cells represent statistical assured values for confidential level 5%. The blank cells means not statistical assured values.

Conclusion

- Urban soils situated lengthways the main streets are: Urbic Regosol, Regosol and Humic Regosol in Bucharest city, Urbic and Mixic Entiantrosol in Iassy city and Urbic Mixic and Urbic–Alluvic Entiantrosol in Baia Mare city.
- Between analyzed heavy metals (Cd, Co, Cr, Cu, Mn, Ni, Pb and Zn), lead is the chemical element that has been found at pollutant level in soils at all three locations.
- Lead concentration in the three municipalities' soils, at medium level content, exceed the maximum allowable limit values by 1.5 times (Bucharest), 1.2 times (Iassy), and 7.9 times (Baia Mare).
- Pedogegeochemic (PAI) and Geochemic (GAI) abundance indexes values are small for Cr, Cu, Mn and Ni (around one value), showing in this way geogenic origin of these chemical elements, while high values of the same abundance indexes for Cd, Pb and Zn reveal a significant anthropogenic contribution.
- Urban soils situated lengthways the main streets of Baia Mare are polluted with lead, zinc, copper and cadmium, while urban soils at similar locations in Bucharest and Iassy are contaminated very severe with lead and zinc, severe with copper, and moderate with cadmium.
- The greatest part of total pollutant chemical elements (Cd, Cu, Pb and Zn) is present in the soluble fraction (soil solution) and potential soluble (exchangeable fraction and organic matter bound fraction).

References

1. Lacatusu R, Lungu M, Kovacsocics B, Breaban I-G, Rusu C, Risnoveanu I (2005) Heavy metals in urban soils of Iassy municipium (Romania), Proceedings of 2nd International Conference "Disaster and Pollution Monitoring", Iassy, 17–19 November 2005, pp. 175–182
2. Manta DS, Angelone M, Bellanca A, Neri R, Sprovieri M (2002) Heavy metals in urban soils: a case study from the city Palermo (Sicily), Italy, Science of Total Environment, 300:229–243
3. Davies BE (1990) Lead in heavy metals in soils, BJ Alloway (ed), Blackie & Son, Glasgow/London
4. Kamal AAM, Eldamaty SE, Faris R (1991) Blood lead level of Cairo traffic policeman, The Science of Total Environment, 105:165–170
5. Monaci F, Moni F, Lanciotti E, Grechi D, Bargagli R (2000) Biomonitoring of airborne metals in urban environments: new tracers of vehicle emission, in place of lead, Environment Pollution, 107:321–327
6. Lacatusu R, Lacatusu A-R, Lungu M, Breaban IG (2008) Macro- and microelements abundance in some urban soils from Romania, Carpathian Journal of Earth and Environmental Sciences, 3:73–81
7. Lacatusu R, Kovacsovics B (1994) Soils heavy metals fractioning method, RNSSS Publications, 28A:187–194 (published in Romanian)
8. Lacatusu R, Ghelase I (1992) On abundance of heavy metals in soil, Environment (Mediul inconjurator), III, 4:45–53 (published in Romanian)
9. Fiedler HJ, Rösler HJ (1988) Spurenelemente in der Umwelt, Ferdinand Enke Verlag, Stuttgart
10. Kloke A (1980) Richtwerte `80. Orientirungs daten fűr tokerierbare Gezamgehalte Ainiger Elemente in Kuturböden, mitt. VDULFA, H. 1–3, 9–11
11. Lacatusu R (1998) Appraising levels of soil contamination and pollution with heavy metals, in Developments for planning the sustainable use of land resources HJ Heineke, W Eckelmann, A-J Thomasson (eds), European Commission, European Soil Bureau, pp. 393–402

Part VI
Urban Climate and Climate Change

Effects of Transport Infrastructure Plans on Climate Change. Application to the Spanish Strategic Transport and Infrastructure Plan 2005–2020 (PEIT)

Elena López, Andrés Monzón, and Panayotis Christidis

Introduction

The inclusion of strategic environmental effects of transport infrastructure plans in assessment methodologies is increasingly required by the research community, decision-makers and relevant stakeholders. One of these strategic effects is the measurement of their contribution to climate change. This contribution is considered critical because of its sustainability implications [1, 2].

First, the climate change phenomenon is directly linked to energy consumption and directly related to greenhouse gas (GHG) emissions. These are strategic environmental aspects of great interest due to both the need to comply with international environmental commitments and the urgency to reduce energy consumption, which has a greater economic component.

In the past few decades there has been an increased concern for assessing the environmental effects of transport and developing mechanisms to report their evolution, such as the periodic 'Transport and Environment Reporting Mechanisms' (TERM) Reports [3]. At the EU level, the transport sector is the primary driver of the growth in total energy consumption, which is likewise directly linked with total emissions [3]. These concerns have also been stressed in the mid-term review of the Transport White Paper [4], which calls for integrated actions in order to achieve a European sustainable transport system.

However, the important efforts devoted to environmental abatement policies have not achieved the expected reduction targets. In fact, the high rate of increase in transport demand is outstripping the rate of improvement in environmental technology for transport [5, 6]. The result has been a significant increase in greenhouse gas (GHG) emissions from transport, which threatens European progress towards

E. López and A. Monzón
TRANSyT-UPM: Centre for Transport Research, Universidad Politécnica de Madrid, Madrid, Spain

P. Christidis
European Commission, Joint Research Centre, Institute for Prospective Technological Studies (JRC-IPTS), Seville, Spain

S. Rauch et al. (eds.), *Highway and Urban Environment*,
Alliance for Global Sustainability Bookseries 17,
DOI 10.1007/978-90-481-3043-6_38, © Springer Science+Business Media B.V. 2010

its international commitments, such as the Kyoto targets and the proposals by the EU Council for further emission reductions for developed countries beyond the Kyoto Protocol period (2008–2012).

Despite the above mentioned importance of including GHG emissions as an impact variable at strategic planning levels, there is a lot of research work still to be done, in order to improve key methodological issues for their evaluation, as the research community has not yet developed a unified assessment procedure. This paper moves one step forward in this direction, with the proposal of a methodology to assess climate change effects of transport infrastructure plans.

Methodology

The methodology is developed with the support of a Geographical Information System (GIS), in the following four stages:

Stage 1: Definition of the study area
This stage includes the delimitation of the study area, which includes both the national territory of the plan under consideration and the cross-border regions in neighboring countries. The level of aggregation and the zonification is also defined in this stage. This is needed to select centroids both for the origins (i) and destinations (j) for the accessibility analysis.

Stage 2: Implementation of the transport and socio-economic system in the GIS
In this stage each of the centroids i of the study area defined above is characterized in terms of its transport and socio-economic data. The transport system is modeled in a vectorial GIS, resulting in an intermodal graph containing the road and rail networks. For each arc on the road network, the length, estimated speed according to type of road and resulting travel time are also recorded (for more details see [7, 8]). For the rail mode, each arc is given a commercial speed according to infrastructure and quality of service characteristics.

Stage 3: Calculation of travel time savings
The approach used to compute travel time savings is based on the calculation of accessibility indicators, which have a wide potential for their application in strategic planning processes [7]. The selected formulation is for this paper that of the location accessibility indicator, which computes average travel time to the set of destinations. This indicator was previously used in similar studies at the Spanish national level [9]. The formulation chosen is included in Eq. 1.

$$L_i = \sum_i \frac{I_{ij} \cdot P_j}{\sum_j P_j} \tag{1}$$

The location indicator (Li) is computed as the average travel time (in minutes) to the set of destinations, using the population of each destination (Pj) as the weighting variable.

The next step after the input database is stored in the GIS consists in the calculation of the road and rail travel times between each i–j pair. These travel times, along with population destination data, are subsequently introduced in Eq. 1 to obtain each node accessibility value. Finally, the accessibility values of each origin centroid are computed and stored in the GIS.

Stage 4: Calculation of climate change effects
The methodology defines a procedure for the assessment of the climate change effects of transport infrastructure plans. The selected indicator to measure climate change effects are total greenhouse gas emissions (GHG), measured in equivalent tons of CO2. CO2 emissions are included as indicators for climate change issues in different environmental indicator list [3].

The single aggregated value of the location indicator defined above is computed and the result is translated into the corresponding increases in travel demand with the use of the range of travel time elasticities [10–12]. The next step consists in transforming the estimated increase in travel demand into the corresponding increase in GHG emissions. This estimation has been carried out with version 2.44 of the TREMOVE model [13]. TREMOVE is a policy assessment model designed to study the effects of different transport and environment policies on the emissions of the transport sector. Model runs are carried out with the data on induced traffic, resulting in the corresponding CO2 emissions. The model estimates the transport demand, the modal split, the vehicle fleets, the emissions of air pollutants and the welfare level under different policy scenarios. All relevant transport modes are modeled, including air transport. Maritime transport is treated in a separate model. TREMOVE models both passenger and freight transport, and covers the period 1995–2020. TREMOVE consists of 21 parallel country models. Each country model consists of three inter-linked 'core' modules: a transport demand module, a vehicle turnover module and an emission and fuel consumption module, to which a welfare cost module and a well-to-tank emissions module is added. Annual CO2 emissions are computed and summed up to calculate total tons of CO2, using the vehicle-kilometers traveled generated by the TREMOVE transport model per mode and the information on the national vehicle fleet (drives, car categories and emission standards).

In the next section, the validity of the above described methodology was tested by applying it to the Spanish Strategic Transport and Infrastructure Plan (PEIT) 2005–2020.

Results and Discussion

Case Study Description

The Spanish Transport and Infrastructure Strategic Plan 2005–2020 includes an ambitious extension of the Spanish high capacity land transport networks. The PEIT network extension includes the construction of 5,000 km of High Capacity

Roads (HCR) and 6,000 km of High-Speed Rail (HSR) lines. The assessment is carried out on the basis of the comparison between the 'construction alternative' (APEIT) and the 'do-nothing alternative' (A0). The accessibility calculations were made using a network accessibility analysis GIS toolbox [14]. The general methodological stages explained in the previous section are given as theoretical background for the case study application which follows next.

Implementation of the Methodology

Stage 1: Definition of the study area
The study area basically comprises the Spanish mainland and its corresponding cross-border regions in neighbouring countries, which include Portugal and the three southern French NUTS-2 (Nomenclature of Territorial Units for Statistics, defined by the Statistical Office of the European Communities, i.e. EUROSTAT) regions.

Stage 2: Implementation of the transport and socio-economic system
In order to calculate accessibility values, a dense intermodal (road and rail) network was modelled with the support of a GIS; the ArcGis software was used in this case. Accessibility values are obtained for each node of the network, which coincide with the nodes of the road network, which are nearly 12,000 km. The first task consisted in modelling the road network of the do-nothing alternative (A0). A vectorial GIS was used, in which the network is modelled as a graph with a set of nodes and arcs. For each arc on the road network, the length, estimated speed according to the type of road (120 km/h for highways, 110 for expressways, 90 for interregional roads, 80 for other roads and 50 for urban roads) and resulting travel time were also recorded. For the rail mode, each arc is given a commercial speed according to both infrastructure and quality of service characteristics. Rail network modelling tasks are significantly more complex than those of the road mode, as it is necessary to include track gauge (Iberian/UIC) data, the location of the stations and frequency of service information in order to calculate travel times, as described in [9], which is not possible to detail in this paper for space reasons.

The population is the selected variable to measure each destination's attractiveness in the accessibility model. In Spain, the selected destination centres correspond to the centroids of the approximately 8,000 municipalities of the Spanish mainland.

Stage 3: Calculation of travel time savings
In this stage, the location accessibility values of each origin centroid i is computed, using Eq. 1. Intermediate calculations include the measurement of each i–j travel time, using minimum-path algorithms embedded in the GIS. The location indicator is therefore used as a proxy for the evaluation of travel time savings, when its results in the PEIT alternative are compared to those of the do-nothing alternative. Hence, a single aggregated value of the location indicator for all Spain has been computed and compared to that of the do-nothing alternative. For clarity reasons, the

results were calculated independently for road and rail modes. Therefore, the analysis of results included below was split in the two corresponding subsections.

Stage 4: GHG calculations

The result of the location accessibility indicator, in percentage change reduction, has been translated into the corresponding increases in travel demand with the use of a range of commonly used travel time elasticities. These values are summarized in Table 1.

These increases in travel demand are included in the TREMOVE model in order to calculate the corresponding increases in GHG emissions. Results obtained for the road and rail modes are summarized in Table 2. The percentage change results obtained by the road and rail alternatives are obviously not directly comparable, as these values are heavily influenced by the emission levels of the do-nothing alternative. A 'global' relative percentage change has therefore been computed, representing the percentage change compared to the sum of road and rail emissions.

Table 1 Travel time savings and estimated induced traffic

	Location indicator (min)			Induced traffic (%)	
Transport mode	Do-nothing alternative	PEIT alternative	Reduction (%)	Minimum	Maximum
Road	156.81	153.18	2.31	1.12	4.62
Rail	325.81	213.86	34.36	58.41	92.72

Table 2 Forecasted induced traffic and corresponding increases in GHG emissions. Do-nothing vs. PEIT alternative. Road and rail modes

			Road	Rail
Traffic (million vkm)	Do-nothing alternative		332 359.16	275.43
	PEIT alternative	Minimum	336 081.58	436.28
		Maximum	347 714.15	530.75
		Mean	340 036.65	370.17
GHG emissions (t CO_2)	Do-nothing alternative		72 513 765.92	234 275.13
	PEIT alternative	Minimum	73 279 001.33	365 755.62
		Maximum	75 670 365.08	442 987.18
		Mean	74 474 683.47	404 371.40
Increase in GHG emissions	Absolute (t CO_2)	Minimum	765 235.41	131 480.49
		Maximum	3 156 599.16	208 712.05
		Mean	1 960 917.55	170 096.27
	Relative[a] (%)	Minimum	1.06	56.12
		Maximum	4.35	89.09
		Mean	2.70	72.61
	Global relative[b] (%)	Minimum	1.05	0.18
		Maximum	4.34	0.29
		Mean	2.70	0.23

[a]Percentage change of each mode emissions of the do-nothing alternative.

[b]Percentage change of total road and rail emissions of the do-nothing alternative.

As Table 2 values show, the road mode do-nothing alternative accounts for over 72.5 million tons of $CO2$. It can be seen how the mean increase in GHG emissions due to the extension of the HCR network included in the PEIT accounts for near 2 million tons of $CO2$, which represents a 2.70% increase compared to both the do-nothing alternative value for the road mode, and total road and rail emissions of the do-nothing alternative. This comparison could be carried out if different road and/or rail alternatives were assessed.

Regarding the rail mode; do-nothing alternative accounts for only 234,000 t. The extension of the HSR network included in the PEIT accounts for nearly 170,000 t of $CO2$, which represents over a 72% increase, in terms of the do-nothing alternative value for the rail mode, whereas it represents only a 0.23% increase of total road and rail emissions of the do-nothing alternative.

The comparison of the absolute increases in GHG emissions between road and rail modes (2 million vs. 170,000 t of $CO2$) gives us an idea of the significant difference in the contribution of the above transport modes to GHG emissions, which is obviously proportional to their corresponding traffic volumes.

Conclusions

This paper defines a methodology for the assessment of climate change effects of transport infrastructure plans, based on the calculation of GHG emissions. The methodology has proven its efficacy in deriving GHG emissions via the calculation of an accessibility indicator, and the resulting induced travel demand that stems from travel time savings. This approach could be of valuable use in cases when a calibrated transport demand model is not available, which is frequently the case at strategic levels.

The application of the methodology has also highlighted the significant differences in the contribution to GHG emissions of road and rail modes, given the severe unbalance in modal split. A modal shift towards rail mode would result in important reductions in GHG emissions, along with other environmental benefits of more balanced modal shares.

Accompanying measures are needed in order to encourage a modal shift from road to rail, in order to take full advantage of network accessibility improvements derived from infrastructure investments. Improvements in punctuality, reliability, perceived comfort and higher frequencies are some of the possible measures that may be effective for this purpose.

In this sense, the integration of road and rail modes in order to measure modal shifts is on our research agenda. Finally, other future research directions we are exploring refer to the application of the methodology to a set of road and rail alternatives, in order to assess the sensitivity of the results to different variables, such as rail commercial speed.

Acknowledgments The results included in this paper are part of the project "Assessment of the effects of transport infrastructure plans, on mobility, the territory and the environment", financed by the Spanish Ministry for Science and Technology and the STEPS project (Scenarios of the Energy and Transport System and their Potential Effects) of the sixth FP of the European Commission. The authors would also like to thank Emilio Ortega and Santiago Mancebo for their SIG work in modeling the accessibility indicators.

References

1. Button KJ, Verhoef E (1998) Transport at the Edge of Mobility and Sustainability. In Transport networks in Europe: concepts, analysis and policies, KJ Button, P Nijkamp, and H Priemus, (Eds.), Edward Elgar Publishing, Cheltenham/Northampton, MA
2. Greene DL, Wegener M (1997) Sustainable transport. Journal of Transport Geography, 5:177–190
3. EEA (2006) Energy and environment in the European Union. Tracking progress towards integration. EEA Report N0 8/2006. European Environment Agency, Copenhagen
4. European Commission (2006) Communication from the Commission to the Council and the European Parliament- Keep Europe Moving. Sustainable mobility for our continent. Mid-term review of the European Commission's 2001 White Paper
5. Monzón A, Nuijten A (2006) Transport strategies under the scarcity of energy supply, The STEPS Consortium, The Netherlands
6. Stead D (2001) Transport intensity in Europe – indicators and trends. Transport Policy, 8:29–46
7. Gutiérrez J, Monzón A (1998) Accessibility, network efficiency, and transport infrastructure planning. Environment and Planning A, 30:1337–1350
8. Martín JC, Gutiérrez J, Román C (2004) Data Envelopment Analysis (DEA) index to measure the accessibility impacts of new infrastructure investments: the case of the high-speed train corridor Madrid-Barcelona-French Border. Regional Studies, 38:697–712
9. López E (2007) Assessment of transport infrastructure plans: a strategic approach integrating efficiency, cohesion and environmental aspects. Ph.D. dissertation, Universidad Politécnica de Madrid, Madrid, 2007 Available at http://www.ad.upm.es
10. Bel G (1997) Changes in travel time across modes and its impact on demand for inter-urban rail travel. Transportation Research E 33:43–52
11. Goodwin PB (1996) Empirical evidence on induced traffic: a review and synthesis. Transportation, 23:35–54
12. Savelberg F, Vogelaar H (1987) Determinants of a northern high-speed railway. Transportation, 14:97–111
13. Transport & Mobility Leuven and KU. Leuven (2006) Final Report on TREMOVE model v 2.44. Service contract for the European Commission, DG Environment, Brussels
14. Mancebo S (2006) AccesstUls.aml (2006-10-26) ® Network accessibility analysis toolbox. Available at http://topografia.montes.upm.es/accesstuls.html

Designing High Traffic Volume Urban Motorway Pavements to Maximize Sustainability

Michael L.J. Maher, Ludomir Uzarowski, and Gary Moore

Introduction

The Kyoto Protocol was adopted in late 1997 to address the problem of global warming by reducing the world's greenhouse gas emissions. It went into effect on February 16, 2005 and as of April 2008, a total of 178 countries and other governmental entities have ratified the agreement. It is estimated that transportation is responsible for over 20% of the world's greenhouse gas emissions. Of this total, some 85% is from road transportation. A wide range of strategies have been put forward to curb the continuing growth in road transport emissions. These include improved vehicle efficiencies, low carbon fuels, and travel reduction. One aspect of road transportation that has not received much attention is the design philosophy for road structural design.

The road design and construction industry needs to become part of the sustainability solution by proactively implementing technologies and construction practices that contribute to achieving the challenging emission reduction goals. In addition to improved asphalt technology, better construction and rehabilitation methods, and optimized pavement selection based on life-cycle thinking, more attention needs to be focused on delivering more sustainable pavements. The sustainable pavement can be defined as one that provides the best balance between economic, environmental and social considerations. In essence, this is a pavement that is safe and durable, has less maintenance demands, longer time between major rehabilitation interventions, and where life cycle costs are minimized.

M.L.J Maher
Golder Associates Ireland, Naas, Co, Kildare, Ireland

L. Uzarowski
Golder Associates Ltd, Whitby, Ontario, Canada

G. Moore
City of Hamilton, Hamilton, Ontario, Canada

S. Rauch et al. (eds.), *Highway and Urban Environment*, Alliance for Global Sustainability Bookseries 17,
DOI 10.1007/978-90-481-3043-6_39, © Springer Science+Business Media B.V. 2010

The effectively designed sustainable pavement should aim to:

- Minimize the use of natural resources
- Reduce energy and fuel consumption during construction and operations
- Minimize greenhouse gas (GHG) emissions
- Limit pollution (air, water, earth, noise, etc.)
- Improve health, safety and risk prevention
- Ensure a high level of user comfort and safety
- Provide long term value for money

Historically, most municipal governments have been forced by political pressure and scarce financial resources into a short-term approach toward pavement design and management. They have also relied heavily of traditional or 'recipe' design methods. This has led to a cycle of acceptance of pavements that lose ride quality very quickly and need major rehabilitation every 18 to 25 years. Even the conventional life cycle cost analysis process frequently fails to fully demonstrate the folly of a short-term design philosophy when it comes to pavements. The twenty-first century realities of excessive energy consumption, dwindling natural resources, environmental impacts of construction and the importance placed by the public on making our roads safer, rarely get factored into the analysis. Nor do the indirect costs sustained by the motorist in maintaining vehicles that are habitually driven on bad roads and the extra fuel consumption as a result of traffic delays from road repair activities get considered. This paper describes a more enlightened approach for pavement design for high volume urban motorways.

Road Transportation and Sustainability

Sustainability principles have application to every stage of road planning, development, construction, operations and maintenance. Some examples include longer lasting pavements, intelligent transport systems, alternative fuels, improved safety features to reduce accidents, and incentives for car pooling. In the past, options for the development of public road networks were largely dictated by economics. In the last 30 years or so, environmental and social impacts have had an increasingly large influence on road design policies and standards. More recently, context sensitive design principles have also begun to influence traditional design approaches [1]. Sustainability when applied to road transportation needs to be broad ranging and must consider a wide range of themes under the broad categories of economic, social and environmental aspects.

Today's economic activity is based on a legacy of facilities and infrastructure. Progress towards sustainability will be a long journey and will only be achieved on a project by project basis and by joint co-operation between owners and their engineers. Progress towards sustainability goals is measured under three main categories: environmental, social and economic. The implementation of this approach is focused on achieving a balance between these categories, rather than achieving one at the expense of another, as illustrated in Fig. 1.

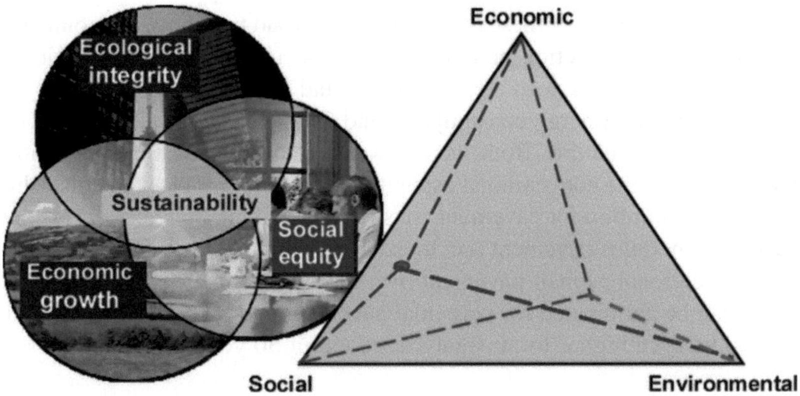

Fig. 1 Sustainability strives for balance between key indicators

Maximizing Sustainability in Urban Pavement Design

The main topic for this paper deals with applying sustainability principles to road structural design. Historically, empirical approaches have been used for flexible pavement design. The accepted life cycle involves periodic routine maintenance and a major rehabilitation treatment every 18 to 25 years. With the rapidly increasing traffic volumes on urban arterial roadways, larger municipalities are looking for ways to extend the effective road service life so as to minimize the disruptions to normal traffic operations and the associated driver delays and inconvenience during road rehabilitation works. Clearly, huge benefits would accrue in terms of sustainability and value for infrastructure investment if the life of flexible pavements could be increased to 50 years or more.

It has been observed that energy consumption and greenhouse gas (GHG) emissions during construction and maintenance activities are significant [2]. However, the emissions from vehicles delayed by lane closures during road works are even larger. Road rehabilitation in urban areas causes serious disruption to traffic and may require the detouring of traffic through densely populated neighborhoods. Long life pavements have obvious advantages that can be quantified in terms of improved environmental quality and reduced GHG emissions. This is especially important in high density urban environments. Additional benefits are the conservation of road aggregate and bituminous resources by reducing the need for repeated removal and replacement of major road elements.

In any analysis of energy consumption and greenhouse gas emissions related to roads, the entire production and construction cycle must be taken into consideration. A study completed in France [3] shows that over a 30 year period, traffic consumes between 10 and 345 times more energy than road construction over the same period, depending on the traffic volume. At first this may suggest that the impact of road design and construction is not a significant contributor to emissions. On the contrary, when one considers that transportation accounts for 26% of all

GHG emissions and 85% of these are generated by road transport, the contributions from design and construction activities cannot be ignored. Road agencies, road designers and contractors need to be aware of and consider these issues in policy and decision-making. Energy consumption and GHG emissions also depend on the type of pavement structure. Both were lower for the long lasting high modulus pavements than for the conventional asphalt pavements [2]. This study also showed that energy consumption for pavement construction and maintenance was about 9% lower for high modulus pavement (i.e. those made with high strength asphalt layers) than for conventional asphalt pavement over a 30 years analysis period. A similar difference can be anticipated for perpetual pavement and it is likely that this difference would be even higher for an analysis period of 50 years.

Perpetual Pavement Design Strategy

The traditional pavement designs developed originally were compared with the initial perpetual pavement design. As part of the feasibility study, life-cycle costs, environmental benefits of the perpetual pavement design, pavement sustainability aspects and public satisfaction were analyzed. Traditionally designed pavements tend to fail in either one of two modes, fatigue cracking or permanent deformation, also referred to as rutting. To significantly extend the life of a traditional pavement, it needs to have two main attributes as follows [4, 5]:

1. Total asphalt thickness of more than 200 mm. It has been shown that flexible pavements with more than 200 mm of hot mix can resist fatigue cracking (bottom-up cracking) regardless of the number of axle load repetitions. The resistance to fatigue cracking is further enhanced by increasing the asphalt cement content of the bottom lift of hot mix asphalt and the air voids in the mix are reduced to about 2–3%. This specially designed layer is described as a 'rich-bottom lift'.
2. The increased hot mix asphalt thickness above the rich-bottom lift comprises high quality, rut-resistant mixes that protect the rich-bottom mix from distress and resist asphalt rutting.

Damage to the pavement is now limited to surface tyre abrasion and loss of friction. This surficial damage can be corrected by surface milling and thin overlay.

Case Study

The Red Hill Valley Parkway (RHVP) is a modern urban Expressway in the City of Hamilton, Ontario, Canada. It is the final leg of a longer Freeway project considered to be the largest municipal road project in Canada with an estimated final total cost of Can$430 Million. An initial traffic volume of about 40,000 vehicles per day and full capacity volumes in excess of 90,000 vehicles per day are expected for this

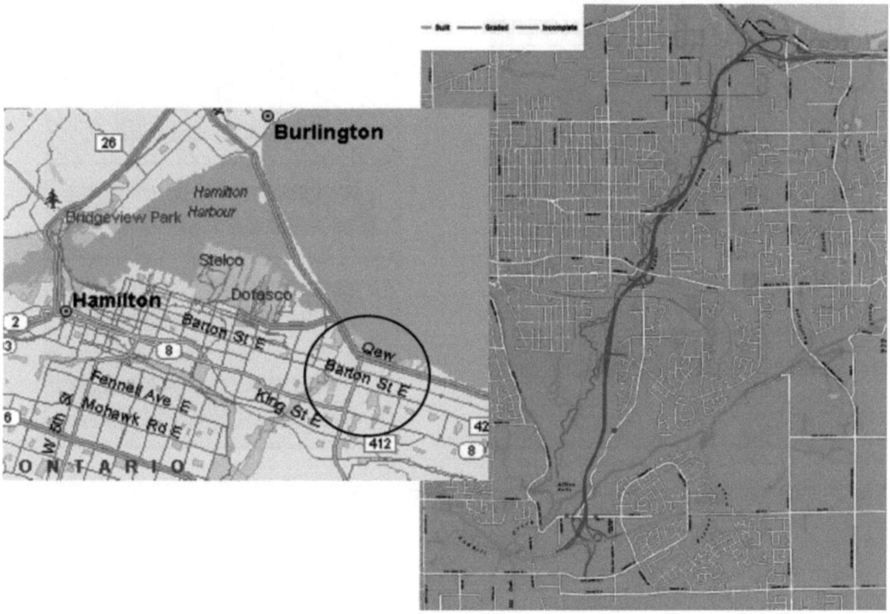

Fig. 2 Location map and route plan for the RHVP

section of the city's crucial transportation artery. The 7.5 km long RHVP is located in an environmentally sensitive area in the City of Hamilton along the Red Hill Creek [4] (Fig. 2). The City of Hamilton decided that, given the projected traffic volumes, the traditional pavement designed for a 20-year life might not be acceptable and that a more radical and sustainable approach was needed. A feasibility study was completed comparing both pavement design alternatives. The RHVP is projected to sustain more than 30 million Equivalent Single Axle Loads (ESAL's) over a 20-year period. The conventional pavement design, based on geotechnical investigations completed between 1999 and 2004, was to support a 20-year traffic loading.

Perpetual pavement designs were then completed for all sections to support the traffic loading over a period of 50 years (90 million ESAL's). The selected pavement designs for both alternatives are shown in Fig. 3. The perpetual pavement incorporates an 80 mm layer of an asphalt-rich mix, which will protect against the initiation of load induced fatigue (bottom-up) cracking.

A life cycle cost analysis was undertaken to compare the perpetual pavement design to the conventional asphalt pavement alternative. The Ministry of Transportation approach was used [6].

The results (just considering the main lane pavements) confirm that while the initial construction costs for the perpetual pavement design is higher, the long term

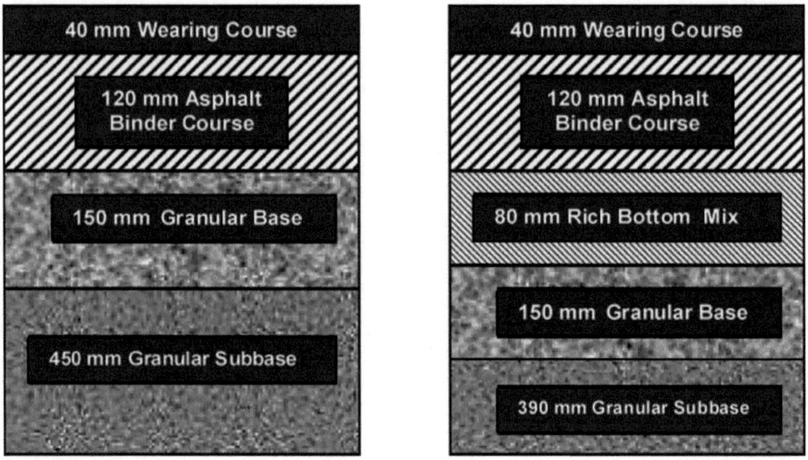

Fig. 3 Comparison of conventional pavement structure (*left*) and perpetual pavement (*right*)

Table 1 Life cycle cost comparison

	Construction cost (million Can$)	Maintenance and rehabilitation (million Can$)	User delay cost (million Can$)	Total cost (present worth million Can$)
Conventional	12.4	4.5	1.3	18.2
Perpetual	13.7	2.3	0.5	16.5

life cycle cost is lower (Table 1). The user-cost estimates include user delay costs only (i.e. delay costs and queuing costs) and not vehicle-operating costs.

Reduced Environmental Impacts

To quantify the environmental benefit of perpetual pavement over a traditional pavement on the RHVP, the Pavement Life-cycle Assessment Tool for Environmental and Economic Effects (PaLATE) software was used. The program was developed by Dr. Arpad Horvath from the University of California, Berkeley as a decision-making tool used to evaluate road construction in terms of life-cycle costing and environmental impact [8]. PaLATE does not take into account vehicle emissions during the service life. Both, perpetual and traditional pavement designs were compared during the initial construction and maintenance and rehabilitation stages. The environmental impact includes energy consumption, CO2, NOx, PM10 and SO2 emissions. The results are summarized in Table 2.

Table 2 Comparison of environmental impact of alternative pavement design strategies

Pavement type	Energy (million MJ)	CO_2 (t)	NO_x (t)	PM_{10} (t)	SO_2 (t)
Traditional	338	19,000	182	151	3,900
Perpetual	230	13,000	123	101	2,600
Difference	108	6,000	59	50	1,300

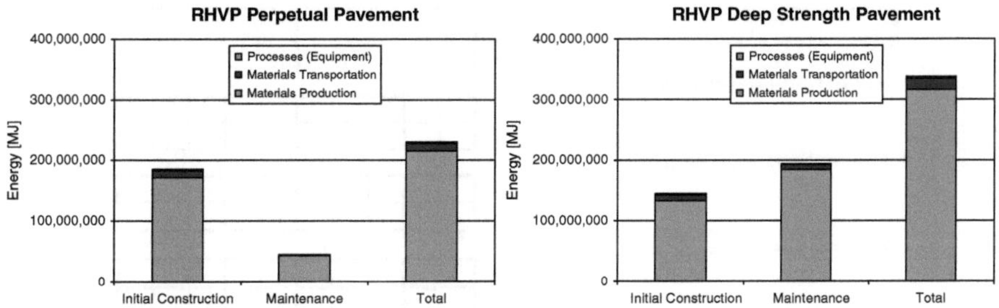

Fig. 4 Life-cycle energy consumption comparisons

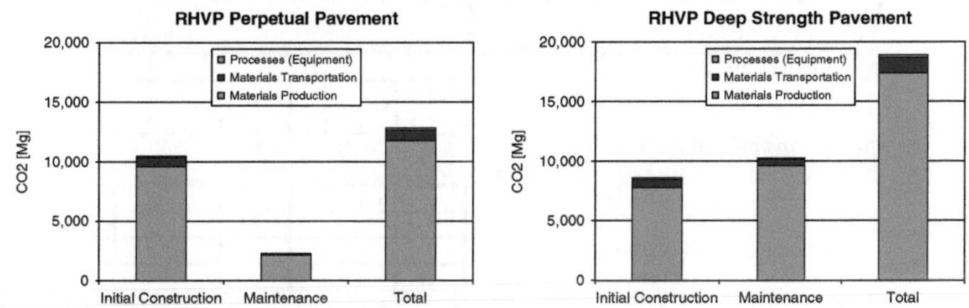

Fig. 5 Life-cycle CO_2 emissions comparison

Over the analysis period of 50 years, the energy consumption and the gas and particulates emissions during the life cycle of the perpetual pavement are reduced by about one third as compared with the conventional pavement structure. Figures 4 and 5 compare the energy consumption and CO2 emission for both alternatives.

Long lasting or perpetual pavements appear to have clear advantages over conventional asphalt pavements for high volume urban motorways. New mix technologies such as warm asphalt mixes, the continued and increased use of pavement recycling are also very promising strategies to achieve more sustainable pavement designs. To assess the sustainability of alternative strategies it is necessary to consider economic, environmental and social impacts. The economic and environmental aspects can be relatively easily quantified, but social opportunities are generally less tangible and so more difficult to integrate in a practicable tool.

To allow some quantification of the sustainability of a particular pavement design option, a simple comparison tool has been developed for the management of construction and demolition waste in land development projects [9] and recently modified [10]. In this process, a series of project-specific indicators are applied to the road project. The output is in the form of a graphical total sustainability chart

ENVIRONMENTAL ASPECT	Option 1					Option 2				
	-2	-1	0	1	2	-2	-1	0	1	2
Use of Natural Resources										
New Aggregates				X						X
Bituminous Materials				X				X		
Atmospheric Emissions										
Air Emissions		X							X	
Dust and Noise		X				X				
Waste Generation										
Excess Material Off-site		X						X		
Value-Added Re-Use			X					X		
Consumed energy										
Consumed energy		X						X		
Transportation impact										
Transportation impact	X								X	
SOCIAL ASPECT										
Health and safety										
Local resident safety			X						X	
Worker safety		X								X
Impact on community										
Work pollution (noise, dust, visual aspects, etc)			X						X	
Duration of works		X							X	
Quality of Life (Frustration, ride quality, etc.)	X									X
Adding value to the road corridor				X						X
Equity										
Training of employees			X					X		
Preferably local job			X					X		
Corporate image										
Corporate image			X							X
Future Maintenance Needs										
On-going local employment					X		X			
ECONOMIC ASPECT										
Economic performance										
Total construction cost				X			X			
Life Cycle Cost		X								X
Local suppliers										
Local suppliers			X				X			
Sustainability										
Durability			X							X
Level of maintenance and repair										
Level of Maintenance and repair			X							X
Technological aspect										
Research and development			X						X	
Technological uncertainties				X			X			

Fig. 6 Sustainability comparison of alternative strategies

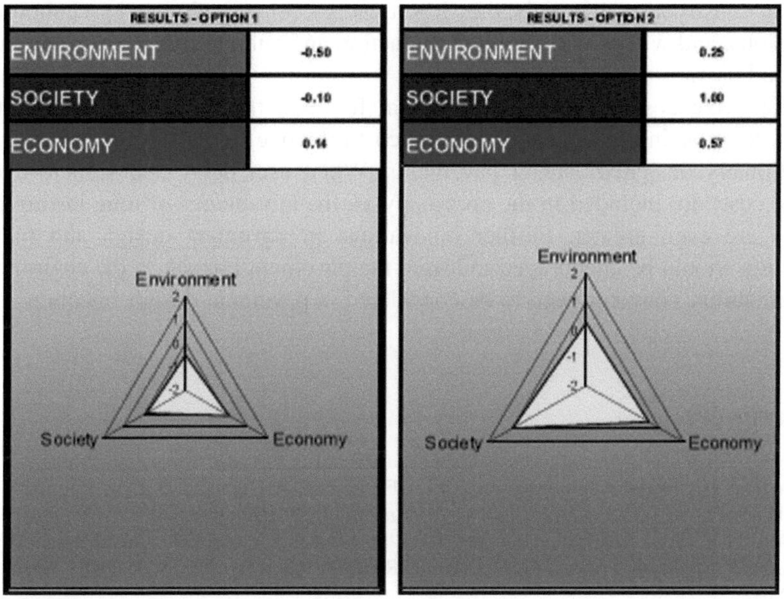

Fig. 7 Sustainability comparison of alternative strategies

that compares the viable alternatives and allows the most sustainable option to be readily identified.

The key indicators for assessment of each scenario can be selected based on what is considered important for a particular agency. While financial considerations will usually dominate, social and environmental criteria also need to be included. The alternatives are rated from −2 to +2 under a series of attributes grouped under environmental social and economic. Figure 6 illustrates the rating grid for the two options. The graphical presentation of the results is presented in Fig. 7. In this instance option 2 (perpetual pavement) obtains the best overall sustainability balance (represented by the larger yellow triangle) and therefore is the preferred solution for the RHVP.

Conclusions

The increasing awareness of the damage to the environment caused by road transportation is increasing demand for more responsible pavement structural designs. Better materials and improved analytical techniques now allow asphalt pavements to be designed to last 50 or more years with minimal maintenance intervention. These long lasting (perpetual) asphalt pavements have lower environmental and social impacts and are more economic over the longer term. They make better use of road construction materials, they lower energy consumption and greenhouse gas emissions, and will provide higher satisfaction and enhanced driving experience to the road users.

Life cycle cost analysis indicates that for high traffic volume urban arterial roads, long lasting pavements are more cost effective over their service lives than traditionally designed asphalt pavements. When user delay and vehicle maintenance costs are included in the cost analysis, the advantages of long lasting pavements are even greater. Further innovations in pavement design and material selection should be encouraged to lessen the pavement impact on the environment. Sustainability concepts need to be considered in pavement design selection.

References

1. Federal Highways Administration (1997) Flexibility in Highway Design, Publication No. FHWA-PD-97-062, Federal Highway Administration, Washington, D.C.
2. Corte J-F (2001) Development and Uses of Hard-Grade Asphalt and of High-Modulus Asphalt Mixes in France, Transportation Research Circular, No. 503, TRB, Washington, D.C.
3. Chappat M, Bilal J (2003) La route écologique du future. Consommation d'énergie & émission de gaz à effet de serre, COLAS Repoer, 40 p
4. Asphalt Pavement Alliance (2002) Perpetual Pavements: A Synthesis, APA 101, Asphalt Pavement Alliance, Lanham, MD
5. Transportation Research Board (2001) Perpetual Bituminous Pavements, Transportation Research Circular Number 503, Washington, D.C.
6. Maher M, Uzarowski L, Moore G, Aurilio V (2006) Sustainable Pavements - Making the Case for Longer Design Lives for Flexible Pavements, Canadian Technical Asphalt Association, Proceedings, 51st Annual Conference, Charlottetown, Prince Edward Island, pp. 43–65
7. Ministry of Transportation of Ontario (2003) Guidelines for the Use of Life Cycle Cost Analysis on MTO Freeways, Ministry of Transportation of Ontario, Materials Engineering and Research Report, Downsview, Ontario
8. Horvath A (2003) Decision-Support Tool for Life-cycle Assessment of Pavements, Transportation Research Board Summer Workshop "Beneficial Use, Sustainability, and Pollution Prevention in Transportation Infrastructure," Portsmouth, NH
9. Clarke-Whistler K, Petersen M (2005) Sustainable Resource Management in Land Development, Proceedings of CABERNET 2005, Belfast
10. Beaudry P (2007) Sustainability Principles as Drivers of Innovation, presented at the 2007 Golder Associates Ltd. Principal's Meeting, Toronto, March 28–30

Fuel Tax Levels Necessary to Achieve the Agreed Reduction Targets of CO_2 Emissions. The Case of Madrid

Daniel de la Hoz, Luis A. Guzmán, Paul Pfaffenbichler, and Simon Shepherd

Introduction

Sustainable urban areas are widely considered a promising target for every city. Different policies are being designed in order to tackle the multifaceted range of transport-related problems in urban agglomerations and therefore contribute significantly to the overall quality of life in cities. The recently published communication "Green Paper - Towards a new culture for urban mobility" [1], clearly says that "European towns and cities are all different, but they face similar challenges and are trying to find common solutions": for making our cities sustainable. This is not a minor task. Over 60% of the population lives in urban areas and 85% of the EU's gross domestic product is created in urban areas [1, 2].

All around Europe, increased traffic, both in the city centre and the metropolitan area is a common phenomenon. Externalities measured in terms of delays, pollution, stress, inequities, etc.; drive our cities into a spiral of degradation. On the other hand, climate change is recognized as an international problem where all are involved. The increase in traffic and the 'stop-go' nature of driving in urban areas implies that cities are becoming a major and growing source of CO2 emissions. Lot of hope has been pinned on technical progress. However, this can be insufficient due to the uncertainty over when new reliable fuel or vehicles technologies will be introduced. For that reason, the focus is more and more shifting to market-driven instruments, like taxation measures, which, apart from creating incentives [3, 4] to develop and use low-emission technologies, can also reduce unsustainable mobility demand [1].

D. de la Hoz and L.A. Guzmán
Centro de Investigación del Transporte, Universidad Politécnica de, Madrid, Spain

P. Pfaffenbichler
Institut für Verkehrsplanung und Verkehrstechnik, TU, Wein

S. Shepherd
Institute for Transport Studies, University of Leeds, Leeds, UK

S. Rauch et al. (eds.), *Highway and Urban Environment*,
Alliance for Global Sustainability Bookseries 17,
DOI 10.1007/978-90-481-3043-6_40, © Springer Science+Business Media B.V. 2010

Fuel taxes were not initially designed for environmental purposes, but their consequences are certainly environmental. The stated motives for fuel taxes vary considerably from one place to another [5]. Nowadays, fuel taxes could play a crucial role to motivate a change, but still it is difficult to implement as policies are shaped by economic interest, and the growing dependence of cars among the population makes it unpopular [6].

There are several economical studies based on the topic of searching for the optimal carbon-tax system to achieve some certain abatement objectives and sustain the macroeconomic welfare [7–10]. Others have propound an integrated economy-climate model under a system dynamics (SD) approach [11, 12]. In this study fuel taxes will be evaluated as an important instrument for the environment and urban sustainability. Using a system dynamic approach for urban mobility, this study shows how the energy taxation policy can contribute to the climate policy. A target looking up process will be developed in order to define the level of a CO2 emissions fuel duty necessary to achieve a specific CO2 reduction target on a regional area. Madrid Regional area was used as case study.

Methodology

System Dynamics

Falling CO2 emission level can be derived from a reduction of fuel consumption (due to a change on driving behavior, optimization or cutting fuel consumption on vehicles, alternative fuels…), a change on the mode of transport to a "greener" one (due to the promotion of public transport, cycling or pedestrian routes, promote car sharing, pricing or market-driven measures, transit oriented design of new developments…), or even avoiding the need to transport (through mixed urban planning, use of telematics…). Due to the complexity and multifaceted of the transport system and its connections – interactions to other subsystems (economic, social, health, environment, territory or activities allocations, etc.), traditional econometric approach may have limitations. Thus, in this case, a system dynamic approach is propounded. A system is defined as a group of entities which act and interact toward the achievement of a specific reason. This system may just represent a part or a subset of an overall system. It is called dynamic if its interactions cause changes over time [13], which helps for evaluating policy measures in the long term where the behavior to be assessed is a consequence of complex interactions [14, 10].

Land Use and Transport Model. MARS Model

The model used here, called MARS (Metropolitan Activity Relocation Simulator) model [15], is based on synergetic principles [16]. MARS is a strategic, interactive land-use and transport interaction (LUTI) model.

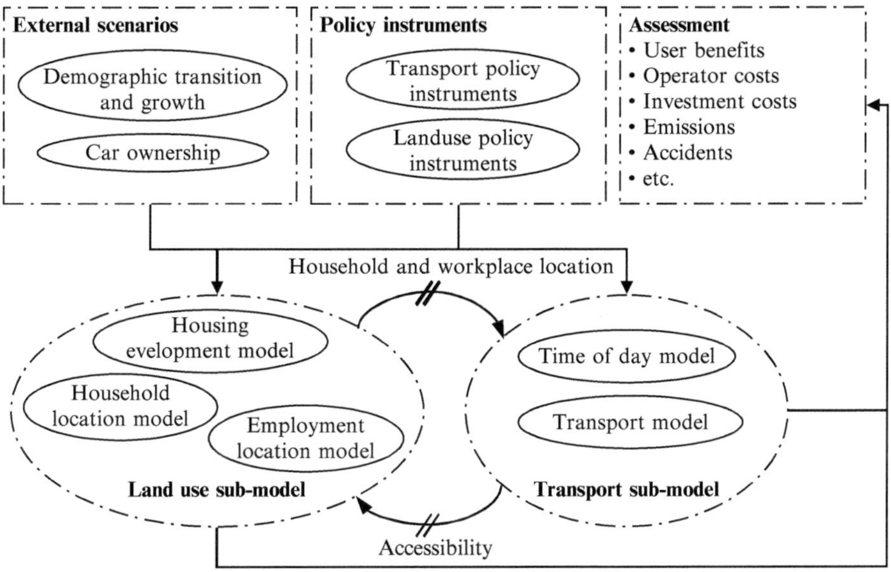

Fig. 1 Basic structure of the MARS sub-models [15]

The MARS model includes a transport model which simulates the travel behavior of the population related to their housing and workplace locations, a housing development model, a household location choice model, a workplace development model, a workplace location choice model, as well as a fuel consumption and emission model. All these models are interconnected with each other and the major interrelations are shown in Fig. 1.

The technique of Causal Loop Diagrams (CLD) is used to explain the major cause and effect relations employed in the model. In the case of car usage, for example, Fig. 2 shows some of the Casual Loops (balancing and reinforcing casual loops) that drive the behavior of commuting trips taken by car from one zone to another. Similar CLDs could be drawn for other modes and for non-commute trips [17]. Changes of the fuel prices to the end users influences the attractiveness to use the car, which will be equilibrated by others CLD as changes on the speed (time) due to a different level of congestion. Changes on accessibility will modify, in a long term, the residential and workplace location which may also impact the attractiveness to car use as well.

Problem Statement – Scenarios

The study was developed by seeking the appropriated CO2 fuel tax level in order to achieve predefined CO2 emissions targets. Under a disparity duty structure based on life-cycle emissions, fuels with higher emissions of pollutants and GHG

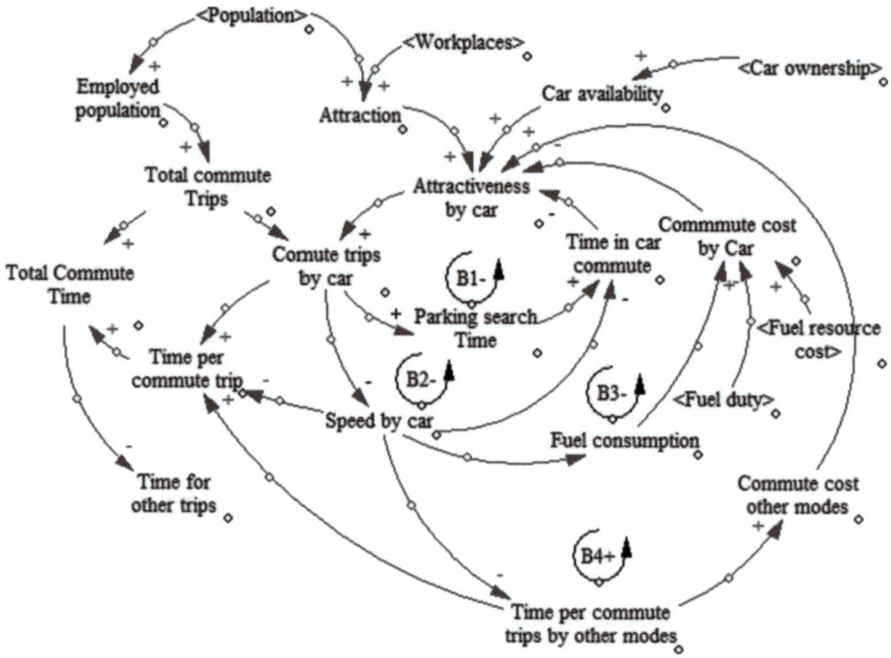

Fig. 2 CLD for the transport model – commute trips by car in MARS using the software VENSIM [24]

would be penalized with higher levels of tax. This tax levels will be clearly be affected by the evolution both of the oil price (Scenarios A = Low Price; B = High Price) and the technology development (Scenarios 1 = Low Technology Investments; Scenarios 2 = High Technology Investments).

By combining the policy dimension with the energy-availability dimension, a scenario framework was set up.

In order to explore the optimal tax policy under different situations, not just from the technological or the oil price situation, but within a strategy of different travel demand measures, an integrated policy measures was put forward in order to seek for a more balanced strategy with the same target (Scenarios 3 = Travel Demand Measures (TDM)). These dimensions were compared with business as usual alternatives (labeled as Scenarios 0), where no policy measures were assumed (Table 1).

Scenarios are summarized in Table 2, based on the figures contemplated in STEPs EU Funded Project [18]. Vehicle fleet stocks were constructed based on ASTRA model results for STEPs Scenarios [19].

CO_2 Emissions Target Adopted

Several assumptions were set up for the fuel tax regulation scenarios. Regarding climate policy, the scenario assumes that all the countries fulfill their individual Kyoto target in accordance with the Burden Sharing Agreement with the targets

Table 1 Scenarios for measure assessment

			Fuel tax regulation CO_2 target		
		Business as usual	Low tech. investment	High tech. investment	TDM
Energy availability	Generally accepted energy supply forecast	A0	A1	A2	A3
	Worst case energy supply forecast	B0	B1	B2	B3

Table 2 Specification for the Baseline Scenarios (based on STEPs EU Co-funded R&D Project [20])

Oil market (annual change in %)		Scenario A	Scenario B
Fuel prices	Gasoline/liter	+1.0% /	+4.0%
	Diesel/liter	+1.0% /	+4.0%
	Electric/unit	+0.0%	+0.0%
	CNG/unit	+1.0%	+4.0%
	Hybrid/unit	+1.0%/	+4.0%
	Hydrogen/unit	+1.0%	+1.0%
Transport energy sub-system (annual change in %)		Scenario 1	Scenario 2
Improving energy efficiency for car	Fuel consumption/ gas. Car	–0.5%	–2.0%
	Fuel consumption/ diesel car	–1.0%	–3.0%
Alternative tech.	Emission factors	–8.1%	–16.0%
Car fleet (growth/share)	Conventional (gas/ diesel)	–1%/72%	–2.1%/55%
	Hybrids:	+12.5%/15%/	+13.5%/20%
	CNG	+10%/10%	+2%/15%
	Electric	+3%/1%	+7%/5%
	Hydrogen	+3%/2%	7.8%/5%
Transport scheme (final % of change)			Scenario 3
PT accessibility		Access time (min)	–30%
PT frequency		Waiting time (min)	+20%
Public transport fares		€	–50%
Public Transport Speed (exclusive lanes)		In-vehicle time (min)	+25%
Traffic calming		Average speed	–10%
Intermodal network		Changing time	–15%
Current parking schemes (inner city)		Short term (€)	Up to 5€
		Long term (€)	Up to 12€

expressed in terms of CO2 emissions. In the case of Spain, 15% over 1990 CO2 emissions level was the commitment. But there are not specific CO2 targets for the transport sector, and even less on a local environment. The role of transportation for reducing GHG emissions needs to be determined, and targets need to be specified.

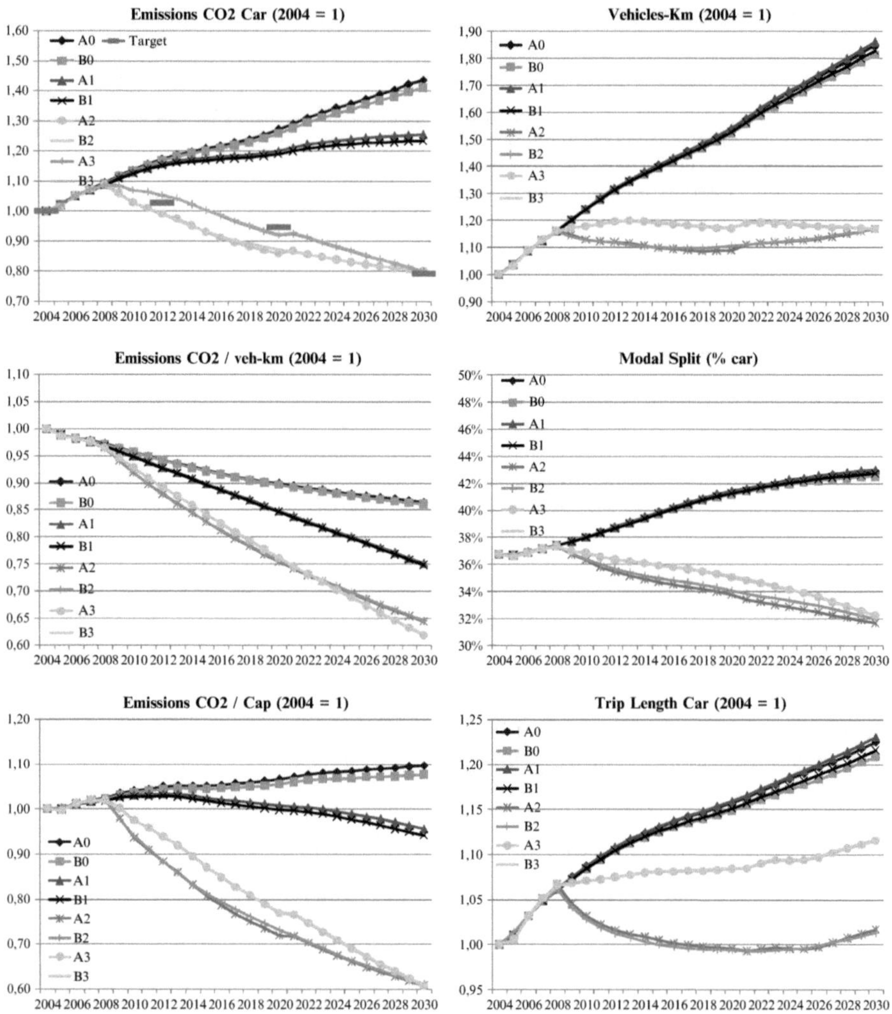

Fig. 3 CO_2 emissions and mobility behavior results from the case study

To determine transport's future contribution it was decided to use the average contribution from transport to emissions in a carbon constrained context. In the case of Spain, the projections to 2012 are that they will increase to 65% over 1990s levels, on diffuse sectors (where transport is). For other sectors (energy and industry) the forecast is just an increment of 37%. The diffuse sectors are responsible for about 46–49% of the GHG emissions. Transport has a weight on the diffuse sectors in the range from 42% on 1990 to 49% on 2004 (that means a 24% of total 2004 GHG) [21]. Taking into account that transport won't be able to follow the stabilization curve that other sectors may do over the coming year, a more relaxing target curve is propounded. However, it is necessary to mention that even the weakest of the targets will require a significant reduction from current emission levels. A target of CO2 emissions levels from 1990 + 71% (2012), + 61 (2020) and + 35 (2030) was settled for the case study.

Results and Discussion for Madrid Case Study

Madrid Region is the biggest urban conglomeration in Spain and third in Europe after London and Paris. Mobility demand in Madrid grows continuously. According to the last mobility survey (2004), numbers of trips in a work day have increased 40% over 1996 ratios. Apart from population growth, trips per person increased 20% on the period 1996–2004 [22, 23], and due to urban sprawl, a further car dependence is generated.

Impacts on CO_2 Emissions and Mobility Behavior

Regarding base line scenarios (A0, B0, A1, B1), it is clearly showed that technology improvements (A1/B1) generate higher potential savings on CO2 emissions in all Fig. 3. Higher fuel prices on vehicle use in scenarios B strengthen this trend.

From the technological investment scenario point of view, (CO2) seems to be efficient but not enough to reach CO2 emissions target. It does, on the other hand, induce a higher mobility in terms of distance and car trips (Scenarios 1).

Once a target of CO2 is settled, an optimization process is developed in order to determinate the fuel tax needed for achieving it (Scenarios A2/B2). Duty levels in these scenarios need to grow on average annual ratios of 18–20% in order to fulfill CO2 emissions requirements, which are economically unfeasible, socially unacceptable and politically unpopular, but environmentally efficient. What this really shows is firstly that a deep change is needed to really achieve the targets propounded for our cities' sustainability. Secondly, strategies for sustainable mobility could not be based on an isolated action.

As expected, the impact on mode share can be viewed in pairs of scenarios. Obviously the demand regulation scenarios (Scenarios 2) have the greatest impact on car use due to the significant increase in costs. Similarly A0/A1 and B0/B1 are grouped together and the relative changes are small within these groupings. What clearly shows is that higher technology scenarios improve environmental impacts, but it does not impact the car dependence of our cities.

Regarding scenarios A2/B2 and from an environmental point of view, high-price car usage appears to be a partial substitute for car dependence and changing behavior. This is needed to be carefully designed in coherence with other measures in order to seek for an optimal operating cost level. Scenarios A3 and B3, indicate that changes are derived by same motivations plus a better alternative in PT usage which does not limit or change mobility demands.

These results show that increasing fuel prices reduce the likelihood of commuting by car and a change on mobility behavior (modal shift, amount of distance traveled). The debate over whether fuel economy standards or fuel taxation is the more appropriate policy instrument to raise fuel economy and reduce CO2 emissions is not appropriated. Both are needed. Results from scenarios A3 and B3 reconfirm the extensively expressed affirmation that individual policy measures are not sufficient to deal with the diverse and multifaceted sustainability concerns associated with transport. In order to achieve improvements in energy intensity, CO2 emissions, congestion and air pollutant emissions, a package of measures is necessary.

Conclusion

As the transport sector is responsible for many long term impacts, such as CO2 emissions or fossil fuel consumption, there is obviously a strong call for bringing together mobility and sustainability. To increase the "environmental friendliness" of energy taxation, a new fuel taxation policy have been set up for achieving a specific CO2 emission target. Clearly, the outlined target represents an effort that would require great changes in the coming years, both regarding the kind of transport and the way that transport is perceived and utilized by individuals and organizations (their mobility and behavior).This is reflected in the level of fuel tax needed (Scenarios 2). Evidence of the effectiveness of fuel taxation as a environmental transport policy in this context are clear, as it was able to cut CO2 emissions growth. But the levels propounded are economically, socially and politically unfeasible. The effectiveness of the measure improves when other, such as technological improvement measures were implemented.

However, there are a truly uncertainty associated with trust on technological improvement. There is a risk that the technology will not deliver the required CO2 reductions. The probability that this may be too late, even if it is really achieved, is a significant element that makes it more urgent to considerer other measures from now. An alternative to a complete dependence on technological evolution is to start implementing schemes which are designed for changing our mobility behavior. Fuel taxation can be part of an integrated package of policy measures. Transparency and logic for destinations of revenues obtained by this policy needs to be understandable and shared by the public.

References

1. COM (2007) Green Paper – Towards a new culture for urban mobility. European Commission. Directorate General for Energy and Transport. 551
2. Eurostat. The Statistical Office of the European Communities
3. European Parliament and Council (2003) On the Promotion of the Use of Biofuels or other Renewable Fuels for Transport (Directive 2003/30/EC)

4. CEU (2005) Directive of the European Parliament and of the Council on the promotion of clean road transport vehicles. Official Journal of the European Union (Directive COM (2005) 634), Council of the European Union
5. Parry I, Small K (2005) Does Britain or the United States have the right gas-oline tax? American Economic Review 95:1276–1289
6. Hammar H, Lofgren A, Sterner T (2004) Political economy obstacles to fuel taxation. Energy Journal 25:1–17 ISSN:0195–6574
7. Nordhaus W (1991) The costs of slowing climate change: a survey. Energy Journal 12:37–65
8. Dean A, Hoeller P (1992) Costs of reducing CO2 emissions: evidence from six global models. OECD Economic Studies No. 19, Winter, pp. 16–47
9. Azar Ch, Schneider SH (2002) Are the economic costs of stabilizing the atmosphere prohibitive? Ecological Economics 42:(1–2):73–80
10. Weber M, Barth V, Hasselmann K (2005) A multi-actor dynamic integrated assessment model (MADIAM) of induced technological change and sustainable economic growth. Ecological Economics 54:306–327
11. Piattelli M, Cuneo MA, Bianchi NP, Soncin G (2002) The control of goods transportation growth by modal share replanning: the role of a carbon tax. System Dynamics Review 18:47–69
12. Kunsch PL, Springael J (2008) European Journal of Operational Research 185:1285–1299. Cited 9 Nov 2006
13. Ogata K (1978) System Dynamics. Eaglewood Cliffs, NJ, Prentice-Hall
14. Abbas KA, Bell MGH (1994) System dynamics applicability to transportation modeling. Transportation Research A, 28A:(5):373–400
15. Pfaffenbichler P (2003) The strategic, dynamic and integrated urban land use and transport model MARS (Metropolitan Activity Relocation Simulator) – Development, testing and application. PhD thesis. University of Technology, Vienna
16. Haken H (1983) Advanced Synergetics - Instability Hierarchies of Self- Organizing Systems and Devices; Springer Series in Synergetics 20, New York, Springer-Verlag
17. Pfaffenbichler P, Emberger G, Shepherd S (2008) The integrated dynamic land use and transport model MARS. Journal of Networks and Spatial Economics (DOI 10.1007/s11067-007-9050-7)
18. STEPs (2006) Scenarios for the Transport Systems and Energy Supplies and their Effects. Transport Strategies Under the Scarcity of Energy Supply
19. ASTRA (2000) Assessment of Transport Strategies, Final Report for the European Commission, 48 p. http://www.iww.uni-karlsruhe.de/ASTRA
20. Monzón A, Nuijten A (2006) Transport strategies under the scarcity of energy supply, DG TREN, European Commission
21. Royal Legislative Decree (2006) REAL DECRETO 1370/2006. Plan Nacional de Asignación de derechos de emisión de gases de efecto invernadero, 2008–2012. Ministerio de la Presidencia
22. CRTM (1996) Encuesta Domiciliaria de Movilidad de Madrid. Consorcio Regional de Transporte de Madrid
23. CRTM (2004) Encuesta Domiciliaria de Movilidad de Madrid. Consorcio Regional de Transporte de Madrid
24. VENSIM, The Ventana Simulation Environment, Ventana Systems, Inc

Emission Inventory for the Road Transport Sector in the Urban Area of Naples: Methodology and Results

Paolo Iodice, Massimo Cardone, Adolfo Senatore, and Mariano Migliaccio

Introduction

The most recent European directives and Italian legislative decrees that have regulated the environmental matter are oriented to planning and determination of the more opportune strategies for health safeguard and for ecosystem protection. It appears obvious, therefore, that the national and regional programming must use appropriate cognitive instruments to estimate the air quality state and the origins of air pollutions in order to support prevention and reorganization decisions.

The inventory of emission sources is a preliminary and fundamental cognitive element for the planning activity and for air quality management; at a European level the inventories must be drawn up with CORINAIR Methodology (Coordination Information Air) [1, 2]. Such inventories constitute a technological, economic and territorial data collection, which concurs to individualize the pollution sources (industrial, civil, transports, etc.), their localization with spatial disaggregation (regions, provinces and towns), the amount and typology of the polluting substance. The amounts of pollutants emitted from various sources in the zones under investigation can be obtained through direct and continuous measures where possible, otherwise through estimates. The direct measure of the emissions, generally, can be carried out only for the principal industrial systems, usually schematized as punctual sources. For all other sources, called diffused sources (small industries, heating systems, natural sources, urban road traffic) and for the linear sources, it must resort to estimations: the emissions are estimated on the basis of opportune activity indicators and fixed emission factors concerning specific emissive activity.

In Europe, the emissions due to road traffic (SNAP sector 07) are almost always an important fraction of the total emissions of a territory, despite extensive measures world-wide to reduce emissions during the last one or two decades. In Italy,

P. Iodice, M. Cardone, A. Senatore, and M. Migliaccio
Dipartimento di Ingegneria Meccanica per l'Energetica– Università degli Studi di Napoli Federico II

S. Rauch et al. (eds.), *Highway and Urban Environment*,
Alliance for Global Sustainability Bookseries 17,
DOI 10.1007/978-90-481-3043-6_41, © Springer Science+Business Media B.V. 2010

for example, traffic emissions account for about 67% of carbon monoxide emissions, 51% of nitrogen oxides emissions and 40% of hydrocarbon emissions (UNECE/EMEP, 2004). Despite their importance, the emissions from road transport with reference to other human and natural activities, are among the most difficult to calculate because they depend on many variables, which are characterized by a particular degree of uncertainty.

For the present application, the COPERT 4 methodology has been employed. It is the most complete and accepted software tool available in Europe for calculating emissions from road traffic. The COPERT model was developed on behalf of the European Environmental Agency to support European countries for their international reporting obligations; in 2003 about 15 European countries were using the COPERT III model for official emission estimates. In this study the COPERT methodology has been used with a bottom-up approach focusing at municipal level instead of national, then including local parameters relating to fleet, driving patterns, medium trips, average vehicle speed and the fuel consumption. Analysis of research results has identified local critical factors in the Naples urban area.

COPERT IV Methodology

Overview

The COPERT methodology is the collection of algorithms and emission factors suggested at European level for the estimation of emissions from the road transport sector (i.e. the whole SNAP 07 of CORINAIR). It is the most commonly used model in Europe for official national inventories of emissions from road traffic. COPERT 4 is the fourth update of the initial methodology developed on the basis of the work of a working group which was set up for this purpose. The current version is a collaborative effort and draws its main elements from several large-scale European activities: the MEET project, the COST 319 action on the Estimation of Emissions from Transport, and the ARTEMIS project

The methodology allows the estimation of the emissions for 230 vehicle categories belonging to the following five main classes: passenger cars, light duty vehicles, heavy duty vehicles, urban buses and coaches, and two wheelers. Vehicles belonging to such main classes are then distinguished according to the fuel type, the EU Directives to which they conform in terms of emissions, the cylinder capacity and other variables. Emissions can be estimated for 36 pollutants, and also for many secondary pollutants like polycyclic aromatics, dioxins and furans and heavy metals contained in the fuel. COPERT calculates the total emissions of exhaust gases by summing emissions from three different sources, namely the thermally stabilized engine operation (hot), the warming-up phase (cold start) and due to evaporation. For volatile organic compounds, expressions for the determination of the evaporative emissions are also given. A methodology has recently been introduced for calculating the emissions of PM from brakes, tyres and pavement wear (this last must not be confused with resuspension). Since vehicle emissions depend on the engine operation (i.e. driving situation), exhaust emissions are calculated as a function of average speed and for three driving conditions: urban, rural and highway.

Uncertainties and Calculation Procedure

The COPERT methodology can be used to calculate a traffic emission inventory both with a top-down approach and with a bottom-up approach. For countries for which the required input data are not available at low level, it seems to be more appropriate to start at NUTS level 0 (national level) and to allocate emissions to other NUTS levels with the help of available surrogate data.

For countries where the required input is available at smaller NUTS level (including for example traffic counting) it has been proposed to make use of this information and to apply a bottom-up approach, building the national total by summing up emissions from the smaller units. An example of this procedure is presented by [3] in calculating emission for the road transport sector in Sardinia (Italy). Emissions in Sardinia were calculated as the sum of contributions from each municipality; the specific elements of each municipality necessary for the estimates were previously analyzed and evaluated. Application of the methodology at higher spatial resolution has to be done only when more detailed data are available from the user. Several input data in applying the methodology can obviously be only estimates and there is a certain degree of uncertainty in estimating these data. The variables necessary to carry out the calculations are listed in Table 1, together with their qualitative uncertainty. Some of the so-called "soft" variables are associated with large uncertainty. The assignment of an initial value from scratch to each soft variable is difficult, however some of them are suggested by COPERT IV for each country, and others can be found in national transportation studies. Other required variables are known with less uncertainty.

Using these variables and the fuel consumption factors of each COPERT IV vehicle category, in terms of grams of fuel per kilometer, the annual fuel consumption for gasoline, diesel and LPG is estimated. The software estimates the fuel consumption and its percentage difference from the actual consumption; if this difference isn't acceptable some soft variables should be modified. When a reliable result is obtained, all the "soft" variables used for this accepted estimate are used as an input to calculate the emissions using the COPERT IV emission factors.

The Emission Inventory

The collection of the necessary input data to calculate the emissions in the urban area of Naples has required the contribution of agencies and institutions. ISTAT (the Italian Institute of Statistics), the Ministry of Transport and Navigation [4],

Table 1 Variables needed to carry out the emission estimations and their uncertainties

Variable	Uncertainty
Actual fuel consumption	L
Traffic on linear sources	M
Annual average mileage for vehicle category	H
Percentage of mileage travelled on each road type	H
Average speed on each road type	H
Average length of a single trip	H
Average temperatures	L
Number of vehicles in each category	L

L = low uncertainty, indicating a variable obtained from continuous measurements; M = medium, indicating a variable obtained from measurements in specific periods extended to the whole year; H = high, indicating estimations

ACI (the Italian Automotive Association) provides the number of registered vehicles and the fleet composition at municipality level [5]. The Italian Association of Oil Companies reports the fuel sold in each province in annual oil market bulletins [6]. APAT (the Italian Agency for Environmental Protection) provides studies on transportation in Italy and information about the average mileage and the average speed of each vehicle category and on each road type [7]. The base year of the emission inventory is 2006. According to the European Directive 98/70/EC fuels must have specific properties in order to protect human health. Therefore the fuel typology used for the emission calculation is the Fuel 2005. Data about the number of vehicles regularly registered have been corrected through percentage reductions (suggested by APAT) to be applied to vehicles PRE EURO, necessary to represent vehicles actually circulating in Italy. The average length of a trip in Italy is 12 km, as suggested by COPERT. The values of others variables, such as the average speed on each road type and the percentage of mileage traveled on each of road type, have been obtained from studies on transportation (APAT). The emissions due to cold starts were calculated considering minimum and maximum temperatures for each month in Naples municipality.

Results

As a result of the elaboration of the acquired data, the composition of circulating vehicle fleet in Naples municipality and the average mileage estimates in terms of kilometers covered by each vehicle category [Vehic*Km] for year 2006 are reported in Table 2. The estimate of the average mileages, as well as the average speeds, has been validated at the end of the iterative procedure, when the error for the estimation of fuel consumption compared with the fuel sold in Naples municipality was been considered acceptable. In order to calculate the fuel sold in Naples municipality,

Table 2 Composition of circulating vehicle fleet and percentage distribution of the vehicle mileage in Naples municipality for year 2006

Category	Vehicles	Vehicles (%)	Vehic*km (%)
Gasoline passenger cars	371,019	51.1	38.3
Diesel passenger cars	116,021	16.0	20.7
GPL passenger cars	18,298	2.5	4.2
Light duty vehicles	27,036	3.7	5.8
Heavy duty vehicles	5,753	0.7	1.4
Buses	845	0.1	1.3
Mopeds	90,000	12.4	14.1
Motorcycles	97,774	13.5	14.2
Total	726,746	100	100

Table 3 Percentage distribution of the gasoline and diesel passenger cars according to the EU Directives with their mileage in Naples municipality for year 2006

EU Directives	GPC (%)	GPC Veic*km (%)	DPC (%)	DPC Veic*km (%)
Euro 0	34.9	20.4	20.4	11.8
Euro 1	16.5	16.8	6.7	6.8
Euro 2	29.7	33.0	18.9	17.9
Euro 3	12	19.0	36.8	43.0
Euro 4	6.8	10.8	17.3	20.5

GPC = gasoline passenger cars; DPC = diesel passenger cars.

the fuel sold in Naples province has been unbundled at municipality level using the number of vehicles registered in Naples municipality and province as spatial proxy.

The most widespread vehicle category is represented by gasoline and Diesel passenger car. These vehicles are then distinguished (Table 3) according to the EU Directives to which they conform. It has been found that in relation to the distribution of the total mileage for the gasoline passenger cars a high percentage is still represented by EURO 0 vehicles. Indeed in the Naples municipality, as in many other areas of southern Italy, the renewal of vehicle fleet is extremely slow compared to other Italian regions, with obvious repercussions on air quality.

Instead in relation to diesel passenger cars, much more favorable results have been observed. This vehicle category is on average younger than the gasoline passenger cars and the largest percentage, around 43%, is represented by EURO 3 cars. In order to determine the estimated total emissions in the municipality of Naples in 2006, also journeys of residents in other municipalities, commuting daily to Naples for work or study, has been considered. The census ISTAT 2001 provides the number of these trips and the means of transport used for each town surrounding Naples (Fig. 1).

Total emissions from road traffic in Naples municipality for 2006 are summarized in Table 4. It is important to stress that the reported COV emissions include

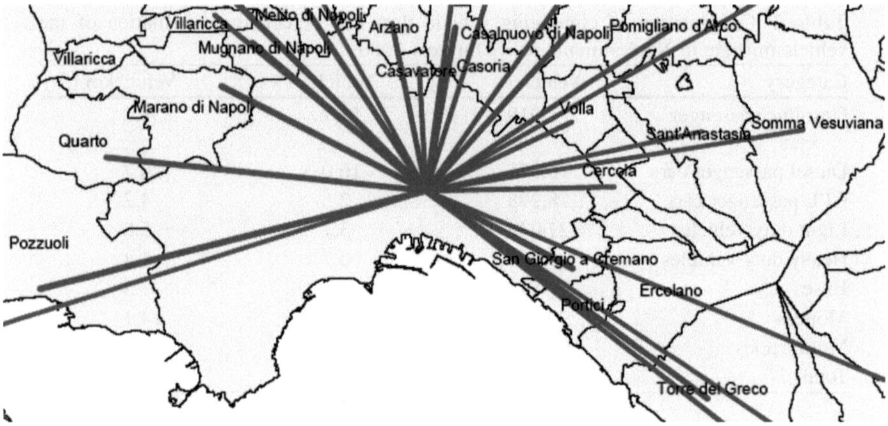

Fig. 1 Daily journeys from vast area toward Naples municipality

Table 4 Estimated total emissions from urban road traffic in Naples during 2006

Pollutant	Emissions (t)	Pollutant	Emissions (t)
CO	16,849	COV	4,340
NO_x	2,105	PM_{10}	180

evaporative emissions and that the PM emissions include brakes, tyres and road pavement wear. The emissions of CO, NOx, COV and PM10 are illustrated for each vehicle class in Fig. 2.

Passenger cars contribute more than 57% to the total road traffic emissions of CO, and about 53% is from gasoline passenger cars (as might be expected, since the combustion is incomplete due to lack of oxygen in engines powered by gasoline).

Previously, it has been observed that the gasoline passenger cars fleet in Naples municipality is relatively old, and old vehicles are generally characterized by higher emissions compared to the most recent ones. In Fig. 3 the carbon monoxide emissions due to gasoline passenger cars have been split according to the EU Directives: EURO 0 passenger cars constitute the largest contribution to the CO emissions.

Figure 2 also shows NOx emissions due to different vehicle classes; urban buses represent the largest contribution in the urban area of Naples (about 26.5%). The predominant share of these NOx emissions is determined by conventional buses (Fig. 4), which are characterized by a high emission factor (about 26 g/km for an average speed of 12 km/h).

The emissions of VOC are also shown in Fig. 2 for different vehicle classes. Again a considerable contribution (more than 28%), is due to gasoline passenger cars, also considering the share of evaporative emissions. About 53% is brought by

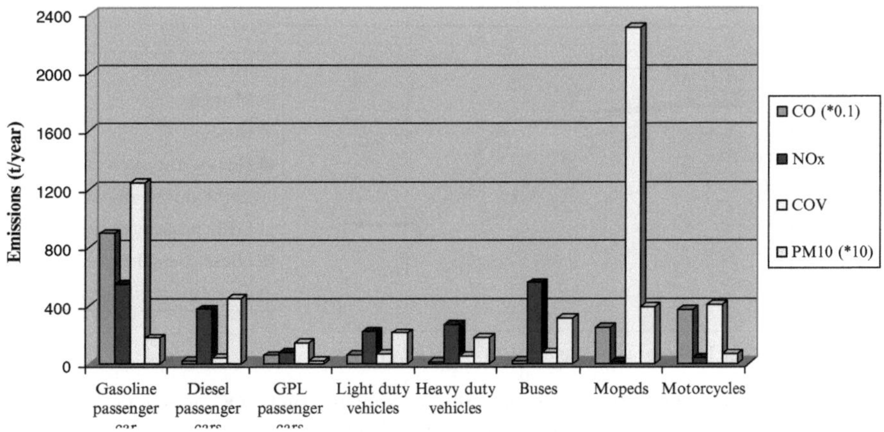

Fig. 2 CO, NO_x, COV and PM_{10} total emissions from different vehicle classes. (CO emission are divided by 10, PM_{10} are multiplied by 10)

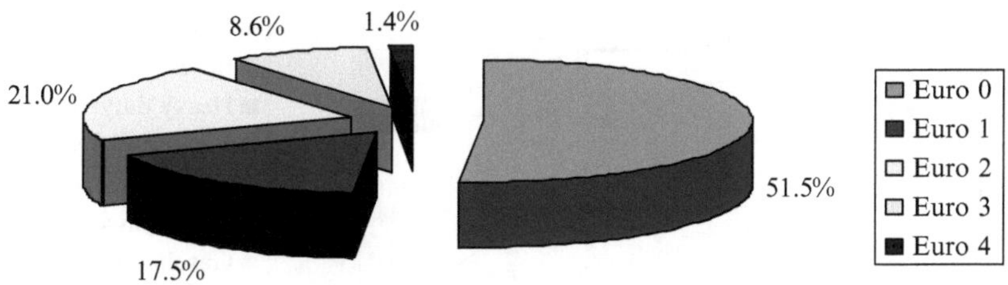

Fig. 3 CO emissions due to gasoline passenger cars disaggregated for EU Directives

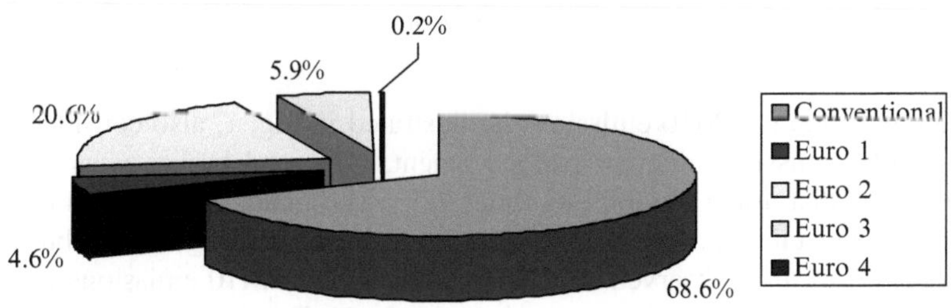

Fig. 4 NO_x emissions due to buses disaggregated for EU Directives

mopeds, despite that the percentage of circulating mopeds in Naples municipality (12.4%) and the average mileage estimates in terms of kilometers per year covered by this vehicle category (14.1%) turned out lower, compared to other categories (Fig. 5). Two-stroke gasoline engines equipped with carburetor that characterize the mopeds will produce a raised emission factor (about 13 g/km in urban environments), justifying such results [8].

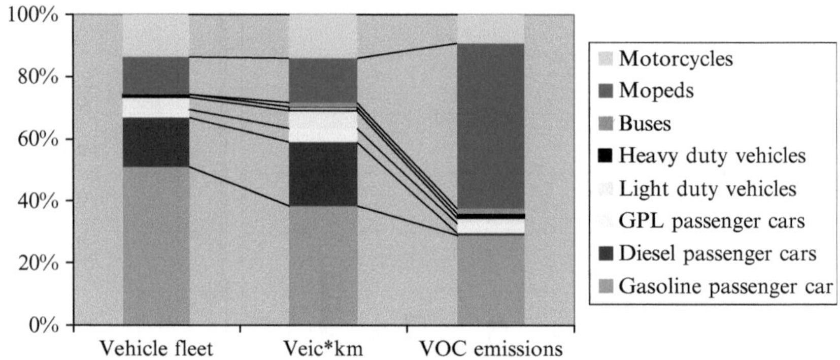

Fig. 5 Circulating vehicle fleet, average mileage estimates, and VOC emissions percentage distribution for each vehicle category in Naples municipality

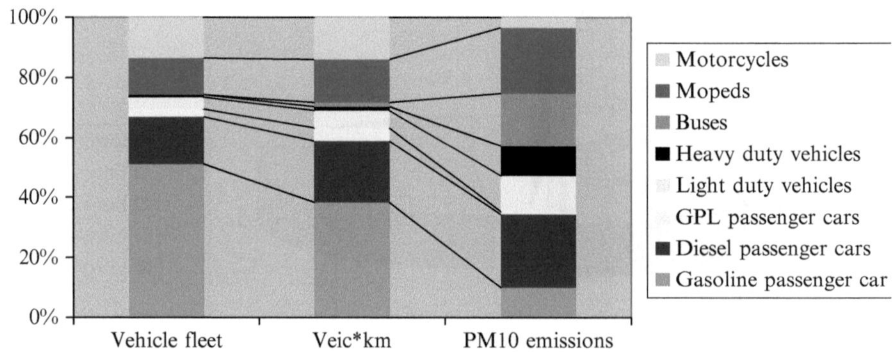

Fig. 6 Circulating vehicle fleet, average mileage estimates, and PM_{10} emissions percentage distribution for each vehicle category in Naples municipality

The repartition of PM10 emissions is illustrated in Fig. 1, also considering the share of PM emissions due to road pavement, tyres and brakes wear for each vehicle class, that affect the total to about 22%. The major contribution to PM10 emissions from tailpipe, however, comes from diesel passenger cars, buses and mopeds (Fig. 6). It is observed that buses weigh upon PM10 emissions for 22%, despite the fact that their percentage in the vehicle fleet is only 0.1% and their average mileage per year does not exceed 1.5%. EURO 0 buses, devoid of modern filtering particulate devices (installed on the latest vehicles), are characterized by emission factors above 1.65 g/km.

Conclusion

The emission inventories from road transport are becoming more and more important in order to accomplish the requirements of the European Directives concerning air quality. In order to estimate the emissions from road transport in Naples municipality, the COPERT IV model has been employed. In this study the COPERT methodology has been used with a bottom-up approach, focusing at municipal level instead of national, then including local parameters relating to fleet, driving patterns, medium trips, the average vehicle speed and the fuel consumption. Besides, it has also considered the journeys of residents in towns surrounding Naples which daily commute to work and study places.

By means of the analysis of research results it has been found that in relation to the gasoline passenger cars, a high percentage is still represented by EURO 0 vehicles, with obvious repercussions on emissions. Indeed, in the Naples municipality, the renewal of the vehicle fleet is extremely slow compared to other Italian regions. In relation to diesel passenger cars, much more favorable results have been observed; in fact this vehicle category is on average younger than the gasoline passenger cars.

These inventories, moreover, must be considered as dynamic instruments since both updating of information as well as data reliability and details are in continuous evolution and improvement.

References

1. CORINAIR (1988) "European Inventory of emissions of pollutants into the atmosphere", Commission of the EC – CORINAIR project, DG XI, 30/3/1988.
2. Eggleston S, Gaudioso D, Gorissen NRC, Samaras Z (1993) "CORINAIR 1990 Emissions from Road Traffic. Methodology and Emission Factors"
3. Bellasio R, Bianconi R, Corda G, Cucca P (2006) – Emission inventory for the road transport sector in Sardinia
4. MINISTERO DEI TRASPORTI (1990–2000): "Conto Nazionale dei Trasporti"
5. ACI (2206) "Annuario Statistico – Automobilismo e Trasporti", Direzione Centrale Coordinamento Centro Studi, Ufficio Statistica, Roma, 2006
6. MINISTERO DELL'INDUSTRIA "Bollettino Petrolifero - vari anni"
7. APAT – Qualità dell'Ambiente Urbano, III Rapporto
8. Migliaccio M (2005) "Sulla efficacia del blocco delle autovetture nei centri urbani per il miglioramento della qualità dell'aria"

Transport Energy Use and Emission Trends in Spain

Pedro J. Pérez-Martínez and Andres Monzón de Cáceres

Introduction

The Spanish transport sector has grown rapidly during the last 15 years and will continue growing in next years. The transport sector is the source of greenhouse gases (GHG) with the highest growth rate, especially due to road transport growth [1]. This paper reviews sector impact on GHG emissions and considers effects of different alternative measures on emissions and subsequent energy consumption.

Reduction of GHG emissions in the transport sector could be achieved: reducing activity, improving energy efficiency of different transport modes and fuels and changing modal share [2]. Measures that can be applied in the transport sector from saving and improvement of energy efficiency are closely linked to sector's nature [3]. These measures include: right application of energy prices, financial and fiscal incentives, reduction of commute trips, transport infrastructure planning and land use, development of low carbon fuels and bigger use of telecommunication technologies.

The Intergovernmental Panel on Climate Change (IPCC) of the United Nations considers that saving and energy efficiency constitute an essential element until technological innovations, current and emergent, could be massively implemented [4]. IPCC stands out main available technologies and commercial practices to mitigate GHG emissions: energy efficient vehicles, hybrid vehicles, clean diesel vehicles, biofuels, and modal shift from road to rail, public transport and no motorized transport modes. Similarly, IPCC reflects technologies and practices potentially commercialized before 2030: second generation biofuels, energy efficient aircrafts, advanced hybrid and electric vehicles with more powerful and reliable batteries. Decrease of transport activity and improvement of energy efficiency are not unique factors driving towards low carbon transport economy. Transport economy is low in carbon if it uses low carbon fuels [5]. It is not only necessary to reduce GHG

P.J. Pérez-Martínez and A.M. de Cáceres
Universidad Politécnica de Madrid, Centro de Investigación del Transporte (TRANSyT), C/Profesor Aranguren s/n 28040, Madrid, España

S. Rauch et al. (eds.), *Highway and Urban Environment*, Alliance for Global Sustainability Bookseries 17,
DOI 10.1007/978-90-481-3043-6_42, © Springer Science+Business Media B.V. 2010

emissions in the final energy consumption, but also in the chain of energy transformations which allow having final energy [6].

Even though there is awareness about CO2 emissions reduction and there is improvement on vehicle efficiency due to new technology of engines and fuels, private vehicle ownership and use continue to grow at an increasing rate together with rising personal incomes and desire to experience faster and more reliable transportation technology. Inefficient modal share and dominance of passenger road transport are key factors of increasing level of vehicle private use and ownership, and associated levels of unsustainability. As personal incomes rise, relative use of private transportation over public transit increases. Similarly, freight transport energy use will continue growing unless there are important reductions in energy intensities of road freight and there is change of modal share.

Transport and Greenhouse Gases

In 1990, transport consumed 39.5% of the total primary energy in Spain and 40.7% in 2004 [7]. In 2004, final energy consumption of the transport sector was slightly more than 38 million tones (tones oil equivalent). Besides being the economic sector with major final energy consumption, transport is the sector with major consumption of fossil fuels (55.2%, 2004). In absolute terms, GHG emissions from transport during this period have grown 66% [8]. At an annual growth rate of 3.7%, emissions are expected to double over 20 years. Emission growth is due mainly to road passengers and freight transport. Road transport alone is responsible for 75% of total sector emissions.

GHG emissions of the transport sector are not explained by either the population growth or the economical growth, because they have lower growth rates. This indicates that production processes in our country have an increasing growth of transport, contrary to EU objectives to generate economical growth with smaller increments of passengers and freight transport flows [9].

On the other hand, annual consumption of 26 million road vehicles is 34,696 million liters of fuel (mainly gasoline and diesel). Of this total, consumption of the car fleet (19.5 million) represents 54%, 2.4 million trucks 33%, vans 10% and buses and motorcycles 3% [10, 11]. It must stand out that consumption of the truck fleet represents 60% of the fuel used by cars. Urban consumption constitutes 22% of the total consumption, of this 73% corresponds to vehicles with diesel engines. These mean consumptions vary based on the type of engine and vehicle. For further details Table 1 can be consulted.

Current transport demand trends and associated GHG emissions in Spain have higher growth rates than in the rest of EU countries (Fig. 1). The mobility of persons and goods grows at a higher rate compared with the mobility of European neighbors. It is observed, in addition, that the growth of transport passengers is greater than the growth of transport freight, when Europe has an opposite trend. These data show the greater importance of the problem in our country, and that in our case, the mobility of persons is still more worrying than the one of goods.

Table 1 Traffic, fleet and fuel consumption by transport mode and fuel type, Spain 2004 [7, 10, 11]

Transport mode	Interurban traffic (10^6 veh-km)	Total fleet (vehicles)	Interurban	Consumption Urban (10^6 L)	Total	Mean consumption (l/100 km)
Gasoline engine						
Motorcycles	1,229	1,612,482	80	43	122	6
Vans	6,979	737,024	942	235	1,177	13
Cars	67,327	12,035,098	6,323	1,783	8,106	9
All	75,535	14,384,604	7,344	2,071	9,415	10
Diesel engine						
Trucks	30,482	2,419,908	9,054	578	9,632	30
Buses	1,424	56,957	400	55	455	28
Vans	14,831	1,592,039	1,763	588	2,351	12
Cars	119,532	7,506,821	8,501	4,367	12,868	7
All	166,269	11,575,725	19,719	5,562	25,281	12

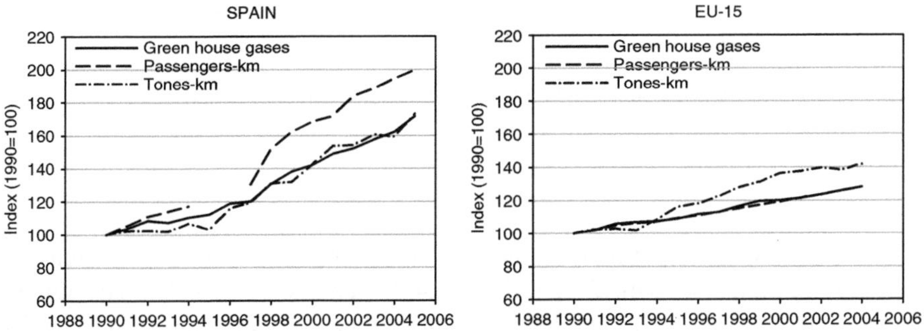

Fig. 1 GHG and transport activity trends in Spain and EU-15, 1990–2004

Therefore, a decided action to supply alternatives to motorized mobility is needed, especially regarding private car use, to reach destinations and goods that guarantee the well-being of the society. Change will not only take place improving and extending the supply and what is needed is a change of mentality in transport modal choice. In this way, travel responsibility is transferred to persons, companies and stakeholders in charge of territorial planning and urban development.

Inefficient Modal Share and Energy Intensity

Former trends are due to an increase of transport activity, important modal imbalance and energy inefficiency: rail transport, in an increasing demand scenario, not only does not grow in absolute terms, but also continues losing market share, both passengers and freight, and has lower values than the European average.

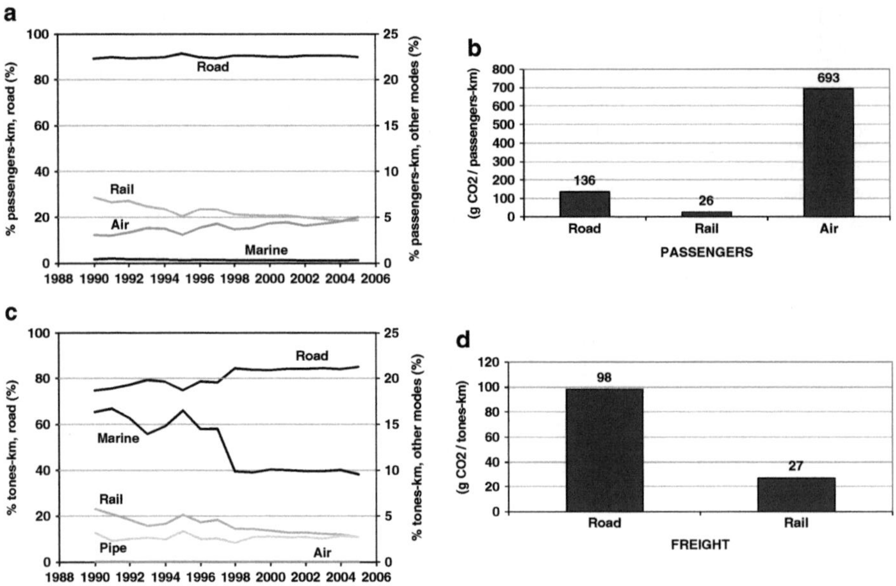

Fig. 2 Modal share (**a**, **c**) and emission intensity (**b**, **d**) by transport mode [7, 8]

In 2005, road transport represented 90% of passengers-kilometers, whereas air transport participated with 5%, rail with 4.7% and marine 0.3%. The higher growth rate between 1995 and 2005 has corresponded to air transport (131.7%). Figure 2a shows a decrease of rail share and a stagnation of marine transport. In 2005, road transport represented 85% of tones–kilometers, whereas marine transport participated with 9.6%, and rail and pipe represented 2.7% each one. The higher growth rate between 1995 and 2005 has corresponded to road transport (72.5%). Figure 2c shows a significant change of rail and cabotage marine towards road transport.

Energy intensity, expressed in megajoules (MJ) per passenger–kilometers or tone–kilometer (in terms of emissions, equivalent grams of CO2), is determined by two factors: energy required to move vehicles and the use of vehicle capacity. Energy intensity and emissions of road transport is five times higher than rail for passengers (Fig. 2b), and four times that of freight (Fig. 2d). For this reason, the decreasing trend of rail transport explains part of the increasing growth of GHG emissions in Spain. It can be stated that Spain is in the worse scenario, because the present situation is one of the most deficient in Europe, and the trend is to make worse, and on an accelerated way. GHG emissions grow at higher rates compare to other EU countries, and in addition the dominant and increasing transport modes are less energy efficient.

Dominant Position of Road Transport

Figure 3 draws some conclusions about reasons of the modal share imbalance shown in the previous section. First graph (Fig. 3a) expresses growths of vehicles and motorization index during the period 1990–2005, higher than 50%. In 2005, the number of vehicles exceeded 25 million and the motorization index was higher than 600 vehicles per 1,000 inhabitants. In Fig. 3b it is interesting to see, that average mileage per vehicle remains constant and equal to 10,700 km/year, so increase of total vehicle-kilometer is due to increase of private mobility using new passenger vehicles. Average mileage per inhabitant has increased 76%, meaning more trips per person and to larger distances. In the lapse of 15 years, there is a change from 3,851 km per inhabitant and year to 6,778 km, and resulting increases of energy consumption and GHG emissions. A British study reveals that the fundamental cause of distance increase per inhabitant by car is the increase of the motorization index [12].

As average consumption per vehicle is practically constant (Fig. 3c), consequently the consumption of energy per inhabitant is increasing parallel to vehicles kilometer per inhabitant. Energy consumption per inhabitant changed from 20 GJ in 1990 to 30 GJ in 2005. Although energy efficiencies of vehicles and fuels have

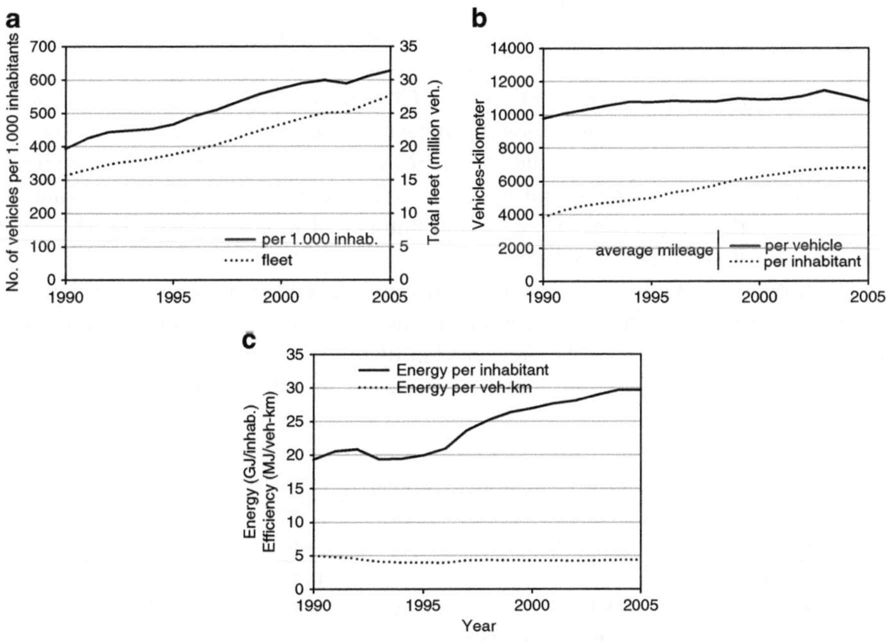

Fig. 3 Road transport trends in Spain: number of road vehicles per 1,000 inhabitants and total number of vehicles (**a**), average vehicle mileage per vehicle and inhabitant (**b**) and average direct energy consumption per inhabitant from road transport, and average fuel efficiency of road vehicles (**c**) [7, 11]

improved, at the same time power and size of new vehicles are increasing. This is the reason why there has been no net saving in average consumption per vehicle-kilometer, around 5 MJ/vehicle-kilometer (0.12 L/km). In terms of CO2 intensity, this corresponds to 119 goe/vehicle-kilometer (equivalent grams of oil per vehicle-kilometer unit).

CO_2 Emissions Trend Scenarios and Potential Mitigation Strategies in the Transport Sector

The situation of the transport sector in Spain, when related to energy saving and CO2 emissions, implies that there is a need to develop a master plan and joining efforts from different areas and fields. It is necessary to define future scenarios and to establish improvement objectives. The current trend scenario (BAU) has an annual emission growth rate of 3.7%. For this reason, a reduction scenario is needed (RED) with the objective to decrease GHG 20% until 2020. This scenario must take place through the fulfillment of the National Plan of Allocation of Emission Rights in 2010 (+37% compared to the 1990 level) [13]. This means a total reduction of emissions, between 2005 and 2020, of 1,145 million tones CO2 compared to the trend scenario.

Reduction of emissions can only be obtained through a clear policy, which includes a set of measures of efficiency in all fields, looking for synergies among them and coordinating a formal action plan. Changes in CO2 emission trends could be achieved by means of the following measures: vehicle and fuel technological (VFT)improvements, freight modal share (FMS) changes, improvement of interurban modal share of passengers traffic (IMT), improvement of urban modal share of passengers traffic, reduction of length and number of motorized trips (UMT) and efficient use of vehicles (EUV).

The potential reduction of these measures will depend on how they are applied, controlled and monitored; consequently no accurate figures are given. However, following the indications of the Action Plan of the Strategy of Energy Efficiency (E4) [14], the objectives could match those listed in Table 2. As it shown, a big

Table 2 Reduction of CO_2 emissions by measures groups, 2004–2020

Emissions		2020 ($MtCO_2$)	Cumulative 2004–2020 ($MtCO_2$)
Trend scenario BAU (1)		193.3	2,500.9
Saving measures	VFT: vehicle and fuel technology	48.7	380.0
	FMS: freight modal share	13.7	106.1
	IMT: interurban modal traffic	8.3	65.0
	UMT: urban modal traffic	15.4	123.1
	EUV: efficient use of vehicles	54.8	470.4
	Total saving (2)	141.0	1,144.6
Reduction scenario RED (1–2)		52.4	1,356.6

effort is necessary, because it would be expected to almost reduce the emission trends in 2020 to the fourth, changing from 193 to 52 million tones CO2. This effort would furthermore be supposed to almost reduce to the half the accumulated emissions during period 2005–2020, gradually increasing the intensification of the measures.

Discussion and Conclusions

Transport in Spain is responsible for almost 31% of CO2 emissions. If the factors that generate the transport activity continue, CO2 emissions will be increased by 157% compared to the level of the Kyoto Protocol in 2020 (base year of reference 1990). Under the Kyoto Protocol, Spain is committed to reduce its greenhouse gases between 2008 and 2012 until reaching 15% increase on 1990 levels. During the period 1990–2020, transport CO2 emissions are expected to increase 196%, towards 193.3 million tones of carbon in 2020.

The reduction of transport GHG emissions will be complicated if factors that have increased emissions in the past are important factors in the increase of future emissions. Although management measures of transport systems have limited reduction potential of emissions, these measures are necessary to reduce other transport externalities: accidents, congestion, noise, and air pollution. These measures are important considering that emissions from other economic sectors have decreased or grown at smaller rate than those of transport.

This paper evaluates different future scenarios corresponding to different emission reduction measures. Between these scenarios and measures, efficient use of vehicles has the greater potential and is the key to drastically reduce carbon emissions along with technological improvements of vehicles and fuels. Efficient use of vehicles represents an appropriate management of the transport system and could reduce base foreseen emissions by 28% in 2020. Improvements in technology of vehicles and fuels continue past trends of energy intensity of passengers and freight transport modes and could reduce base emissions by 25% in 2020. Changes in interurban and urban modal shares of passengers and freight could reduce CO2 emissions by 19%, leading to a stabilization of emissions by 52.4 million tones of carbon in 2020: any additional significant reduction of CO2 emissions would need introduction of great scale additional measures such as low carbon fuels, fiscal measures and e-work.

References

1. Pérez-Martínez PJ (2007) Mobility and environment in Spain. In: Morrison G, Rauch S (eds) Highway and Urban Environment, Proceedings of the 8th Highway and Urban Environment Symposium. Springer Verlag, Dordrecht, pp 35–43
2. Schipper L, Scholl L, Price L (1997) Energy use and carbon emissions from freight in 10 industrialized countries: an analysis of trends from 1973 to 1992. Tranpn Res D 2:57–76

3. Rodenburg CA, Ubbels B, Nijkamp P (2002) Policy scenarios for achieving sustainable transportation in Europe. Transport Rev 22:449–472
4. Kahn Ribeiro S, Kobayashi S, Beuthe M, Gasca J, Greene D, Lee DS, Muromachi Y, Newton PJ, Plotkin S, Sperling D, Wit R, Zhou PJ (2007) Transport and its infrastructure. In: Metz B, Davidson OR, Bosch PR, Dave R, Meyer y LA (eds) Climate Change 2007: Mitigation. Contribution of Working Group III to the Fourth Assessment Report of the Intergovernmental Panel on Climate Change. Cambridge University Press, Cambridge, United Kingdom and New York, NY, USA, pp 323–385
5. Johansson B (1995) Strategies for reducing emissions of air pollutants from the Swedish transportation sector. Transport Res A 29:371–385
6. Van Wee B, Janse P, Van Den Brink R (2005) Comparing energy use and environmental performance of land transport modes. Transport Rev 25:3–24
7. MFO Ministerio de fomento (2006) Los transportes y los servicios postales. Madrid, Spain, 359 pp
8. MMA Ministerio de Medio Ambiente (2006) Inventario de Gases de Efecto Invernadero de España-Edición 2006 (serie 1990–2005), sumario de resultados. Madrid, Spain, 28 pp
9. EEA Environmental European Agency (2006) Transport and environment: facing a dilemma. TERM 2005. Copenhague, Denmark, 52 pp
10. ME Ministerio de Economía (2006) Informe de Recaudación Tributaria. Madrid, Spain, 162 pp
11. MI Ministerio del Interior (2006) Anuario Estadístico General 2005. Madrid, Spain, 112 pp
12. Kwon T H, Preston J (2005) Driving Forces behind the Growth of Per-capita Car Driving Distance in the UK, 1970 to 2000. Transport Reviews 25:467–490
13. MP Ministerio de la Presidencia (2006) Plan Nacional de Asignación de derechos de emisión de gases de efecto invernadero, 2008–2012. Madrid, Spain, 119 pp
14. IDAE Instituto diversificación y ahorro energía (2006) Plan de Acción 2008–2012 de la Estrategia de Ahorro y Eficiencia Energética en España E4. Madrid, Spain, 212 pp

Index

D

U

V

W

MIX
Papier aus verantwortungsvollen Quellen
Paper from responsible sources
FSC® C105338

If you have any concerns about our products,
you can contact us on
ProductSafety@springernature.com

In case Publisher is established outside the EU,
the EU authorized representative is:
Springer Nature Customer Service Center GmbH
Europaplatz 3, 69115 Heidelberg, Germany

Printed by Libri Plureos GmbH
in Hamburg, Germany